Dept of Plant Pathology
139 Funchess Hall
Auburn University, AL 36849

CRC Series in Naturally Occurring Pesticides
Series Editor-in-Chief
N. Bhushan Mandava

Handbook of Natural Pesticides: Methods
Volume I: Theory, Practice, and Detection
Volume II: Isolation and Identification
Editor
N. Bhushan Mandava

Handbook of Natural Pesticides
Volume III: Insect Growth Regulators
Volume IV: Pheromones
Editors
E. David Morgan
N. Bhushan Mandava

Future Volumes
Handbook of Natural Pesticides

Insect Attractants, Deterrents, and Defensive Secretions
Editors
E. David Morgan
N. Bhushan Mandava

Plant Growth Regulators
Editor
N. Bhushan Mandava

Microbial Insecticides
Editor
Carlo M. Ignoffo

CRC Handbook of Natural Pesticides

Volume V
Microbial Insecticides
Part A
Entomogenous Protozoa and Fungi

Editor

Carlo M. Ignoffo, Ph.D.
Laboratory Director
U.S. Department of Agriculture
Agricultural Research Service
Biological Control of Insects Research Laboratory
Columbia, Missouri

CRC Series in Naturally Occurring Pesticides

Series Editor-in-Chief
N. Bhushan Mandava, Ph.D.

CRC Press, Inc.
Boca Raton, Florida

Library of Congress Cataloging-in-Publication Data
(Revised for volume 5)

CRC handbook of natural pesticides.

(CRC series in naturally occurring pesticides)
Vol. 5 has title: Handbook of natural pesticides.
Includes bibliographies and indexes.
   Contents: v. 1. Theory, practice, and detection —
v. 2. Isolation and identification —   — v. 5.
Microbial insecticides/editor, Carlo M. Ignoffo
(pt. A)
   1. Natural pesticides — Handbooks, manuals, etc.
I. Mandava, N. Bhushan, 1934-   . II. Morgan,
E. David (Eric David) III. Title: Handbook of natural
pesticides. IV. Series.
SB951.145.N37C73 1985   632'.95   84-12092

Library of Congress Cataloging-in-Publication Data
ISBN 0-8493-3651-1 (v. 1)

This book represents information obtained from authentic and highly regarded sources. Reprinted material is quoted with permission, and sources are indicated. A wide variety of references are listed. Every reasonable effort has been made to give reliable data and information, but the author and the publisher cannot assume responsibility for the validity of all materials or for the consequences of their use.

All rights reserved. This book, or any parts thereof, may not be reproduced in any form without written consent from the publisher.

Direct all inquiries to CRC Press, Inc., 2000 Corporate Blvd., N.W., Boca Raton, Florida, 33431.

© 1988 by CRC Press, Inc.

International Standard Book Number 0-8493-3660-0 (V. 5A)

Library of Congress Card Number 84-12092
Printed in the United States

CRC Handbook Series in Naturally Occurring Pesticides

INTRODUCTION

The United States has been blessed with high quality, dependable supplies of low cost food and fiber, but few people are aware of the never-ending battle that makes this possible. There are at present approximately 1,100,000 species of animals, many of them very simple forms, and 350,000 species of plants that currently inhabit the planet earth. In the U.S. there are an estimated 10,000 species of insects and related acarinids which at sometime or other cause significant agricultural damage. Of these, about 200 species are serious pests which require control or suppression every year. World-wide, the total number of insect pests is about ten times greater. The annual losses of crops, livestock, agricultural products, and forests caused by insect pests in the U.S. have been estimated to aggregate about 12% of the total crop production and to represent a value of about $4 billion (1984 dollars). On a world-wide basis, the insect pests annually damage or destroy about 15% of total potential crop production, with a value of more than $35 billion, enough food to feed more than the population of a country like India. Thus, both the losses caused by pests and the costs of their control are considerably high. Insect control is a complex problem for there are more than 200 insects that are or have been subsisting on our main crops, livestock, forests, and aquatic resources. Today, in the U.S., conventional insecticides are needed to control more than half of the insect problems affecting agriculture and public health. If the use of pesticides were to be completely banned, crop losses would soar and food prices would also increase dramatically.

About 1 billion pounds of pesticides are used annually in the U.S. for pest control. The benefits of pesticides have been estimated at about $4/$1 cost. In other words, chemical pest control in U.S. crop production costs an estimated $2.2 billion and yields a gross return of $8.7 billion annually.

Another contributing factor for increased crop production is the effective control of weeds, nematodes, and plant diseases. Crop losses due to unwanted weed species are very high. Of the total losses caused by pests, weeds alone count for about 10% of the agricultural production losses valued at more than $12 billion annually. Farmers spend more than $6.2 billion each year to control weeds. Today, nearly all major crops grown in the U.S. are treated with herbicides. As in insect pest and weed control programs, several chemicals are used in the disease programs. Chemical compounds (e.g., fungicides, bactericides, nematicides, and viracides) that are toxic to pathogens are used for controlling plant diseases. Several million dollars are spent annually by American farmers to control the diseases of major crops such as cotton and soybeans.

Another aspect for improved crop efficiency and production is the use of plant growth regulators. These chemicals that regulate the growth and development of plants are used by farmers in the U.S. on a modest scale. The annual sale of growth regulators is about $130 million. The plant growth regulator market is made up of two distinct entities — growth regulators and harvest aids. Growth regulators are used to increase crop yield or quality. Harvest aids are used at the end of the crop cycle. For instance, harvest aids defoliate cotton before picking or desiccate potatoes before digging.

The use of modern pesticides has accounted for astonishing gains in agricultural production as the pesticides have reduced the hidden toll exacted by the aggregate attack of insect pests, weeds, and diseases, and also improved the health of humans and livestock as they control parasites and other microorganisms. However, the same chemicals have allegedly posed some serious problems to health and environmental safety, because of their high toxicity and severe persistence, and have become a grave public concern in the last 2 decades. Since the general public is very much concerned about their hazards, the U.S. Environmental

Protection Agency enforced strong regulations for use, application, and handling of the pesticides. Moreover, such toxic pesticides as DDT, 2,4,5-T and toxaphene were either completely banned or approved for limited use. They were, however, replaced with less dangerous chemicals for insect control. Newer approaches for pest control are continuously sought, and several of them look very promising.

According to a recent study by the National Academy of Sciences, pesticides of several kinds will be widely used in the foreseeable future. However, newer selective and biodegradable compounds must replace older highly toxic persistent chemicals. The pest control methods that are being tested or used on different insects and weeds include: (1) use of natural predators, parasites, and pathogens, (2) breeding of resistant varieties of species, (3) genetic sterilization techniques, (4) use of mating and feeding attractants, (5) use of traps, (6) development of hormones to interfere with life cycles, (7) improvement of cultural practices, and (8) development of better biodegradable insecticides and growth regulators that will effectively combat the target species without doing damage to beneficial insects, wildlife, or man. Many leads are now available, such as the hormone mimics of the insect juvenile and molting hormones. Synthetic pyretheroids are now replacing the conventional insecticides. These insecticides, which are a synthesized version of the extract of the pyrethrum flower, are much more attractive biologically than the traditional insecticides. Thus, the application rates are much lower in some cases, one tenth the rates of more traditional insecticides such as organophosphorus pesticides. The pyrethroids are found to be very specific for killing insects and apparently exhibit no negative effects on plants, livestock, or humans. The use of these compounds is now widely accepted for use on cotton, field corn, soybean, and vegetable crops.

For the long term, integrated pest management (IPM) will have tremendous impact on pest control for crop improvement and efficiency. Under this concept, all types of pest control — cultural, chemical, inbred, and biological — are integrated to control all types of pests and weeds. The chemical control includes all of the traditional pesticides. Cultural controls consist of cultivation, crop rotation, optimum planting dates, and sanitation. Inbred plant resistance involves the use of varieties and hybrids that are resistant to certain pests. Finally, the biological control involves encouraging natural predators, parasites, and microbials. Under this system, pest-detection scouts measure pest populations and determine the best time for applying pesticides. If properly practiced, IPM could reduce pesticide use up to 75% on some crops.

The naturally occurring pesticides appear to have a prominent role for the development of future commercial pesticides not only for agricultural crop productivity but also for the safety of the environment and public health. They are produced by plants, insects, and several microorganisms, which utilize them for survival and maintenance of defense mechanisms, as well as for growth and development. They are easily biodegradable, often times species-specific and also sometimes less toxic (or nontoxic) on other non-target organisms or species, an important consideration for alternate approaches of pest control. Several of the compounds, especially those produced by crop plants and other organisms, are consumed by humans and livestock, and yet appear to have no detrimental effects. They appear to be safe and will not contaminate the environment. Hence, they will be readily accepted for use in pest control by the public and the regulatory agencies. These natural compounds occur in nature only in trace amounts and require very low dosage for pesticide use. It is hoped that the knowledge gained by studying these compounds is helpful for the development of new pest control methods such as their use for interference with hormonal life cycles and trapping insects with pheromones, and also for the development of safe and biodegradable chemicals (e.g., pyrethroid insecticides). Undoubtedly, the costs are very high as compared to the presently used pesticides. But hopefully, these costs would be compensated for by the benefits derived through these natural pesticides from the lower volume of pesticide use and reduction of risks. Furthermore, the indirect or external costs resulting from pesticide poisoning, fatalities, livestock losses, and increased control expenses (due to the destruction of natural enemies and beneficial insects as well as the environmental contamination and pollution from chlorinated, organophosphorus, and carbamate pesticides) could be assessed against benefits vs. risks. The development and use of such naturally occurring chemicals could become an integral part of IPM strategies.

As long as they remain endogenously, several of the natural products presented in this handbook series serve as hormones, growth regulators, and sensory compounds for growth, development, and reproduction of insects, plants, and microorganisms. Others are useful for defense or attack against other species or organisms. Once these chemicals or their analogs and derivatives are applied by external means to the same (where produced) or different species, they come under the label "pesticides" because they contaminate the environment. Therefore, they are subject to regulatory requirements, in the same way the other pesticides are handled before they are used commercially. However, it is anticipated that the naturally occurring pesticides would easily meet the regulatory and environmental requirements for their safe and effective use in pest control programs.

A vast body of literature has been accumulated on natural pesticides during the last 2 or 3 decades; we have been assembling this information in these handbooks. We have limited our attempts to chemical and a few biological aspects concerned with biochemistry and physiology. Wherever possible, we tried to focus attention on the application of these compounds for pesticidal use. We hope that the first two volumes which dealt with theory and practice served as introductory volumes and will be useful to everyone interested in learning about the current technology that is being adapted from compound identification to the field trials. The subsequent volumes deal with the chemical, biochemical, and physiological aspects of naturally occurring compounds, grouped under such titles as insect growth regulators, plant growth regulators, etc.

In a handbook series of this type with diversified subjects dealing with plant, insect, and microbial compounds, it is very difficult to achieve either uniformity or complete coverage while putting the subject matter together. This goal was achieved to a large extent with the understanding and full cooperation of chapter contributors who deserve my sincere appreciation.

The editors of the individual handbooks relentlessly sought to meet the deadlines and, more importantly, to bring a balanced coverage of the subject matter, but, however, that seems to be an unattainable goal. Therefore, they bear full responsibility for any pitfalls and deficiencies. We invite comments and criticisms from readers and users as they will greatly help to update future editions. It is hoped that these handbooks will serve as a source book for chemists, biochemists, physiologists, and other biologists alike — those engaged in active research as well as those interested in different areas of natural products that affect the growth and development of plants, insects, and other organisms.

The editors wish to acknowledge their sincere thanks to the members of the Advisory Board for their helpful suggestions and comments. Their appreciation is extended to the publishing staff, especially Amy Skallerup, Melanie Mortellaro, and Sandy Pearlman for their ready cooperation and unlimited support from the initiation to the completion of this project.

N. Bhushan Mandava
Editor-in-Chief

FOREWORD

Pests of crops and livestock annually account for multi-billion dollar losses in agricultural productivity and costs of control. Insects alone are responsible for more than 50% of these losses.

For the past 40 years the principal weapons used against these troublesome insects have been chemical insecticides. The majority of such materials used during this period have been synthetic organic chemicals discovered, synthesized, developed, and marketed by commercial industry. In recent years, environmental concerns, regulatory restraints, and problems of pest resistance to insecticides have combined to reduce the number of materials available for use in agriculture. Replacement materials reaching the marketplace have been relatively few due to increased costs of development and the general lack of knowledge about new classes of chemicals having selective insecticidal activity.

In response to these trends, it is gratifying to note that scientists in both the public and private sectors have given significant attention to the discovery and evaluation of natural products as fertile sources of new insecticidal agents. Not only are these materials directly useful as insect control agents, but they also serve as models for new classes of chemicals with novel modes of action to attack selective target sites in pest species. Such new control agents may also be less susceptible to the cross resistance difficulties encountered with most classes of currently used synthetic pesticide chemicals to which insects have developed immunity.

Natural products originating in plants, animals, and microorganisms are providing a vast source of bioactive substances. The rapid development and application of powerful analytical instrumentation, such as mass spectrometry, nuclear magnetic resonance spectroscopy, gas chromatography, high performance liquid chromatography, immuno- and other bioassays, have greatly facilitated the identification of miniscule amounts of active biological chemicals isolated from natural sources. These new scientific approaches including biotechnology and tools are addressed and reviewed extensively in these volumes.

Some excellent examples of success in this research involve the discovery of insect growth regulators, especially the so-called juvenoids, which are responsible for control of insect metamorphosis, reproduction, and behavior. Pheromones which play essential roles in insect communication, feeding, and sexual behavior represent another important class of natural products holding great promise for new pest insect control technology. Another exciting approach is the development, commercialization, and use of safe, effective, naturally occurring microorganisms that have been formulated into microbial insecticides for the suppression and control of insect pests. All of these are discussed in detail in Volumes dealing with insects.

It is hoped that the scientific information provided in these volumes will serve researchers in industry, government, and academia, and stimulate them to continue to seek even more useful natural materials that produce effective, safe, and environmentally acceptable materials for use against insect pests affecting agriculture and mankind.

Orville G. Bentley
Assistant Secretary
Science and Education
U.S. Department of Agriculture

PREFACE

Our current generation realized a dream initially envisioned and prestructured by our predecessors. There has always been an expectation that disease-producing microorganisms of insects (i.e., entomopathogens) could be used to control insect pests. This expectation can be found in the descriptions of insect maladies from early Greek and Roman literature to the first conceptualization by Agostino Bassi (1834) that a microorganism (a fungus, *Beauveria bassiana*) can cause a disease (in silkworms).

Although Pasteur and LeConte both suggested (ca. 1874) that microorganisms might be used to control insect pests, the first concerted attempt to actually use an entomopathogen (a fungus, *Metarhizium anisopliae*) to control an insect pest (wheat cockchafer and sugar beet curculio) was demonstrated by E. Metchnikoff and I. Krassilstchik (ca. 1879). No major advancement occurred thereafter until the 1940s when R. T. White and S. R. Dutky demonstrated that a milky-disease bacterium (*Bacillus popilliae*) could be mass produced and effectively used to control grubs of the Japanese beetle. Significant major advancements, however, have occurred within the last three decades in the development, registration, commercialization, and use of a microbial insecticide (i.e., insect control agents formulated from microorganisms or their products). At least one of each major type of microorganism was developed and registered as a commercial microbial insecticide during this period. These registrations include the commercialization of a bacterium, *B. thuringiensis* (1961) and the first-time labeling and commercialization of a virus, *Baculovirus heliothis* (1974); a protozoan, *Nosema locustae* (1980); and a fungus, *Hirsutella thompsonii* (1981). If number of trade name products is considered then there are about three dozen commercial microbial pesticides available today. As one specific example...varieties of *B. thuringiensis* are used to formulate at least a dozen different commercial products for the control of either mosquitoes, beetles, caterpillars, spider mites, or lygus bugs.

The characterization and documentation of these significant advancements in the development and use of entomopathogens are the objectives of this two-part treatise on Microbial Insecticides. Scientists and others interested in an extensive review and bibliography will find these handbooks ideal both as a source of information and as a reference source. Authoritative researchers in their respective specialities have followed a common outline and format to bring consistent, thorough coverage to each presentation. Part A deals with entomogenous protozoa and fungi; Part B treats the entomogenous bacteria and viruses. Each presentation begins with an historical synopsis of the group and concludes with a glance into the future. Between these two extremes are discussions of potential candidates for development into microbial insecticides; specificity and virulence; stability and persistence in the environment; methods of production; and specific examples of uses of microorganisms as microbial insecticides. The text includes synoptic tables and is illustrated with photographs and drawings that provide excellent pictorial examples of each group of microorganism.

Carlo M. Ignoffo
Columbia, Missouri

ACKNOWLEDGMENT

The editor expresses his heartfelt appreciation to all the contributing authors for their patience and diligence. The editor also expresses his sincere gratitude to Flori Ignoffo, Dolores Reddick, and the many who assisted in any way in either the preparation, editing, reviewing, or proofing of the texts. Other specific acknowledgments by name are included by the authors within each section. Lastly we express out thanks to the staff of CRC Press, Inc. for all their untiring efforts.

DEDICATION

These volumes are dedicated to all scientists and their support groups, be they technical, familial, or spiritual, that have contributed to the development of microbial insecticides.

THE EDITOR-IN-CHIEF

N. Bhushan Mandava, holds B.S., M.S., and Ph.D. degrees in chemistry and has published over 150 papers including two patents, several monographs and reviews, and books in the areas of pesticides and plant growth regulators and other natural products as well as analytical instrumentation. As editorial advisor, he has edited three special issues on countercurrent chromatography for the *Journal of Liquid Chromatography*. He is now a consultant in pesticides and drugs. Formerly, he was associated with the U.S. Department of Agriculture and the Environmental Protection Agency as Senior Chemist. He has been active in several professional organizations, was President of the Chemical Society of Washington, and serves as Councilor of the American Chemical Society.

THE EDITOR

Carlo M. Ignoffo, Ph.D. is an Insect Pathologist and Director of the Biological Control of Insects Research Laboratory, U.S. Department of Agriculture, Agricultural Research Service, Columbia, Missouri, and also holds an appointment as Adjunct Professor, Department of Entomology, University of Missouri-Columbia.

Dr. Ignoffo received his B.S. degree in Zoology from Northern Illinois University in 1950 and his M.S. and Ph.D. in Entomology in 1954 and 1957, respectively, from the University of Minnesota. He served in the U.S. Army, during the period 1954—1956 (Chemical Corps) conducting research on vectors of agents infectious to man and animals. Since the completion of his doctorate in 1957, he has been involved in teaching, research, and research leadership as Associate Professor of Entomology-Biology from 1957—1959 (Iowa Wesleyan College); a Research Entomologists/Pathologist from 1959—1965 (Entomology Research Division, USDA); a Research Pathologist and Administrator in industry from 1965—1971 (Bioferm/International Minerals and Chemical Corporation); and as a Research Leader/Laboratory Director (in his current appointment) since 1971.

Dr. Ignoffo is a charter member of the Society for Invertebrate Pathology and is or was a member of the American Association for the Advancement of Science, Entomological Society of America, American Institute of Biological Sciences, International Organization for Biological Control, and the Tissue Culture Association of America. He served as a technical advisor and research consultant to industry as well as International and National policy-making groups concerned with the development, production, use, and safety of microbial insecticides.

Research accomplishments are documented by authorship or coauthorship of more than 230 articles in scientific journals, including book chapters and review articles. He is a patentee in the field and was responsible for the isolation, commercialization, and registration of the world's first viral commercial pesticide. Dr. Ignoffo's life-time research focus has been in the development and use of microbial insecticides, especially viral insecticides. His current major research interest is in the characterization and manipulation of the genome of insect viruses in order to increase their ease of production, effectiveness, and persistence as viral insecticides.

ADVISORY BOARD

Clayton C. Beegle, Ph.D.
Research Entomologist
U.S. Department of Agriculture
Cotton Insects Research
Brownsville, Texas

Wayne M. Brooks, Ph.D.
Professor
Department of Entomology
North Carolina State University
Raleigh, North Carolina

Clayton W. McCoy, Ph.D.
Professor
Citrus Research and Education Center
Institute of Food and Agricultural
 Sciences
University of Florida
Lake Alfred, Florida

CONTRIBUTORS

Drion G. Boucias, Ph.D.
Associate Professor
Department of Entomology
Institute of Food and Agricultural
  Sciences
University of Florida
Gainesville, Florida

Wayne M. Brooks, Ph.D.
Professor
Department of Entomology
North Carolina State University
Raleigh, North Carolina

Clayton W. McCoy, Ph.D.
Professor
Citrus Research and Education Center
Institute of Food and Agricultural
  Sciences
University of Florida
Lake Alfred, Florida

Robert A. Samson, Ph.D.
Professor of Mycology
Centraalbureau voor Schimmelcultures
Baarn, The Netherlands

TABLE OF CONTENTS

Entomogenous Protozoa 1
Wayne M. Brooks

Entomogenous Fungi 151
Clayton W. McCoy, Robert A. Samson, and Drion G. Boucias

Index 237

ENTOMOGENOUS PROTOZOA

Wayne M. Brooks

INTRODUCTION

The protozoa are a diverse and heterogeneous group of microorganisms. Many are associated with insects in relationships ranging from commensalistic to pathogenic. Those protozoa pathogenic for insects will be discussed with emphasis on species with demonstrated or promising potential as microbial insecticides. Unlike many of the entomopathogenic bacteria, viruses, or fungi, few of the entomogenous protozoa are highly virulent or fast acting. Most species produce chronic infections characterized by a general debilitation of the host. Thus, protozoa are most considered as candidates for long-term application or introduction programs and offer little potential as short-term, quick-acting microbial insecticides.[1-5]

The potential for utilizing protozoa as microbial insecticides must be visualized in the context of their general characteristics as pathogens of insects. Infection is usually initiated with peroral ingestion of spores, cysts, or other stages by a larval stage of the host. Some species also are commonly transmitted on the surface of the egg (transovum) or within the egg (transovarian) of their host. In special cases, some protozoa are transmitted mechanically via the ovipositional activities of hymenopterous parasites. Infections are generally sublethal and chronic in nature and pathognomonic signs or symptoms are seldom exhibited by infected hosts. Most infections are characterized by such nonspecific signs and symptoms of disease as sluggishness, irregular growth, loss of appetite, malformed larvae, pupae, or adults, or adults with reduced vigor, fecundity, and longevity. The period of infection may only be slightly shorter than the normal life span of the host, and in some cases larval or pupal periods are lengthened. While infected hosts may die prematurely, the actual cause of death is difficult to determine. Competition for nutrients may be important where the extracellular phase of protozoan development is extensive. However, this possibility as well as the possibility that some protozoa produce toxins has not been thoroughly investigated. Death of the host is most likely the result of mechanical destruction of cells by developing stages of the protozoan.

With few exceptions, most protozoa must be produced in vivo using whole-organism technology. This is inherently more difficult and expensive than is the production of bacteria and fungi using fermentation technology. Long-term storage of infective stages of protozoa is difficult and persistence of unprotected stages, under field conditions, is of short duration. While the host range of only a few species has been examined extensively, it has become increasingly evident that most species are not as host specific as was once generally considered. The host spectrum of a few species, however, is sufficiently broad to encourage their production as microbial insecticides. Limited assessment of infectivity to nontarget organisms and the obvious need for increased research in the areas of formulation and application technology mark the infant state of use of protozoa as microbial insecticides. Additional comprehensive and general discussions of protozoa as pathogens of insects are presented in earlier reviews by Paillot,[6] Steinhaus,[7,8] Aoki,[9] Lipa,[10,11] Weiser,[12,13] and Brooks.[14]

HISTORICAL BACKGROUND

The entomogenous protozoa played a prominent role in the development of the field of insect pathology although most of the limited efforts to utilize protozoa as microbial control agents have been conducted since the early 1950s. Reports in the early 1800s of observations and descriptions of gregarines associated with insects are discussed briefly by Steinhaus[15] and similar early reports of entomogenous flagellates are presented by Wallace.[16] More significantly, Pasteur[18] (along with the earlier observations of Bassi[17] on the muscardine disease of silkworms caused by a fungus) helped establish insect pathology as a science through his classical study of a protozoan infection (pebrine) of the silkworm *Bombyx mori*. Pasteur showed that the peculiar microscopic corpuscles (spores) seen by earlier students of silkworm diseases were the cause of the disease and made observations that led to a method for obtaining disease-free silkworms. While the true nature of the parasite as a protozoan was unknown to Pasteur or to Naegeli[19] who named the parasite *Nosema bombycis*, subsequent observations by Balbiani[20] and Stempell[21] established the identity of the parasite as a protozoan in the order Microsporida. It is also significant that Pasteur[22] made one of the first definite suggestions that microorganisms might be used to control insects. He suggested the use of ''les corpuscles'' of pebrine against the grape phylloxera, a pest threatening grape production in France at the time. More detailed accounts of Pasteur's contributions to insect pathology are presented by Steinhaus.[8,15]

Many protozoa were described from insects and other invertebrates in the early 1900s and some, such as *Ameson pulvis* from the green crab *Carcinus maenas*, were suggested as possible biological control agents.[23] However, the first apparent attempt to use a protozoan as a microbial control agent was carried out by Taylor and King.[24] Utilizing cyst-containing feces collected from grasshoppers infected with the amoeba *Malameba locustae*, they applied the amoeba in a mixture with bran and molasses along roads and fences. A low incidence of infection was found in grasshoppers after 8 weeks but insufficient data were collected to determine if any control was achieved. Serious attempts were not made until the early 1950s to use protozoa as microbial control agents, perhaps influenced in part by the chronic nature of most protozoan infections and the often erratic results in tests involving some entomogenous fungi and bacteria. Limited attempts utilizing various microsporidia were carried out in the U.S. by Hall[25] and Zimmack et al.[26] and in Czechoslovakia by Weiser and Veber.[27,28] The first extensive tests to utilize protozoa as microbial control agents were carried out by McLaughlin and his associates in the mid 1960s[29-33] and involved the neogregarine *Mattesia grandis* and the microsporidium *Nosema gasti* against the cotton boll weevil, *Anthonomus grandis*, in the southern U.S. More recently, the mass production and successful use of the microsporidium *Nosema locustae* against grasshoppers on rangelands[34-38] led to registration of this protozoan in 1980 by the U.S. Environmental Protection Agency. This was the first protozoan commercially produced as a microbial insecticide in the U.S. Earlier reviews on protozoa as microbial control agents have been presented by McLaughlin,[1] Pramer and Al-Rabiai,[3] Tanada,[39] Brooks,[4] Henry,[40] Wilson,[41] and Canning.[5]

TAXONOMY

As a heterogeneous group of essentially single-celled, eukaryotic organisms, the protozoa have been usually placed together as a matter of convenience rather than as a natural grouping of related organisms. They have been treated classically as a single phylum and traditionally divided into two subphyla with five classes based on locomotory organelles. However, the classification of the protozoa has undergone considerable revision at the suprafamilial level in recent years.[42,43] In the most recent report of the Committee on Systematics and Evolution of the Society of Protozoologists,[43] the protozoa are treated as a subkingdom and seven phyla are recognized, five of which contain entomogenous protozoa. A synopsis of this classification scheme showing the major taxa that include entomogenous protozoa is presented in Table 1. In the subsequent discussion, emphasis will be placed on those taxa which include species pathogenic for insects. Useful keys to most of the common genera of entomogenous protozoa are provided by Poinar and Thomas.[44,44a]

Table 1
CLASSIFICATION SCHEME OF ENTOMOGENOUS PROTOZOA

Taxa[a]	Representative genera
Phylum Sarcomastigophora	
Subphylum Mastigophora	
Class Zoomastigophorea	
Order Kinetoplastida	*Herpetomonas, Crithidia, Leptomonas*
Order Retortamonadida	*Retortamonas*
Order Diplomonadida	*Octomitus*
Order Oxymonadida	*Oxymonas, Pyrosonympha*
Order Trichomonadida	*Trichomonas, Devescovina*
Order Hypermastigida	*Trichonympha, Joenia*
Subphylum Sarcodina	
Superclass Rhizopoda	
Class Lobosea	
Subclass Gymnamoebia	
Order Amoebida	*Malpighiella, Malameba, Malpighamoeba*
Phylum Apicomplexa	
Class Sporozoea	
Subclass Gregarinia	
Order Eugregarinida	*Gregarina, Ascogregarina*
Order Neogregarinida	*Mattesia, Farinocystis, Ophryocystis*
Subclass Coccidia	
Order Eucoccidiida	*Adelina, Legerella, Barrouxia*
Phylum Microspora	
Class Microsporea	
Order Minisporida	*Chytridiopsis, Hessea*
Order Microsporida	*Nosema, Pleistophora, Amblyospora*
Phylum Ascetospora	
Class Stellatosporea	
Order Balanosporida	*Haplosporidium*
Phylum Ciliophora	
Class Kinetofragminophorea	
Subclass Vestibulifera	
Order Trichostomatida	*Balantidium*
Subclass Suctoria	
Order Suctorida	*Discophrya, Rynchophrya*
Class Oligohymenophorea	
Subclass Hymenostomatia	
Order Hymenostomatida	*Tetrahymena, Lambornella*
Subclass Peritrichia	
Order Peritrichida	*Epistylis, Opercularia*
Class Polyhymenophorea	
Subclass Spirotrichia	
Order Heterotrichida	*Nyctotherus*

[a] Compiled primarily from Reference 43.

The phylum Sarcomastigophora includes those protozoa referred to as flagellates and amoebae which generally possess flagella and/or pseudopodia (sometimes both types of organelles are present) and only a single type of nucleus. They typically do not form spores. Flagellates (subphylum Mastigophora) typically possess one or more flagella and usually reproduce asexually by longitudinal binary fission. The entomogenous flagellates are included in the class Zoomastigophorea characterized by the lack of chromatophores and being predominantly parasitic. The largely mutualistic and commensalistic flagellates which occur in the intestinal track of termites, roaches, and a few other insects are included in the orders

Retortamonadida, Diplomonadida, Oxymonadida, Trichomonadida, and Hypermastigida. Most of the entomogenous, pathogenic flagellates are included in the order Kinetoplastida, family Trypanosomatidae, and are generally referred to as trypanosomatids. These flagellates are usually elongate and slender in shape and possess a single flagellum either free or attached to the body by an undulating membrane. The flagellum arises within a reservoir and is associated with a contractile vacuole and a Feulgen-positive kinetoplast. While a large number of trypanosomatids are associated with insects, relatively few are pathogenic. The strictly entomogenous trypanosomatids are included in the genera *Herpetomonas*, *Crithidia*, *Blastocrithidia*, and *Rhynchoidomonas*. Many of the flagellates of the genus *Leptomonas* are also entomogenous, and other species are also known from other invertebrates including protozoa, nematodes, and mollusks. The digenetic flagellates of the genera *Leishmania* and *Trypanosoma* have both an invertebrate and vertebrate host while those in the genus *Phytomonas* have a plant and invertebrate host. A list of representative species of the entomogenous trypanosomatids is included in Table 2. An extensive list of the entomogenous trypanosomatids from insects and a review of the systematics of the group was presented by Wallace.[16] A more recent review on the biology of these flagellates from arthropods has also been published.[45]

Amoebae (subphylum Sarcodina) possess pseudopodia, a naked body, and reproduce asexually by binary or multiple fission, often in cysts. Flagella, when present, are usually restricted to developmental stages and none are present in the entomogenous amoebae. Nearly all of the entomogenous amoebae occur in the families Amoebidae and Endamoebidae, order Amoebida characterized by a single nucleus and lobose pseudopodia. The cytoplasm is generally divided into a granular endoplasm filled with inclusions and a hyaline ectoplasm. The strictly entomogenous amoebae belong to the family Amoebidae and are represented by the genera *Malameba*, *Malpighamoeba*, and *Malpighiella*. Among the parasitic Endamoebidae, a few entomogenous species are known in the genera *Entamoeba*, *Endamoeba*, *Endolimax*, and *Dobellina*. These amoebae are considered as commensals of the alimentary tract of various insects, primarily cockroaches. However, Purrini and Halperin[76a] recently described an *Endamoeba* sp. in the gut of the bark beetle, *Orthotomicus erosus*, that they suggested was harmful despite the lack of external symptoms in heavily infected adults. Representative species of entomogenous amoebae are shown in Table 2, only two of which, *Malameba locustae* and *Malpighamoeba mellificae*, produce significant pathological effects in insects.

As the causative agent of amoebic disease of the honey bee, *Apis mellifera*, *M. mellificae* is not a candidate microbial control agent, since the honey bee is a beneficial insect. However, *Malameba locustae* has been shown to be infectious for a wide range of grasshopper species with sufficient virulence to be considered as a potential microbial control agent. The taxonomic status of this species and the other entomogenous amoebae is in need of re-evaluation and probable revision. Most authorities, however, have continued to accept the placement of the strictly entomogenous genera *Malameba*, *Malpighamoeba*, and *Malpighiella* within the family Amoebidae, although Harry and Finlayson[79] suggested that *Malameba locustae* may belong to the family Schizopyrenidae based on preliminary observations on ultrastructure. Purrini[80] recently described another entomogenous amoeba, *Malameba scolyti*, from the bark beetles *Dryocoetes autographus* and *Hylurgops palliatus*.

The phylum Apicomplexa contains those protozoa typically referred to as sporozoa and are characterized by an apical complex (visible with the electron microscope) present at some stage which generally consists of polar ring(s), rhoptries, micronemes, and conoid and subpellicular microtubules. Microspores are generally also present at some stage, cilia are absent, sexuality is by syngamy, and all species are parasitic. The entomogenous sporozoa belong to the class Sporozoa characterized by a well developed apical complex, sexual and asexual reproduction, and spores or oocysts containing infective sporozoites produced by

Table 2
A LIST OF THE PRINCIPAL GROUPS AND GENERA OF ENTOMOGENOUS PROTOZOA, ALONG WITH REPRESENTATIVE SPECIES AND SELECTED REFERENCES[a]

Group	Genus	Representative species	Selected Ref.[b]
Trypanosomatids	Blastocrithidia	caliroae	46
		gerridis	47, 48
		leptocoridis	49, 50
		raabei	51
	Crithidia	acanthocephali	52
		fasciculata	16, 53, 54
		luciliae	55, 56
		oncopelti	57, 58
		tabani	59
	Herpetomonas	ludwigi	60, 61
		muscarum	16, 62, 63
		swainei	64
	Leptomonas	ctenocephali	65, 66
		oncopelti	57, 67
		pyrrhocoris	68—70
		serpens	71
		seymouri	72
	Rhynchoidomonas	drosophilae	73, 74
		luciliae	75
		siphunculinae	76
Amoebae	Amoeba	chironomi	81
	Dobellina	mesnili	82, 83
	Malameba	locustae	24, 77, 79, 84, 85
		scolyti	80
	Malpighamoeba	mellificae	86, 87
	Malpighiella	refringens	88
Eugregarines	Ascogregarina	armigerei	109
		barretti	110
		brachyceri	110a
		chagasi	111
		clarki	112
		culicis	109, 110, 113
		galliardi	114
		lanyuensis	109
		legeri	115
		mackiei	116
		taiwanensis	109
		tripteroidesi	117
Neogregarines	Caulleryella	apiochaetae	118
		pipientis	119
	Coelogregarina	ephestiae	106, 107
		orchopiae	106, 120
	Farinocystis	tribolii	121
	Gigaductus	agoni	122
		anchi	103, 123
		steropi	124
	Lipocystis	polyspora	125
	Lipotropha	macrospora	126
		microspora	126
	Lymphotropha	tribolii	102
	Machadoella	triatomae	127
	Mattesia	dispora	128

Table 2 (continued)
A LIST OF THE PRINCIPAL GROUPS AND GENERA OF ENTOMOGENOUS PROTOZOA, ALONG WITH REPRESENTATIVE SPECIES AND SELECTED REFERENCES[a]

Group	Genus	Representative species	Selected Ref.[b]
		grandis	129—131
		trogodermae	106, 132
	Menzbieria	chalcographi	133
	Orphryocystis	dendroctoni	134
		elektroscirrha	135
	Schizocystis	gregarinoides	136
		legeri	137
	Syncystis	mirabilis	138
	Tipulocystis	maximae	60
Coccidia	Adelina	cryptocerci	139
		sericesthis	140
	Barrouxia	bellostomatis	141
		schneideri	142
	Chagasella	alydi	143
	Ithania	wenrichi	144
	Legerella	hydropori	145
		parva	146
	Rasajeyna	nannyla	147
Microsporidia[c]	Amblyospora	californica	154, 155
		inimica	155, 156
		minuta	155, 157
		opacita	155, 158
	Auraspora	canningae	159
	Bohuslavia	asterias	159a, 159b
		simulii	159b, 159c
	Burenella	dimorpha	160, 161
	Buxtehudea	scaniae	162
	Campanulospora	deliae	162a
	Caudospora	pennsylvanica	163, 164
		polymorpha	165, 166
		simulii	167, 168
	Chapmanium	cirritus	155
		dispersus	155a
	Chytriodiopsis	socius	169, 170.
		typographi	134, 171
	Cougourdella	polycentropi	172
		rhyacophilae	173
	Culicospora	magna	149, 157, 174
	Culicosporella	lunata	149, 175
	Cylindrospora	chironomi	149c
		fasciculata	149d
	Cystosporogenes	operophterae	175a
	Duboscqia	chironomi	176
		coptotermi	177
		legeri	178, 179
	Episeptum	inversum	179a
	Evlachovaia	chironomi	149c
	Golbergia	spinosa	149, 180
	Gurleya	aeschnae	181
		chironomi	182
		legeri	183
	Hazardia	milleri	149, 184

Table 2 (continued)
A LIST OF THE PRINCIPAL GROUPS AND GENERA OF ENTOMOGENOUS PROTOZOA, ALONG WITH REPRESENTATIVE SPECIES AND SELECTED REFERENCES[a]

Group	Genus	Representative species	Selected Ref.[b]
	Helmichia	aggregata	185
	Hessea	squamosa	186a
	Hirsutusporos	austrosimulii	186d
	Hyalinocysta	chapmani	155
		expilatoria	186b
	Issia	globulifera	186c
		trichopterae	149, 167
	Janacekia	debaisieuxi	167a, 167b
	Jirovecia	brevicauda	149, 187
	Mitoplistophora	angularis	188
	Microsporidium[d]	goeldichironomi	189
		hyphantriae	28, 152, 190
	Neoperezia	chironomi	191
	Nosema	algerae	192—194
		apis	195—198
		bombycis	19, 199, 200
		fumiferanae	201, 202
		gasti	203, 204
		locustae	205—207
		pyrausta	208—213
	Octosporea	carloschagasi	214
		muscaedomesticae	215, 216a
	Orthosoma	operophterae	216b, 216c
	Parathelohania	anophelis	155, 157, 217
		legeri	155, 217—219
		obesa	155, 157, 217, 220
	Pegmatheca	simulii	155
	Pilosporella	chapmani	155
		fishi	155
	Pleistophora	californica	221
		kudoi	222, 223
		schubergi	224—227
	Polydispyrenia	simulii	227a, 250
	Resiomeria	odonatae	149d
	Semenovaia	chironomi	149c
	Stempellia	mutabilis	228, 229
	Striatospora	chironomi	149c
	Systenostrema	tabani	155
	Telomyxa	glugeiformis	228, 230
		orae	230a
	Thelohania[e]	fibrata	231, 232
		pristiphorae	233
		pyriformis	157, 220
	Toxoglugea	chironomi	234, 235
		variabilis	236
		vibro	237, 238
	Trichoduboscqia	epeori	239—241
	Tuzetia	ecdyonuri	167a, 242
		schneider	167a, 228
	Unikaryon	bouixi	243a
		minutum	243

Table 2 (continued)
A LIST OF THE PRINCIPAL GROUPS AND GENERA OF ENTOMOGENOUS PROTOZOA, ALONG WITH REPRESENTATIVE SPECIES AND SELECTED REFERENCES[a]

Group	Genus	Representative species	Selected Ref.[b]
	Vairimorpha	necatrix	244, 245
		plodiae	246—248
	Vavraia	culicis	149, 167, 249, 250
	Weiseria	laurenti	251, 252
		sommermanae	164
Ciliates	Lambornella	clarki	258, 259
		stegomyiae	261
	Tetrahymena	chironomi	257, 262
		rotunda	260a
		dimorpha	260b
		sialidos	260i

[a] Exclusive of the commensalistic flagellates primarily associated with termites and blattids, and the predominantly commensalistic eugregarines of coleopterans, orthopterans, dipterans, and other insects.

[b] The selected references generally refer to papers dealing with the description and/or life cycle of the species or in some cases to a major taxonomic treatise which indicates the current status of the species.

[c] Except for the new general and species described recently, an annotated list of the species of microsporidia is included in the comprehensive treatise of Sprague.[152] Many species are also reviewed in the monograph of Weiser.[212]

[d] A collective group without established attributes in which identifiable species can be placed provisionally because their generic status is uncertain.[148]

[e] According to Hazard and Oldacre,[155] this genus is represented by species found in decapod crustaceans. The entomogenous species included here are considered as being of doubtful status.

sporogony. A majority of the described species of entomogenous sporozoa are included in the subclass Gregarinia and are referred to as gregarines. They are characterized by mature gamonts which are extracellular and relatively large, attachment organelles (the mucron or epimerite) may be present, gametes generally similar (isogamous), gamonts which undergo syzygy, the formation of oocysts by zygotes within gametocysts, and a life cycle characteristically consisting of gametogony and sporogony. Of the approximately 1400 known species,[89] about 93% belong to the order Eugregarinida, most of which are harmless commensals of the digestive tracts of insects. The gregarines in this order are characterized by a life cycle involving gametogony and sporogony but not merogony (i.e., there is no asexual reproduction). Mobile species move by gliding or undulation of longitudinal ridges of the body surface. As this group of entomogenous protozoa is exceptionally large in number (1300 species in 195 genera and 29 families) and is largely commensalistic or of low virulency to insects, no attempt will be made to delineate the families, genera, or species involved. Along with the early taxonomic treatments on eugregarines by Watson[90] and Kamm,[91] checklists and revisions of many genera and species of this order have been provided in a recent series of papers by Levine.[92-97] Grassé[98] provides a classification scheme of the higher taxa, and a recent general discussion of gregarines is provided by Manwell.[99]

Among the eugregarines considered to be somewhat pathogenic to their hosts, the species of the genus *Ascogregarina* have received the most attention. These eugregarines (see Table 2) are parasites of mosquitoes and other insects,[94] and at least one species, *A. culicis*, has been evaluated as a potential microbial control agent for mosquitoes.

The gregarines of the order Neogregarinida are usually referred to as neogregarines and are characterized by one or more merogonic schizogonies in addition to the gametogonic and sporogonic schizogonies characterizing the life cycle of the eugregarines. The merogony is considered to have been acquired secondarily and to compensate for the low number of spores produced during sporogony. While several classifications of the Neogregarinida have been proposed, the most recent are those of Grassé[98] and Weiser.[100,101] In Grassé's system, five families are proposed based on the morphology of trophozoites and meronts and their habitat. Weiser divided the group into two families based on the number of merogonic schizogonies. There is less difference in the two systems at the generic level although Weiser recognizes 16 genera in contrast to only 11 by Grassé. As shown in Table 2, there are presently 14 genera which contain entomogenous neogregarines. *Lymphotropha* is a recently described genus,[102] and the genus *Gigaductus* was transferred from the Eugregarinida to Neogregarinida by Ormières.[103] Although the genus *Coelogregarina* is considered by Weiser[104,105] to be synonymous with *Mattesia*, Canning[106] agreed with Ghélélovitch[107] in his original separation of this genus from that of *Mattesia*. A key to most of these genera is provided by Weiser and Briggs,[108] and most are briefly discussed by Weiser.[12,13] Many neogregarines are pathogenic to their hosts and two species, *Mattesia grandis* and *M. trogodermae*, have received considerable attention as microbial control agents.

The sporozoa of the subclass Coccidia have mature gamonts which are small, typically intracellular, and lack a mucron or epimerite. Gametes are characteristically anisogamous and the life cycle involves merogomy, gamogony, and sporogony. Most coccidia are parasites of vertebrates and some species are transmitted by insect vectors. A few species are entomogenous, their occurrence in insects being considered as accidental.[12] The entomogenous coccidia occur in the order Eucoccidiida with a merogonic schizogony. In addition to such vertebrate pathogenic genera as *Haemoproteus* and *Plasmodium* which are vectored by insects, a few strictly entomogenous species are known in several genera of the suborders Adeleina and Eimeriina (Table 2).

The phylum Microspora is characterized by unicellular spores with a polar tube (polar filament) or its rudiment, the absence of mitochondria, and an intracellular parasitic habit. Microsporidia are the most important protozoan pathogens of insects and include a large number of entomogenous species. Many cause significant pathology to a wide range of economically important pest and beneficial insects. Thus, most of the protozoa with promising or demonstrated potential as microbial insecticides are included in this group and will be the focus of attention in the balance of this review. An excellent historical review of the evolution of microsporidian classification was presented by Sprague.[148] Sprague[148] and Weiser[149] recently proposed, almost simultaneously, new taxonomic systems for the microsporidia which are contrasted at the suprageneric level in Table 3, along with the more recent proposal of Issi.[149c]* Although the Committee on Systematics and Evolution of the Society of Protozoologists[43] essentially follows the system of Sprague[148] at the suprafamilial level, Hazard et al.[151] presented an excellent discussion of problems involved in explaining why they believe a stable and comprehensive system for classification of the microsporidia cannot be established at this time. Also cognizant of the unsatisfactory status of microsporidian systematics at the suprageneric level, Larsson[149e] recently presented character-state trees based on ultrastructural and developmental characteristics which he proposes to serve as a focal point for an alternative classification system of the microsporidia. And, while the desirability of a new classification system is readily apparent, most microsporidiologists appear to be following the system of Sprague[148] or have confined their taxonomic considerations to the generic level. Thus, Hazard et al.[151] present a key to the genera containing

---

* Further recent modifications by Sprague[150] and Weiser[149a] are also included in Table 3. However, the suggested changes in the order Chytridiopsida by Weiser[149b] are not shown as they were also not included in his most recent taxonomic proposal.[149a]

Table 3
CLASSIFICATION SYSTEMS FOR THE MICROSPORIDIA

System I[a]	System II[b]	System III[c]
Phylum Microspora	Phylum Microspora	Phylum Microsporidia
Class Rudimicrosporea	Class Metchnikovellidea	Class Microsporidea
Order Metchnikovellida	Order Metchnikovellida	Subclass Metchnikovellidea
Family Metchnikovellidae	Family Metchnikovellidae	Order Metchnikovellida
Class Microsporea	Order Chytridiopsida	Family Metchnikovellidae
Order Minisporida	Family Chytridiopsidae	Subclass Chytridiopsidea
Family Chytridiopsidae		Order Chytridiopsida
Family Hesseidae	Order Hesseida	Family Chytridiopsidae
Family Burkeidae	Family Hesseidae	Family Buxtehudeidae
Family Buxtehudeidae		
	Class Microsporidea	Subclass Cylindrosporidea
Order Microsporida	Order Pleistophoridida	Order Cylindrosporida
Suborder Pansporoblastina	Family Pleistophoridae	Family Striatosporidae
Family Pleistophoridae	Family Thelohaniidae	Family Cylindrosporidae
Family Pseudopleistophoridae	Family Amblyosporidae	
Family Duboscqiidae	Family Culicosporidae	Subclass Nosematidea
Family Thelohaniidae		Order Culicosporida
Family Burenellidae	Order Nosematidida	Family Culicosporidae
Family Amblyosporidae	Family Nosematidae	Family Golbergiidae
Family Culicosporidae	Family Mrazekidae	
Family Gurleyidae		Order Glugeida
Family Telomyxidae		Family Pereziidae
Family Tuzetiidae		Family Glugeidae
		Family Thelohaniidae
Suborder Apansporoblastina		
Family Glugeidae		Order Nosematida
Family Spraguidae		Family Amblyosporidae
Family Unikaryonidae		Family Burenellidae
Family Pereziidae		Family Spraguidae
Family Cougourdellidae		Family Nosematidae
Family Caudosporidae		Subfamily Nosematinae

Subfamily Pseudopleistophorinae

Family Nosematidae
Family Mrazekiidae

a According to Sprague.[148,150]
b According to Weiser.[149,149a]
c According to Issi.[149c]

entomogenous microsporidia, a list of which, along with more recently described genera, is also included in Table 2. An annotated list of the microsporidia is included in the comprehensive treatise of Sprague.[152] Most of the entomogenous microsporidia occur in the order Microsporida of Sprague,[148] the class Microsporididea of Weiser,[149] or the subclass Nosematidea of Issi[149c] and are the classical forms where there is a tendency toward maximum development of spore organelles (polar tubes and polaroplasts) and the presence or absence of a sporophorous vesicle. The majority of the species, considered as potential microbial control agents, occur in the genus *Nosema*. However, since this genus is a very large and heterogeneous group,[152,153] it is likely that many species in future taxonomic works will be assigned to other genera.

The haplosporidia are a group of inadequately known protozoa whose classification has undergone almost as much revision as have the microsporidia with which they were once grouped. Sprague[253] recently reviewed the evolution of haplosporidian classification and proposed their elevation to the phylum rank. This action has been accepted by the Committee on Systematics and Evolution of the Society of Protozoologists[43] and the phylum Ascetospora includes those protozoa whose spores are multicellular (or possibly unicellular), with one or more sporoplasms, without polar capsules or filaments, and are all parasitic. The spores of the more typical haplosporidia from mollusks or various marine or fresh water invertebrates (genera *Haplosporidium* and *Minchinia*) are characterized by a stage in sporogenesis that resembles an acorn in its cupule and an anterior orifice covered externally by an operculum.

Most of the relatively few species that have been described from insects are inadequately known, particularly from an ultrastructural standpoint. *Nephridiophaga blattellae* is probably the best known of the entomogenous species generally considered to be a haplosporidium. The assignment of this species to the Haplosporidia, however, was based only on the fact that the spores are of a simple type without a polar tube. Sprague[253] indicates that none of the positive characters that are distinctive for typical haplosporidia were demonstrated in *N. blattellae* and suggests that the family Nephridiophagidae (genera *Nephridiophaga* and *Physcosporidium*) be rejected from the Ascetospora along with the genus *Coleospora*. In an earlier revision of the genus *Haplosporidium*, Sprague[254] had previously rejected the entomogenous species *H. bayeri* and *H. typographi* from the genus since neither species apparently has spores with a lid or operculum. Weiser[134] reassigned *H. typographi* to the microsporidian genus *Chytridiopsis*. The taxonomic status of two other more recently described species, *H. tipulae* by Huger[255] and *H. simulii* by Beaudoin and Wills,[256] is also questionable and ultrastructural studies on both species are needed. Thus, it appears questionable as to whether or not there are any true haplosporidia associated with insects, and more research on protozoa generally considered as entomogenous haplosporidia is needed.

Ciliates (phylum Ciliophora) are characterized by the presence of simple cilia or compound ciliary organelles at some stage in their life cycle, a subpellicular infraciliature, transverse binary fission, sexuality involving conjugation, autogamy and cytogamy, and generally the presence of two types of nuclei. Most ciliates are free-living and relatively few species are pathogenic for insects. Some ciliates (*Balantidium* and *Nyctotherus*) occur commonly as commensals in the gut of cockroaches and some suctorians (order Suctorida) and peritrichs (order Peritrichida) occur as epibionts attached to the cuticular exoskeleton of various species of aquatic insects. While such associations are of little consequence from a detrimental standpoint, a few ciliates of the order Hymenostomatida cause lethal infections in various aquatic insects. The ciliates in this group are characterized by uniform body ciliature and a ventral and well-defined buccal cavity with ciliature composed of a single undulating membrane and an adoral zone of three membranelles. Entomogenous species are known in the genera *Tetrahymena* and *Lambornella* (Table 2), some of which may be potentially useful for control of mosquitoes or black flies. Although *Lambornella stegomyiae* was previously considered as a member of the genus *Tetrahymena*,[257] a recent study,[258] demonstrating the definitive presence of cuticular "invasion" cysts in the life cycle of a similar species, (described subsequently by Corliss and Coats[259] as *L. clarki*) resulted in the resurrection of the genus *Lambornella*.[259] The well known ciliate *Tetrahymena pyriformis* has been reported as an accidental parasite of various invertebrates,[257] and *T. chironomi* was described by Corliss[257] as causing lethal infections in *Chironomus plumosus*. In more recent reports, *T. rotunda*[260a] and *T. dimorpha*[260b] were described as parasites of black flies in the genus *Simulium*, and *T. sialidos* was described from larvae of the sialid, *Sialis lutaria*.[260i]

Although their definitive affinity with any established phylum of the Protozoa remains to be demonstrated, the helicosporidia are an interesting and anomalous group of entomogenous organisms considered to have some potential as microbial control agents.[263] *Helicosporidium parasiticum* was originally described by Keilin[264] from ceratopogonid flies as an unusual type of parasitic protist and he suggested its temporary inclusion with the Sporozoa. Kudo[265] placed it as a separate order, the Helicosporidia, in the class Cnidosporidia; subsequently Weiser[266,267] suggested that it should be transferred from the Protozoa and proposed its affiliation with the primitive Ascomycetes. Another helicosporidium was isolated recently from nitidulid beetles.[263] And, although initially considered to be conspecific with *H. parasiticum*,[263,268] this parasite is now thought to be a new, undescribed species of *Helicosporidium*.[269] Lindegren and Hoffman[269] also concluded that the *Helicosporidium* species is not an ascomycete and that the presence of Golgi bodies and apparent mitotic divisions of its nucleus suggest, again, an affinity to the Protozoa. Helicosporidia have also been found in several species of mosquitoes,[270-272] and a preliminary evaluation of a species isolated from mosquitoes in Thailand was recently made by Hembree.[274] While the status of the helicosporidia as protozoa needs to be confirmed, the wide host range of these parasites[263,264,266,273-275b] indicates that these organisms should be further explored as potential microbial control agents. This is further indicated by the recent isolation of another *Helicosporidium* sp. from ditch water that was found to be infectious to several species of mosquitoes and whose spores were resistant to freezing and desiccation.[275c]

POTENTIAL CANDIDATE MICROBIAL INSECTICIDES

In this section emphasis will be placed on specific protozoa (Tables 4 and 5) with demonstrated or promising potential as microbial control agents. The selection of specific candidates for discussion, however, is somewhat arbitrary since only one protozoan, *Nosema locustae*, has been registered in the U.S. by the Environmental Protection Agency as a microbial insecticide.[276] The other species selected are included because they have been tested under field conditions with varying degrees of success or have exhibited some potential for control following laboratory tests. These candidates will also be the focus of specific attention in later sections of this chapter.

Malameba locustae (King and Taylor, 1936)

This amoeba was originally described by King and Taylor[77] as *Malpighamoeba locustae* from the Malpighian tubules of the grasshoppers *Melanoplus differentialis*, *M. mexicanus*, and *M. femurrubrum*. In a subsequent publication,[24] however, they proposed a new genus, *Malameba*, for the amoeba since it differed considerably from the honey bee amoeba *Malpighamoeba mellificae*. Despite recent investigations on the bionomics and control of this amoeba, no real effort has been made to re-examine the taxonomic status of the species.*

---

* Another continuing and unresolved problem with this genus is its spelling. Although Taylor and King[24] originally spelled it as *Malameba*, many subsequent authors have referred to it as *Malamoeba*. In this text, the original spelling as presented by Taylor and King will be followed in accordance with Article 32(a) of the International Code of Zoological Nomenclature.[78]

Table 4
CANDIDATE PROTOZOA AS MICROBIAL CONTROL AGENTS

Phylum	Class	Order	Species
Sarcomastigophora	Lobosea	Amoebida	*Malameba locustae*
Apicomplexa	Sporozoea	Eugregarinida	*Ascogregarina culicis*
		Neogregarinida	*Mattesia grandis*
			M. trogodermae
Microspora	Microsporea	Microsporida	*Nosema algerae*
			N. fumiferanae
			N. gasti
			N. locustae
			N. pyrausta
			Pleistophora schubergi
			Vairimorpha necatrix
			Vavraia culicis

Description

Cysts — According to King and Taylor[77] and Henry,[84] fresh cysts (Figure 1) are oval or uniformly ellipsoidal in shape with a thick, hyaline wall that is highly refractive to light under phase microscopy. A variable number of refractive globules also are usually present. Cysts are uninucleate and there may be a prominent vacuole and one or two rodlike inclusions in the cytosome.[77] Fresh cysts average 7 × 12 μm in size and differences in cyst measurements (Table 6) have been attributed to variations in measuring techniques or to the host species involved.[84] Ultrastructurally, the cyst wall is thick and multilayered in nature.[79,277] According to Hanrahan[277] the cyst contents are highly condensed with few details detectable in mature cysts. In the early stages of cyst formation, there is a ring of dark-staining material just inside the plasmalemma composed of densely packed organelles such as ribosomes, mitochondria, and Golgi apparatus. Centrally, the area contains a nucleus and characteristic, membrane-bound granules said to be lipoid in nature by Hanrahan[277] but referred to as granules of glycogen by Harry and Finlayson.[79]

Trophozoites — The trophozoites (Figure 1) of *Malameba locustae* were originally described as being uninucleate, 5 to 10 μm in diameter, and with highly refractile globules in hyaline protoplasm.[24] Henry[84] also indicated that the trophozoites (4 to 12 μm in diameter) contain granules in the transparent cytoplasm and are generally spherical in shape with irregular to elongate forms being common. While most investigators have described only one type of trophozoite in the life cycle of *M. locustae,* Evans and Elias[85] differentiated two types of trophozoites in *Locusta migratoria*. Primary trophozoites, occurring in the ceca and midgut epithelia, were described as typically spherical, 6 to 14 μm in size (average 10.4 μm), with clear cytoplasm, and a large prominent karyosome located centrally. Secondary trophozoites, occurring in the lumen of Malpighian tubules, are slightly smaller in size (7 to 8 × 5 to 6 μm) with a diffusely granular cytoplasm and an indistinct dispersed karyosome. Harry and Finlayson[79] also recognized primary and secondary trophozoites in the life cycle of *M. locustae* in *Schistocerca gregaria*. In her study on the ultrastructure of *M. locustae,* Hanrahan[277] described trophozoites from the midgut epithelium (the primary trophozoites of Evans and Elias[85]) as irregular in shape and containing a single, circular nucleus with a prominent karyosome located centrally or laterally. The cytoplasm contains distinct granular endoplasm and clear ectoplasm. Rough endoplasmic reticulum occurs near the nucleus and may form a complete circle around the nucleus. Large mitochondria, small vacuoles, and a contractile vacuole are present in the cytoplasm along with ribosomes and other granules. Secondary trophozoites, occurring in the lumen of the Malpighian tubules, are similar to

Table 5
SUMMARY OF THE PRINCIPAL HOST-PATHOGEN CHARACTERISTICS OF CANDIDATE PROTOZOA AS MICROBIAL CONTROL AGENTS

Microbial agent	Primary host(s)	Infective stage	Transmission Per os	Transmission Trans-ovarian	Site of infection	Gross pathology	General Ref.
Malameba locustae	Acrididae	Cysts-trophozoites	x		Midgut, gastric ceca, and Malpighian tubules	Molting difficulties, prolonged nymphal instars, sluggishness, anorexia, marked comatose condition with tetanic twitches of legs prior to death; in sublethal infections, reproductive failure	14, 84
Ascogregarina culicis	Culicidae	Oocyst-sporozoites	x		Midgut and Malpighian tubules	Sublethal infections where adult longevity may be affected	296
Mattesia grandis	*Anthonomus grandis*	Spore-sporozoite	x		Fat body	Molting difficulties, deformed pupae or adults, premature death as larva, pupa, or adult; in sublethal infection, reduced fecundity and longevity	130
M. trogodermae	*Trogoderma* spp.	Spore-sporozoite	x		Fat body	Death usually in larval stage, preceded by body distention, sluggish movement, and white fecal exudate often in long strands	106, 132
Nosema algerae	Culicidae	Spore-sporoplasm	x		Varies with host species, from systemic in highly susceptible species to specific site	In highly susceptible species, death usually occurs in the larval or pupal stage; in less susceptible species, adult vigor, longevity, and fecundity are reduced	192, 319

Table 5 (continued)
SUMMARY OF THE PRINCIPAL HOST-PATHOGEN CHARACTERISTICS OF CANDIDATE PROTOZOA AS MICROBIAL CONTROL AGENTS

Microbial agent	Primary host(s)	Infective stage	Transmission Per os	Transmission Trans-ovarian	Site of infection	Gross pathology	General Ref.
					(e.g., nervous system) only in less susceptible species		
N. fumiferanae	Choristoneura fumiferana	Spore-sporoplasm	x	x	Primarily midgut cells; may become systemic	Infection usually sublethal; retarded larval and pupal development; reduced pupal weight and adult longevity and fecundity	320
N. gasti	Anthonomus grandis	Spore-sporoplasm	x	x	Systemic in most host tissues	Chronic infection, apparently with little larval mortality; reduced adult longevity and fecundity; heavily infected weevils may die as pupae or adults	203
N. locustae	Acrididae	Spore-sporoplasm	x		Primarily fat body cells	Prolonged nymphal development, lethargic movements, reduced feeding and adult fecundity; increased cannibalism and death rates among group-reared grasshoppers	321
N. pyrausta	Ostrinia nubilalis	Spore-sporoplasm	x	x	Malpighian tubules, silk gland, reproductive tissues, occasionally other tissues	Prolonged larval development; adult fecundity and longevity reduced along with egg fertility; premature mortality highest when infected transovarially	211, 322, 323, 323a

Pleistophora schubergi	Lepidoptera	Spore-sporoplasm	x	Midgut cells	Prolonged larval development; reduced larval size, vigor, and feeding; death usually in larval or pupal stage; in sublethal infections, pupae and adult may be deformed and cocoons may not be formed; adult longevity shortened	224, 225, 324, 325
Vairimorpha necatrix	Lepidoptera	Spore-sporoplasm	x	Primarily fat body cells	Reduced larval movement and feeding; high larval mortality with few infected larvae able to pupate; infected pupae unable to enclose to adults	245, 247, 326
Vavraia culicis	Culicidae	Spore-sporoplasm	x	Malpighian tubules, fat body cells, and intestine	Infection sublethal with low larval mortality, reduced adult emergence, fecundity, and longevity	250, 327, 328

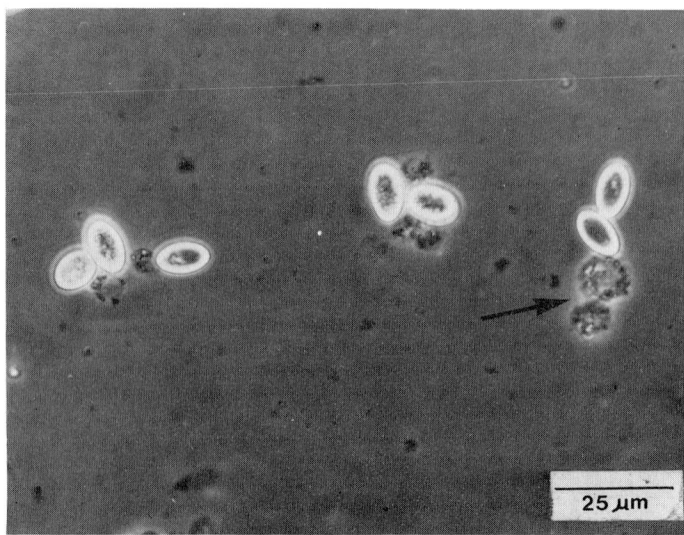

FIGURE 1. Trophozoites (arrow) and cysts of *Malameba locustae*. (Courtesy of J. E. Henry, U.S. Department of Agriculture, Agricultural Research Service, Bozeman, Mont.)

Table 6
COMPARISON OF CYST DIMENSIONS OF *MALAMEBA LOCUSTAE*[a]

Order	Host	Preparation	Length Mean	Length Range	Width Mean	Width Range	Ref.
Orthoptera	*Chortiocetes terminifera*	Fresh	12.6		7.5		282
	Locusta migratoria	Stained		9.0—11.0		5.0—6.0	85
		Fresh		12.5—16.4		7.5—9.3	281
	Locustana pardalina	Fresh	12.0		8.0		279
	Melanoplus bivattatus	Fresh	12.6		7.2		84
		Stained	11.5		6.7		
	M. differentialis	Unknown	9.6	8.5—10.0	5.5	4.6—6.2	77
		Fresh	12.3		7.6		84
		Stained	11.3		7.2		
	M. sanguinipes	Fresh	12.6		7.9		84
		Stained	10.6		6.1		
	Pardillana limbata	Fixed	10.8		6.8		287
	Pteronemobius sp.	Fixed	12.1		7.58		
	Schistocerca gregaria	Stained	9.0	7.0—11.5	5.5	4.5—6.5	284
		Fresh	12.1	8.8—13.6	7.4	7.4—9.2	
Thysanura	*Lepisma saccharina*	Stained	8.4	5.0—9.5	5.5	4.5—6.5	284

[a] Measurements in micrometers.

the primary trophozoites[79,277] except that no distinct ectoplasm is found in the secondary forms. These trophozoites may contain one or two nuclei, numerous granules, mitochondria, contractile vacuoles, and rough endoplasmic reticulum associated with the nucleus. Food vacuoles containing pieces of the brush border of the Malpighian tubules were also described by Harry and Finlayson.[79]

Bionomics

Life cycle — According to most workers, *Malameba locustae* develops extracellularly in the lumen of the Malpighian tubules of grasshoppers and to a lesser extent intracellularly in the epithelial cells of the gastric ceca and midgut.[24,77,84] In more recent studies on the life cycle of this amoeba, Evans and Elias[85] and Harry and Finlayson[79] demonstrated the existence of primary and secondary trophozoites although their accounts differ somewhat in tracing the course of infection. Upon ingestion, the cysts begin to break down almost immediately in the foregut to release the primary trophozoites in the midgut within a few hours after ingestion. The trophozoites feed intracellularly in epithelial cells of the gastric ceca or midgut and eventually give rise to the secondary trophozoites by binary or multiple fission. According to Evans and Elias,[85] infection of the Malpighian tubules is effected by the migration of secondary trophozoites from the epithelial cells to the underlying connective and muscle tissue of the gut muscularis where contact and invasion occur with the anteriorly directed Malpighian tubules. However, in comprehensive histopathological observations on the pathogenesis of infection with *M. locustae* in *Schistocerca gregaria*, Harry and Finlayson[79] could find no evidence that the secondary trophozoites entered Malpighian tubules via the gut epithelium and hemocoel, a conclusion also supported by the observations of Hanrahan.[277] Secondary trophozoites are apparently released from degenerating epithelial cells of the midgut or ceca and enter the lumen of the Malpighian tubules when the gut is empty, just prior to or after molting. According to Harry and Finlayson,[79] the secondary trophozoites feed exclusively as extracellular parasites by engulfing portions of the brush border of tubule cells. However, Hanrahan[277] found no evidence of amoebae attacking the microvilli or the epithelial cells. Cyst production within Malpighian tubules generally occurs in 14 to 18 days, although Henry[84] reported that production was related to the severity of infection with some cysts being found in tubules and fecal material of heavily infected grasshoppers within 9 days after infection.

Natural occurrence — Amoebic disease of grasshoppers is known primarily as a problem of laboratory cultured grasshoppers. Although the source of the particular laboratory colony has not always been indicated, infected grasshoppers have been reported in colonies maintained in Iowa,[77] Montana,[84,278] South Africa,[277,279-281] Australia,[282] England,[79,283] Sweden,[284] and Poland.[284a] Data on natural infections of *M. locustae* in field-collected grasshoppers are less common, being limited to reports by Taylor and King[24] in Iowa; Henry[84] in Montana and California; Lea[285] and Venter[280] in South Africa; Fowler et al.[286a] and Henry et al.[286b] in West Africa; and Ernst and Baker[287] in Australia. These reports indicate that natural infections in grasshoppers are extremely low, although Venter[280] reported that a natural epizootic of *M. locustae* in the brown locust *Locustana pardalina* prevented the development of a second generation of grasshoppers in South Africa when first-generation adults failed to sexually mature.

Ascogregarina culicis Ross, 1898

This eugregarine was first discovered in mosquitoes by Ross[113] in India who named it *Gregarina culicidis* but subsequently used the name *Gregarina culicis* in later publications.[288,289] His initial observations on its life cycle were confirmed by Manson,[290] and the first detailed description of the species and its life history were provided by Wenyon[291] who assigned it to the genus *Lankesteria*. Vávra[110] provided an in-depth taxonomic review of the species of this genus, discussed the poorly defined nature of the genus, and reviewed various proposals that have been made to classify this gregarine among the Eugregarinida. More recently, Levine[94] followed the earlier proposal of Grassé[98] to establish a new genus, *Ascocystis* for those species of *Lankesteria* known from insects and designated *Ascocystis culicis* as the type species of the genus. *Ascocystis* was shown subsequently to be a junior homonym and the new name *Ascogregarina* was proposed for the genus.[292] Other species presently recognized in this genus are listed in Table 2.

Description and Life Cycle

It is not possible to assess the exact identity of some species of *Ascogregarina* from mosquitoes cited as *A. culicis*;[110] and, in at least two cases,[293,294] the eugregarine identified as *A. culicis* is more likely *A. barretti*.[110] Since the original description by Ross[288] is limited, the following account of *A. culicis*, isolated from *Aedes aegypti*, is based on the more recent and detailed observations of Wenyon,[291] Walsh and Callaway,[295] Vávra,[110] McCray et al.,[296] and Lien and Levine.[109]

Cephalins — Arising from the intracellular development of the sporozoites within cells of the larval midgut, the aseptate cephalins are elongate and broadly triangular in shape with a centrally located nucleus and a well developed mucron. Cephalins from ruptured host cells are generally 50 μm or shorter and often are attached to host cell debris for a period of time. The surface of the cephalin is relatively smooth, and the cytoplasm is transparent and contains paraglycogen granules.

Ultrastructurally,[110,295] the pellicle is highly folded with the ectoplasm protruding deeply into each fold. The folds are perpendicular to the body surface in free cephalins. The endoplasm is filled with paraglycogen granules, golgi complexes, endoplasmic reticulum, mitochondria, vacuoles, and amorphous inclusions. The well developed mucron is umbrella or mushroom-like in shape, measuring about 3×1.5 μm. It is evaginated from the anterior end of the cephalin and is connected by a short, heavy stalk. Numerous fine tubules and fibers run from the mucron membrane down to the body. The mucron is left behind upon rupture of the host cells and the freed cephalin may remain attached to cellular debris by means of the conical anterior part of the body represented by the prior attachment point of the mucron.

Gamonts — With rupture of host cells, the released cephalins remain attached to cellular debris or may exist free in the gut lumen of the larval mosquito where they increase in size to become gamonts. The gamonts are generally elongate, and more or less triangular anteriorly with the nucleus still in a central location. Longitudinal striations are conspicuous in the pellicle. Gamonts vary greatly in length, ranging up to over 300 μm in length,[296] but most are only 50 to 150 μm long.[110] Migrations of the gamonts to the lumen of the Malpighian tubules occur when the mosquito pupates and subsequent development is extracellular. The gamonts pair, at first anteriorly, and then laterally, and a hyaline membrane is secreted around each pair of gamonts to form a gametocyst.

Gametocysts — Found primarily in the distal ends of the Malpighian tubules, the gametocysts are spherical in shape and range from 60 to 100 μm in diameter. Each gamont within a gametocyst differentiates into gametes, the nuclei of which differ in size. With conjugation, the zygotes become oocysts (sporocysts) within which eight sporozoites are formed.

Oocysts — The oocysts (Figure 2) are ovoid to lanceolate in shape with truncate ends and variable in size, ranging from 10 to 12×6 to 7 μm. Along with the elongate sporozoites, several residual granules usually occur in a tight cluster near one end of the oocyst. Formation of oocysts generally occurs late in the pupal stage of the mosquito and with adult emergence, the walls of the gametocysts rupture to release large numbers of free oocysts in the lumen of the Malpighian tubules. These pass out of the tubules into the gut and are excreted by the adult.

Sporozoites — Few specific observations on sporozoites of *Ascogregarina culicis* have been presented prior to the report of Sheffield et al.[297] They found that oocyst excystation occurs primarily in the posterior third of the midgut. Newly released sporozoites are 9.5 to 10 μm in length and taper posteriorly with a blunt anterior end. Ultrastructurally, the sporozoite is enclosed by a typical pellicle and the anterior end has a conoid with two associated rings. A polar ring serves as the termination of the subpellicular microtubules and a flask-shaped structure is also situated internal and posterior to the conoid, primarily

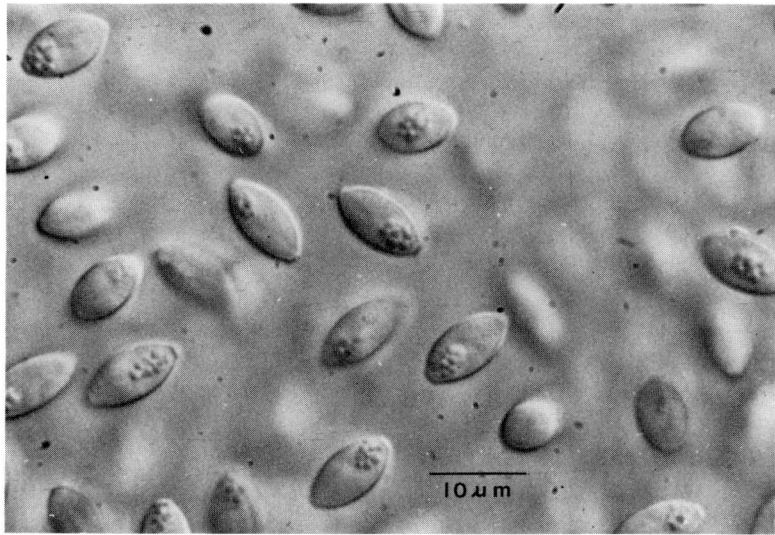

FIGURE 2. Oocysts of *Ascogregarina culicis*, Nomarski interference contrast. (Courtesy of E. D. Rowton.)

in extracellular sporozoites. A micropyle is also present at the anterior end of the sporozoite. The nucleus is apparently located near the posterior end. Subsequent intracellular development of the sporozoites gives rise to the cephalins.

Natural Occurrence

Ascogregarina culicis is a common and cosmopolitan parasite in natural populations of the yellow fever mosquito *Aedes aegypti*, being recorded from India,[113,288,289,298] Brazil,[299] Iraq,[66,291] Africa,[300-302] Philippines,[303] Singapore,[304] Taiwan,[109] Russia,[305,305a] and the southeastern U.S.[295,296,306-311] Eugregarines identified as *A. culicis* have also been reported in other species of *Aedes* from China,[312,313] India,[314] Africa,[315] Philippines,[316] Korea,[313a] Russia,[305a] and Western Samoa.[317] It is not possible, according to Vávra,[110] to confirm the exact identity of some species of *Ascogregarina* cited above because of the lack of structural data. He also concluded that the species studied by Ganapati and Tate[293] and Kramar[294] from *Aedes geniculatus* are likely identical with *A. barretti*.

Few data on natural prevalence of infection in mosquitoes are available in most of the early literature. However, Feng[312] reported collecting larvae and pupae of *Ae. koreicus* in China in which 1 of 19 adults, 0 of 7 larvae, and 7 of 26 pupae were infected with *A. culicis*. Later Feng[313] also found *A. culicis* in 2 of 43 *Armigeres obturbans* and 1 of 6 *Aedes albopictus*. More recent extensive surveys in the southeastern U.S.[307-310] revealed the common occurrence of *Ascogregarina culicis* in areas where *Ae. aegypti* is well established with 25 to 76%[309] and 32 to 86%[308] of the mosquitoes being infected at sites that were positive for the presence of *A. culicis*. The number of collection sites positive for infection by *A. culicis* was also found to increase seasonally but the magnitude and frequency of seasonal infection by *A. culicis* did not coincide with the seasonal population peak for *Ae. aegypti*.[309] The species also was found to infect about 40% of over 3000 *Ae. polynesiensis* collected from natural populations from Western Samoa.[317] More recently, Laird[318] reported that 60% of *Ae. polynesiensis* larvae from the Nukunono village islet of the Tokelau Islands were infected with a species of *Ascogregarina* which was similar morphologically to *A. culicis*.

Mattesia grandis McLaughlin, 1965

This neogregarine was originally described by McLaughlin[129] from the cotton boll weevil, *Anthonomus grandis*.

Description and Life Cycle[129,131]

Sporozoites — Eight sporozoites are produced in each spore (oocyst), generally after dissociation of the spores from the gametocysts and after death of the host. They are gregarinoid in shape (vermiform, bluntly rounded anterior, and sharply pointed posteriorly) with a large nucleus containing an eccentric endosome. The nucleus is located in the broader, anterior end of the sporozoite. The eight sporozoites occur alternately head to tail within the spore and tend to be appressed to the side of the spore wall. They are capable, when in a free state, of slow but forceful movement by bending. In wet-mount preparations, naturally released sporozoites measure about 2.6×6.8 μm, and they are slightly smaller when extruded from spores by pressure (Table 7).

Limited observations on the ultrastructure of sporozoites indicate that the anterior end has a rostrum with a lattice-work pattern formed by spiral fibers of the conoid. A wide collar-like structure is present below the limit between the conoid-containing anterior rostrum and the broadened part of the sporozoite body. A polar ring is located below the rostrum to which the subpellicular microtubules are attached.

In *Anthonomus grandis,* sporozoites are released in the midgut from the spores upon ingestion and invade through the intestinal epithelium to develop initially as micronuclear meronts in adjacent fat body cells.

Micronuclear meronts — The micronuclear meronts, arising from the intracellular development of sporozoites in fat body cells, are initially irregular to spherical in shape and contain one or more small nuclei. As the meronts grow, the nuclei divide and mature meronts (Figure 3A) may contain from 30 to 200 nuclei. Meronts vary in size, ranging up to 20 to 30 μm long (Table 7). Host cells eventually rupture and the meronts continue to undergo merogony extracellularly. The nuclei migrate to the peripheral regions of the meront body where micronuclear merozoites are formed by the constriction of cytoplasm around each nucleus. In wet-mount preparations, late-stage meronts appear spherical or irregular in shape with the merozoites projecting outward from the periphery in various stages from small buds to elongated, vermiform shapes.

Micronuclear merozoites — Formed by peripheral budding from micronuclear meronts (Figure 3B), the motile, elongate merozoites are slightly broader anteriorly with a nucleus located anterior of center. Merozoites with micronuclei apparently continue the cycle of micronuclear merogony while other microzoites with enlarged, macronuclei apparently begin macronuclear merogony. Micronuclear meronts and merozoites are the predominant stage in infected hosts from the 1st through the 5th day postinfection.[130]

Macronuclear meronts — These meronts initiate the second (or macronuclear) merogony and are formed by the intracellular development of micronuclear merozoites whose nuclei enlarge. With subsequent nuclear division, meronts with up to 40 to 60 macronuclei are formed. These macronuclear meronts (Figure 3C) are ovoid to irregularly shaped and range up to 20 μm in diameter (Table 7). Macronuclear merozoites are eventually formed by cytoplasmic budding at the periphery of the meront.

Ultrastructurally the macronuclear meront is enveloped by three membranes, the innermost of which carries on pinocytosis. In young meronts with only one to four nuclei, a specialized area of the cell membrane (the conoidal complex) forms a mucron but this structure is absent in late stage meronts which form the merozoites. The cytoplasm contains one or more nuclei each of which has a double membrane and a large, central nucleolus. Mitochondria with vesicular cristae are located in the cytoplasm along with lamellae of the endoplasmic reticulum, numerous ribosomes, and irregularly shaped cisternae with amorphous contents in the area close to the mucron.

Macronuclear meronts occur as early as 3 days postinfection and are the dominant stage present on days 5 to 7.[130]

Macronuclear merozoites — Late stage, extracellular meronts give rise to macronuclear merozoites by budding in a similar manner as the micronuclear merozoites are formed. The

Table 7
MORPHOMETRICS OF *MATTESIA GRANDIS*, A NEOGREGARINE OF THE COTTON BOLL WEEVIL, *ANTHONOMUS GRANDIS*[a,129]

Developmental stage	Preparation	Length Mean	Length Range	Width Mean	Width Range	Nucleus Mean	Nucleus Range
Sporozoites							
Pressure extruded	Fresh	6.7	5.5—8.4	—	—	—	—
Natural emergence	Fresh	6.8	5.5—8.7	—	—	—	—
Micronuclear meronts							
Young, single nucleus	Fixed	—	—	3.2	2.1—4.9	1.8	1.0—2.5
Large, beginning to form merozoites	Fresh	—	—	13.5	7.8—22.6	1.4	0.9—1.8
Large, with well-formed merozoites	Fresh	—	—	13.8	7.8—22.6	—	—
Micronuclear merozoites	Fresh	12.2	8.2—16.6	1.8	1.3—2.7	1.1	0.8—1.5
Macronuclear meronts							
Young, single nucleus	Fixed	—	—	4.5	3.0—6.0	2.6	1.6—3.6
Mid stage, several nuclei	Fixed	—	—	7.9	5.8—11.4	2.4	1.0—4.2
Late stage, merozoites forming	Fresh	—	—	13.7	9.7—20.6	3.0	2.0—4.1
Macronuclear merozoites	Fresh	10.6	8.5—13.9	3.5	2.2—4.4	2.2	1.5—3.1
Gamonts (unpaired)	Fixed	—	—	4.8	3.2—6.7	1.8	1.3—2.6
	Fresh	—	—	6.7	4.4—8.8	2.5	1.8—4.0
Gametocysts							
Gamonts uninucleate	Fixed	—	—	5.6	4.5—6.8	1.7	1.2—2.3
	Fresh	—	—	5.4	5.0—6.8	1.6	1.0—2.3
Gametes formed	Fresh	—	—	11.2	9.1—13.3	—	—
Zygotes formed	Fresh	—	—	10.8	9.4—12.2	—	—
Spores formed	Fixed	10.2	9.0—10.9	8.5	4.9—11.3	—	—
	Fresh	12.5	11.0—15.7	14.0	12.0—15.9	—	—
Spores (oocysts)							
Free from gametocyst	Fixed	9.3	7.9—12.1	5.6	4.5—7.0	—	—
	Fresh	11.8	8.5—14.2	7.1	3.8—9.4	—	—

[a] Measurements in micrometers.

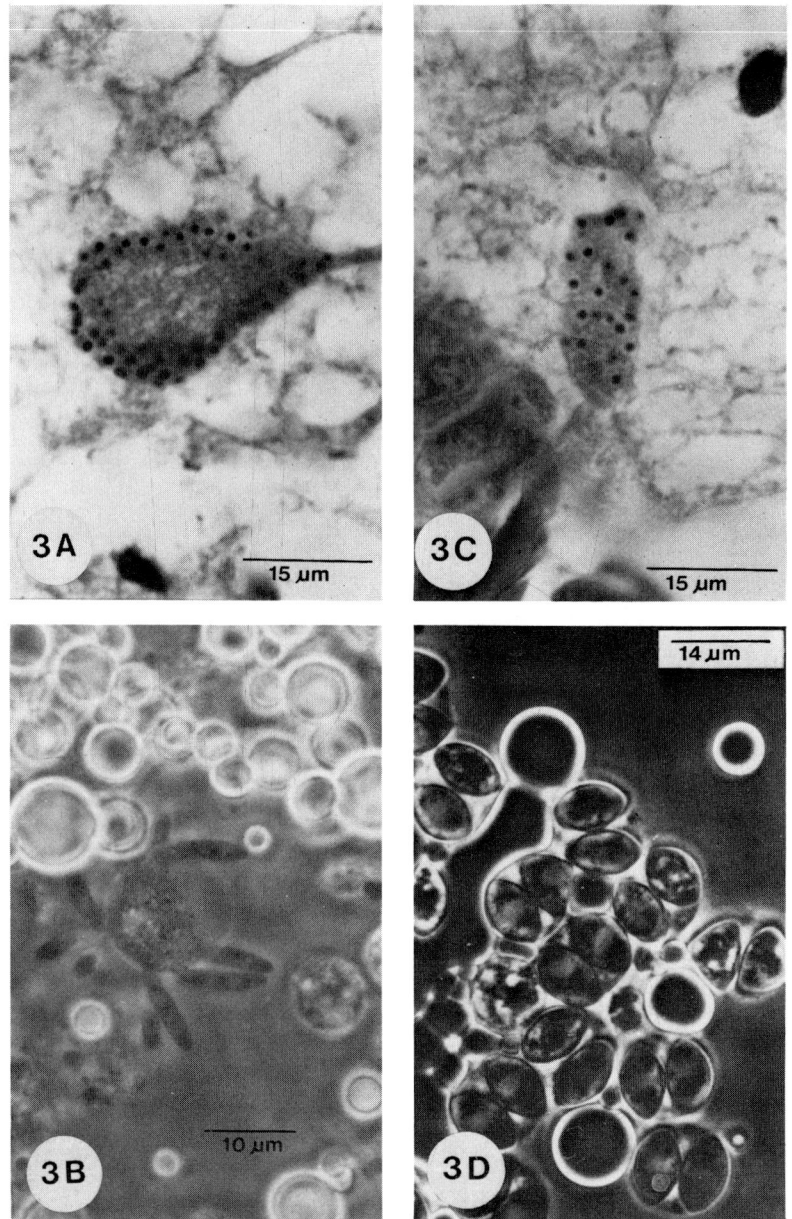

FIGURE 3. Life-cycle stages of *Mattesia grandis*. (A) Micronuclear meront; (B) formation of micronuclear merozoites by budding; (C) macronuclear meront; (D) gametocysts, each containing two mature spores or oocysts. (Courtesy of Society of Protozoology and R. E. McLaughlin, U.S. Department of Agriculture, Agricultural Research Service, State College, Miss.)

macronuclear merozoites are ovoid to pyriform in shape, variable in length (Table 7) (shorter but wider than micronuclear merozites), and with nuclei that are distinctly macronuclear in size. The cytoplasm is lacy or net-like in appearance and the nucleus is located anterior of center. Attachment to the mother meront is by the posterior end. The macronuclear merozoites are less motile than the more elongate micronuclear merozoites and their occurrence generally marks the end of the multiplication phase of the infection. Most of the merozoites apparently round up after their release from the mother meront to form gamonts.

Observations on the ultrastructure of merozoite formation reveal that merozoite buds are formed from macronucleate meronts which lack a mucron by an infolding of the meront membrane. The conoidal complex is formed on the apex of the merozoite bud and the nucleus later migrates into the bud. The merozoite bud is finally separated by constriction from the main body of the meront by a complete membrane similar to that covering the rest of the body. Ultrastructurally the macronuclear merozoites are broadly pear-shaped and sharply pointed at the posterior end. A conical rostrum containing the conoid is present at the anterior pole of the merozoite beneath which lies the polar ring, subpellicular microtubules, and several club-shaped organelles (rhoptries) that make up the apical complex or conoidal complex characteristic of sporozoa. While most of the macronuclear merozoites change into gamonts, ultrastructural observations also revealed that the rostrum of some merozoites becomes retracted and the conoidal complex changes into a mucron — suggesting their continued development as meronts.

Gamonts — Most of the macronuclear merozoites round up to become gamonts after their release from mother meronts and may increase in size before pairing by contact at their apical region to begin gametogony. Prior to the formation of the gametocyst membrane, the pairs of gamonts (copula) are spherical in shape with the halves divided by cell membranes in close apposition. The conoidal complex of both gamonts is also in close apposition and disappears with the formation of the gametocyst membrane. Each gamont has a large nucleus with a dark, prominent nucleus, a considerable amount of rough endoplasmic reticulum, numerous mitochondria with vesicular cristae, and vacuolar spaces. Gamonts are frequently present in infected hosts by the 5th and 6th day postinfection and the resultant gametocysts are the predominant stage present on days 6 and 7.[130] Dimensions of gamonts and gametocysts are presented in Table 7.

Gametocysts and gametes — Following formation of the double gametocyst membrane around each copula, the nucleus of each gamont divides to produce two nuclei of equal size. One nucleus of each gamont divides again and six nuclei of equal size are present in the gametocyst. The second nucleus of each gamont also divides subsequently but the two pairs of nuclear material remain in a delayed state of telophase. This nuclear material from each gamont apparently pairs and two residual bodies are formed. The other four nuclei (gametes) copulate in pairs to produce two zygotes. Ultrastructurally each gamete has a single-layered membrane, one nucleus, several mitochondria, and many lamellae of endoplasmic reticulum.

Spores (oocysts) — The zygotes are initially spherical in shape, enveloped in a double membrane, and contain a single nucleus. Their formation marks the beginning of sporogony and ultimately eight sporozoites are produced within each spore. Each zygote becomes a sporoblast with cytoplasmic condensation and the thickening of the zygote wall. The residual bodies disappear during this phase of development and are seldom detectable in sporoblasts or the mature spores. As the spores near maturity, the gametocyst becomes distended and is broader than long with the gametocyst membrane tightly stretched over the poles of the two spores. The spores, while flattened on the adjacent side in sectioned gametocysts, are usually curved evenly from pole to pole in wet-mount preparations (Figure 3D). The spore wall is of uniform thickness on the sides but possesses a thickened cap at the poles which is filled with a plug of material closing the pore. Limited observations ultrastructurally revealed the spore (oocyst) has a thick coat of two layers that is otherwise structureless. Spore dimensions are presented in Table 7. As mentioned previously, the development of the sporozoites does not occur until after the dissolution of the gametocyst membrane and liberation of the spores with death of the host. Zygotes are apparent on the 7th and 8th day postinfection and spores are the predominant forms thereafter. Huge quantities of spores are usually the only stage present in late stage infections or in moribund hosts; the entire life cycle of the neogregarine occurs in the fat body.[130]

Natural Occurrence

No data on the natural occurrence of *Mattesia grandis* in *Anthonomus grandis* have been obtained. McLaughlin[129] suspected that the pathogen originated from a wild strain of the cotton boll weevil from Mexico which spread subsequently to several laboratory-maintained strains of the weevil in Mississippi.

Mattesia trogodermae Canning, 1964

This species was described by Canning[106] from the khapra beetle, *Trogoderma granarium*. Additional details on the life cycle and various stages of the species are included in unpublished theses by Nara[132] and Pounds.[329]

Description and Life Cycle[106]

Sporozoites — The sporozoites are liberated from spores (oocysts) in the gut of the host larva and are found initially in fat body cells adjacent to Malpighian tubules. Thus, Canning[106] concluded that the sporozoites probably enter the fat body through the Malpighian tubules rather than the gut wall. However, from an extensive histopathological investigation of the pathogenesis of *M. trogodermae* in *Trogoderma glabrum*, Pounds[329] concluded that the sporozoites probably migrate through the midgut wall to infect the fat body. Sporozoites were detected among the microvilli of the midgut epithelial cells but he could not confirm their presence either within or between cells of the midgut. He also found the heaviest foci of infection in fat body cells adjacent to the midgut. While Canning[106] did not specifically describe sporozoites, photographs of live sporozoites by Nara[132] reveal their typical vermiform shape with a rounded anterior end and a sharply pointed posterior end.

Micronuclear meronts — The sporozoites develop intracellularly into micronuclear meronts (Figure 4a) which may contain up to 60 nuclei and measure up to 26 μm in diameter. The nuclei are 1.5 μm in diameter and each contains a spherical, eccentric endosome. The nuclei become peripheral in distribution prior to merozoite formation.

Micronuclear merozoites — Uninucleate globular bodies form at the surface of the meront and usually elongate to form a rosette of merozoites (Figure 4b). The merozoites (Figure 4c) measure 3 to 4 μm in diameter or 5 to 8 × 2.5 to 3 μm depending on the number produced and the state of elongation. They are actively motile and are the principal means of spreading the infection throughout the host fat body. According to Nara,[132] numerous rosettes of micronuclear merozoites are found in *T. glabrum* 5 days postinoculation and a few may be present as late as 13 days when sporogony is essentially completed.

Macronuclear meronts — As a result of micronuclear merogony, the micronuclear merozoites again develop intracellularly to give rise to irregularly spherical meronts with enlarged nuclei (Figure 4d). These macronuclear meronts contain up to 12 nuclei when mature and vary in size up to 20 × 15 μm. Macronuclear merogony results in the production of macronuclear merozoites (Figure 4e).

Macronuclear merozoites — These are motile and measure 4 to 5 μm in diameter. Although not described in detail by Canning,[106] Nara[132] also found that the macronuclear merozoites are formed by peripheral budding from the mother meront and rosettes of macronuclear merozoites are common in *Trogoderma glabrum* between 5 and 9 days postinoculation. She described the merozoites as broadly pear-shaped with a maximum width at about $1/3$ of the body length. The body tapers sharply at the posterior pole.

Gamonts — The macronuclear merozoites round up (Figure 4f) and associate in pairs with the originally spherical gamonts flattening themselves to each other as two hemispheres (Figure 4g).

Gametocysts — A membrane is secreted around each pair of gamonts to form a gametocyst (Figure 4h) varying in diameter from 6 to 8.5 μm. During gametogony, four gametes and

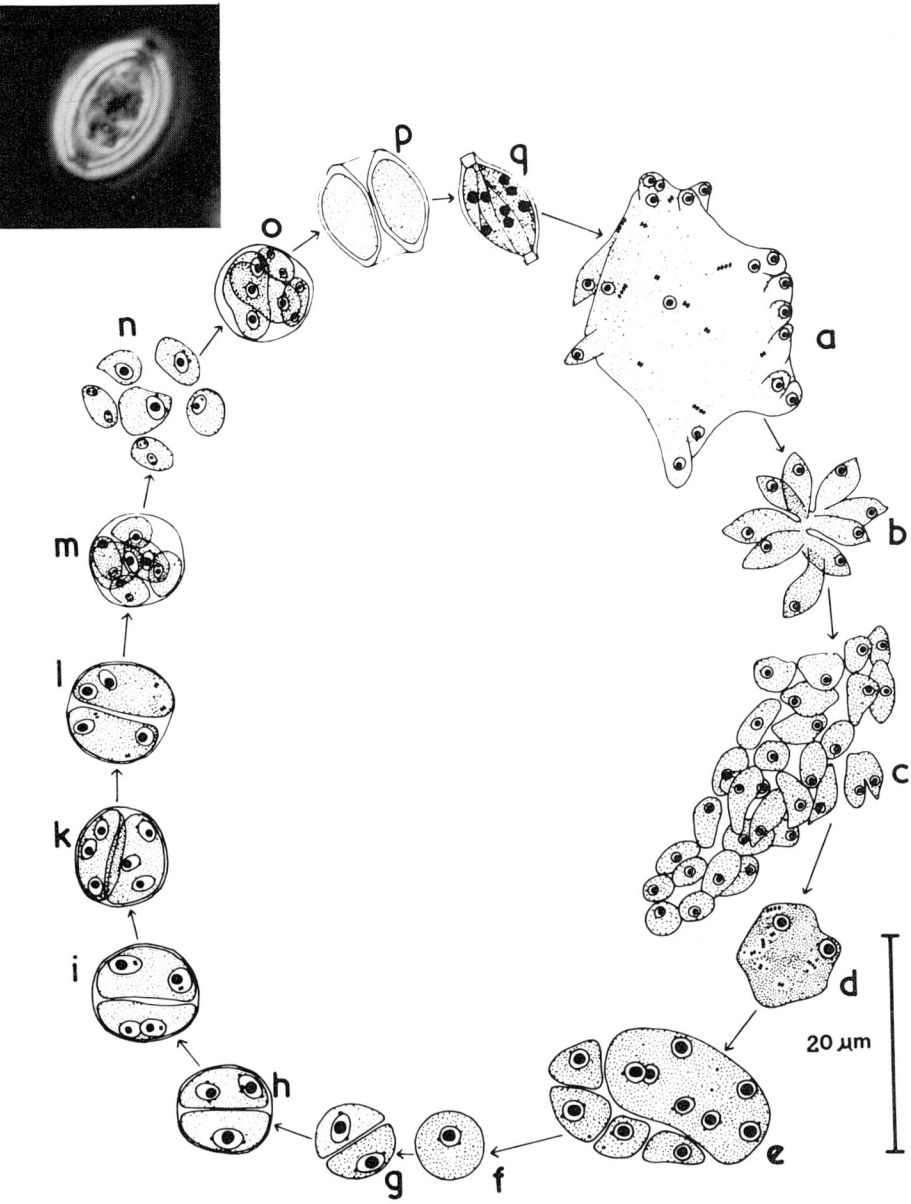

FIGURE 4. Life cycle of *Mattesia trogodermae*. (a) Micronuclear meront; (b) rosette formation of micronuclear merozoites; (c) separation of micronuclear merozoites; (d) young macronuclear meront; (e) separation of macronuclear merozoites; (f) gametocyte; (g) association of gametocytes; (h to l) nuclear division within gametocyst; (m) cytoplasmic division into four gametes and two residual cells; (n) gametes and residual cells of (m) separated out; (o) fusion of gametes; (p) formulation of spore wall around zygotes; (q) mature spores with eight sporozoites and polar plugs. (Courtesy of Academic Press and E. U. Canning). (Insert: mature spore or oocyst, phase, wet-mount preparation; courtesy of W. E. Burkholder, U.S. Department of Agriculture, Agricultural Research Service, Madison, Wis.)

two binucleate residual bodies are formed (Figure 4h to n). The gametes fuse in pairs to form zygotes while the residual bodies disintegrate (Figure 4o). In *T. glabrum*, gametogony begins by the 8th day postinoculation and sporogony is complete within 2 weeks.[132]

Spores (oocysts) — The zygotes develop walls and two lemon-shaped spores (Figure 4p) are formed within the gametocyst. Their shape is usually asymmetrical within the gametocyst

Table 8
PREVALENCE OF INFECTION OF *MATTESIA TROGODERMAE* IN SPECIES OF *TROGODERMA* FROM CALIFORNIA[330]

Year	No. of locations with *M. trogodermae*	No. of insects examined	Infected with *M. trogodermae* (%)
1965	22	435	65
1966	208	3114	76.2
1967	105	1212	89.5
1968	125	885	89.7

Table 9
PREVALENCE OF INFECTION OF *MATTESIA TROGODERMAE* IN *TROGODERMA* SPECIES IN CALIFORNIA FROM 1965—1968[330]

Host species	No. samples	No. insects examined	No. infected	Infected (%)
T. glabrum	23	122	108	88.5
T. grassmani	2	12	12	100
T. inclusum	52	369	156	42.3
T. ornatum	1	10	4	40
T. simplex	517	3735	3202	85.7
T. sternale	133	818	621	75.9
T. variabile	90	580	434	74.8

due to the flattening of their adjoining surfaces. Eight sporozoites (Figure 4q) are formed within each spore as a result of sporogony and occur side by side within the spore or in a coiled state in older spores. Each pole of the spore is filled with a plug of material which is soluble in the digestive juices of the host. Fresh spores (Figure 4 inset) measure 13 × 7.7 µm while fixed spores are slightly smaller, measuring 11 × 6 µm.

Natural Occurrence

Although originally described from a laboratory culture of *Trogoderma granarium* from England,[106] *Mattesia trogodermae* has been found to be common and widely distributed in several other species of *Trogoderma* from California (Tables 8 and 9).[330] Earlier reports of a neogregarine in laboratory colonies of various *Trogoderma* species from Wisconsin[331,332] and Kansas[331,333] are likely attributable to *M. trogodermae*.[132,334] Thus, *M. trogodermae* is probably a cosmopolitan species, occurring as a common pathogen of laboratory colonies and natural infestations of various stored product pests in the genus *Trogoderma*.[335]

Nosema algerae Vávra and Undeen, 1970

This species was described by Vávra and Undeen[192] from a laboratory colony of *Anopheles stephensi* maintained in Illinois. Although this strain of *An. stephensi* had been obtained from the U.S. Public Health Service in Georgia, it had originally been established from material received from the London School of Tropical Medicine and Hygiene in England.[336] A similar microsporidium was also discovered in various species of anopheline mosquitoes maintained in colonies at the University of Maryland and Walter Reed Army Institute of Research in Washington, D.C.,[337] and also from *An. stephensi* maintained in Florida.[336] All of these strains had a common ancestry, i.e., the London School of Tropical Medicine and Hygiene. About this same time, Canning and Hulls[193] described a *Nosema* species from *An. gambiae* maintained in a laboratory colony in Tanzania, East Africa, which they concluded was also identical with *N. algerae*. Although Fox and Weiser[338] had earlier reported a microsporidium from *An. gambiae* and *An. melas* from Liberia, West Africa, as *Nosema stegomyiae*, Canning and Hulls[193] concluded that the microspordium involved was actually *N. algerae*. While not all taxonomic specialists of the microsporidia agree that the microsporidium studied by Fox and Weiser as *N. stegomyiae* should be synonymized as *N. algerae*, it is generally agreed that the microsporidia of various laboratory colonies of *Anopheles* are identical[336] and should be referred to as *N. algerae*.[152,193,194]

In their original description of *N. algerae*, Vávra and Undeen[192] were unable to identify definitely the early merogonic stages of the life cycle and considered all of the vegetative stages they observed to be sporonts or sporoblasts. However, Canning and Hulls[193] and Canning and Sinden[194] were able to find characteristics that enabled them to distinguish between meronts and sporonts. Early meronts were also observed subsequently by Undeen[339] in pig kidney cell cultures and sporoplasms were found by Avery and Anthony[340] in cells of *Anopheles albimanus*. Thus, it appears that some of the stage described by Vávra and Undeen[192] may have been meronts and may account for some of the variation in their description of the life cycle stages as compared with Canning and Hulls[193] and Canning and Sinden.[194]

Description and Life Cycle

Meronts[193,194] — Uninucleate meronts (Figure 5b) are rare and most often are found in the first few days of infection. They appear round in shape with a large nucleus. Most meronts are bi- or tetranucleate (Figure 5c, d) and range in shape from nearly spherical to elongate. The nuclei occur as a diplokaryon with the two closely adjacent nuclei having straight sides where the nuclear membranes are in close apposition. Cytoplasmic fission or cytokinesis of the tetranucleate meronts (Figure 5d, f) gives rise to two binucleate forms (Figure 5g), each with a diplokaryon. Occasionally, octonucleate meronts (Figure 5e) are formed which separate first into tetranucleate forms (Figure 5f) and then into two binucleate forms (Figure 5g). Binucleate meronts may continue to undergo merogony or may transform into sporonts.

Ultrastructurally, meronts show little cytoplasmic differentiation, with rough endoplasmic reticulum, ribosomes, and occasional expanded vesicles being the only structures present. The diplokaryotic nuclei are also readily evident. Meronts are characterized ultrastructurally by their surface structure. A layer of electron-dense material (which later gives rise to the exospore) occurs between the plasma membrane and closely packed tubular structures and also occupies the space between the tubules. Thus, in cross sections of the surface the exospore material appears alternately with spherical sections of the tubules above a continuous layer of the exospore material.

In a recent ultrastructural study on the early developmental stages of *Nosema algerae*, Avery and Anthony[340] found that sporoplasms, which become the meronts, are characterized by whorled vesicles in the cytoplasm and fibrous protrusions on portions of the limiting membrane. Sporoplasms are also binucleate, although the nuclei are not always in diplokaryotic arrangement. Dimensions of the meronts and other vegetative stages are presented in Table 10.

Sporonts[192-194,340] — The sporonts (Figure 5h) are typically binucleate but the occurrence of rare, uninucleate forms has been noted.[193] Sporonts appear identical to the meronts when stained with Giemsa and can only be distinguished from them at the light microscope level

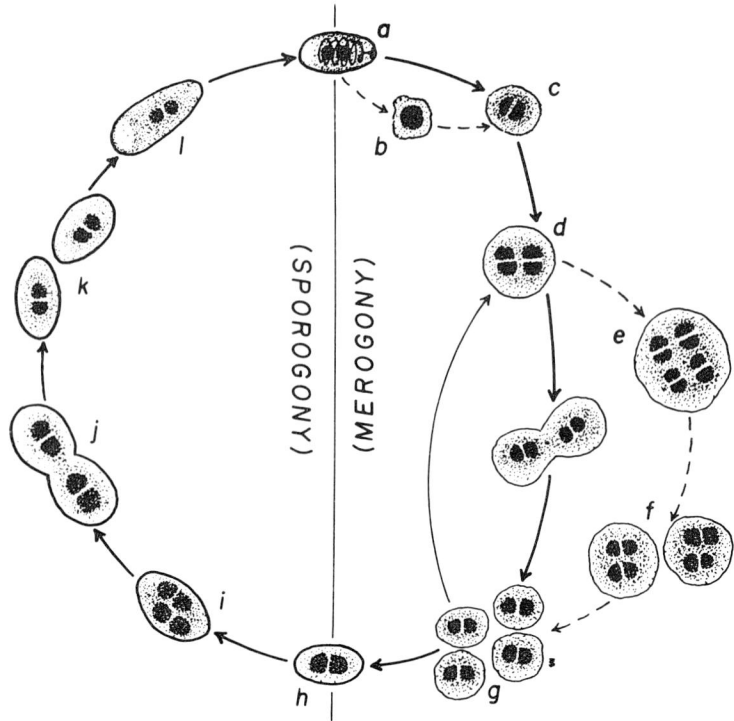

FIGURE 5. Schematic drawing of the life cycle of *Nosema algerae*. (a) Spore; (b) uninucleate meront; (c,g) binucleate meronts; (d,f) tetranucleate meronts; (e) octonucleate meront; (h) binucleate sporont; (i) tetranucleate sporont; (j) sporont undergoing cytokinesis; (k,l) binucleate sporoblasts.

Table 10
DIMENSIONS OF THE LIFE CYCLE STAGES OF *NOSEMA ALGERAE*

Stage	Size[a] (μm)	Shape	Ref.
Sporoplasms	$\bar{x} = 1.28 \pm 0.15 \times 1.91 \pm 0.29$		340
Meronts			
Uninucleate	3.4 × 3.6	Round	193, 194
Binucleate	3.3—3.6 × 5.0	Spherical to elongate	193, 194
Tetranucleate	4.0 × 6.6	Ovoid	193
Sporonts	2.1—2.3 × 4.0—5.3	Elongate	193
	4—5 × 5	Round to ovoid	192
Sporoblasts	2.5 × 4.5	Elongate	193, 194
	2.5 × 5.0	Slender ovoid	192
Spores			
Fresh (from larvae)	2.4—3.2 × 3.7—5.2	Ovoid with one end pointed	192
	$\bar{x} = 2.7 \pm 0.16 \times 4.3 \pm 0.29$		
(from adults)	2.3—3.9 × 3.7—5.4		192
	$\bar{x} = 2.8 \pm 0.20 \times 4.4 \pm 0.36$		
	1.8—2.5 × 3.0—4.3	Ellipsoidal with rounded poles	193
	$\bar{x} = 2.2 \pm 0.01 \times 3.7 \pm 0.32$		
Fixed	ca. 2 × 3.5		192
	$\bar{x} = 2.1 \pm 0.02 \times 3.1 \pm 0.32$		193
Polar tubes	51 ± 9.0		193
	ca. 70		192

[a] Values represent the mean, range, and/or mean ± SD.

by their more elongate shape. In electron micrographs the surface of the sporont possesses a continuous layer of exospore material that renders the tubules (seen in the meronts) invisible. They also contain rough endoplasmic reticulum and ribosomes.

Sporonts are formed from binucleate meronts but evidence for the occurrence of autogamy is conflicting. Vávra and Undeen[192] indicated that the occurrence of diplokarya in all the known stages of the life cycle of *N. algerae* and the retention of their integrity in these stages suggest that autogamy does not occur. However, Canning and Sinden[194] cite as evidence of autogamy the presence of a sporont nucleus in telophase and a diplokaryon with persisting centriolar plaques and fragments of microtubules. At any rate, nuclear division within the sporont gives rise to a tetranucleate sporont (Figure 5i) which divides (Figure 5j) to produce two sporoblasts (Figure 5k), each of which develops into a spore (Figure 5a, l).

Sporoblasts — Representing the prespore stage, sporoblasts have thickened walls; and in well-fixed material, the initial stages in the development of the polar tube are usually evident.[194,340] Sporoblasts are also characterized by cytoplasmic and nuclear condensation.

Spores — Fresh spores (Figure 6A) have been described as ovoid with one pole more pointed than the other[192] and also as typically ellipsoidal with rounded poles.[193] Spore dimensions are shown in Table 10. Posterior vacuoles have been seen in some fresh spores, and extruded polar tubes may have up to 12 to 14 undulations and vary in length up to 70 μm (Table 10). In Giemsa-stained preparations (Figure 6B) the spore nuclei stain deep violet.[192]

Although the ultrastructural features of the spore were not well demonstrated in their sectioned material, Vávra and Undeen[192] suggested that the spore contained eight to nine coils of the polar tube. Other authors have reported from 11 to 14 coils of the polar tube, with 11 to 12 as the most frequent.[193,194,341] Ultrastructurally the mature spore of *Nosema algerae* (Figure 6C) is typical of other species of *Nosema*, containing a polar tube, polar sac, polaroplast, ribosomes, diplokaryotic nuclei, and a two-layer wall composed of a transparent endospore and electron dense exospore. As seen in scanning electron micrographs (Figure 6D), the surface of the spore is smooth and nonornamented. While sometimes seen in fresh spores, Canning and Sinden[194] found no ultrastructural evidence of a posterior vacuole in mature spores.

Natural Occurrence

As already indicated, *N. algerae* has been found primarily as a pathogen in various laboratory colonies of anopheline mosquitoes. However, it has also been recorded, usually at very low incidences, from *Anopheles gambiae* from Liberia[338] and Nigeria,[342] *An. stephensi* from India,[343] *An. albimanus* from the west coast of El Salvador, Central America,[319] and in *Culex sitiens* from Cairns, Australia.[344] Similar and possibly identical species to *N. algerae* have also been recovered from *An. stephensi* collected in Pakistan.[345,346]

Nosema fumiferanae Thomson, 1955

This species was originally described by Thomson[201] as a member of the genus *Perezia* based on the disporous nature of the species and the lack of host cell hypertrophy. Weiser[212] classified it in the genus *Glugea*. In an unpublished thesis, Wilson[202] recognized its affinity with the genus *Nosema*, and it was first cited in print as *N. fumiferanae* by Percy.[347] Its placement in the genus *Nosema* was confirmed by Sprague[152] in his extensive treatise on the systematics of the microsporidia.

As a member of the genus *Nosema*, the stages in the life cycle of *N. fumiferanae* are similar to those described for *N. algerae* and many other species included in this genus. In fact, Nordin and Maddox[348] could not distinguish this species from the type species, *Nosema bombycis,* and several other species described from various lepidopterous hosts. Thus, its identity rests primarily on its association as an endemic pathogen of the spruce budworm, *Choristoneura fumiferana.*

32 CRC Handbook of Natural Pesticides

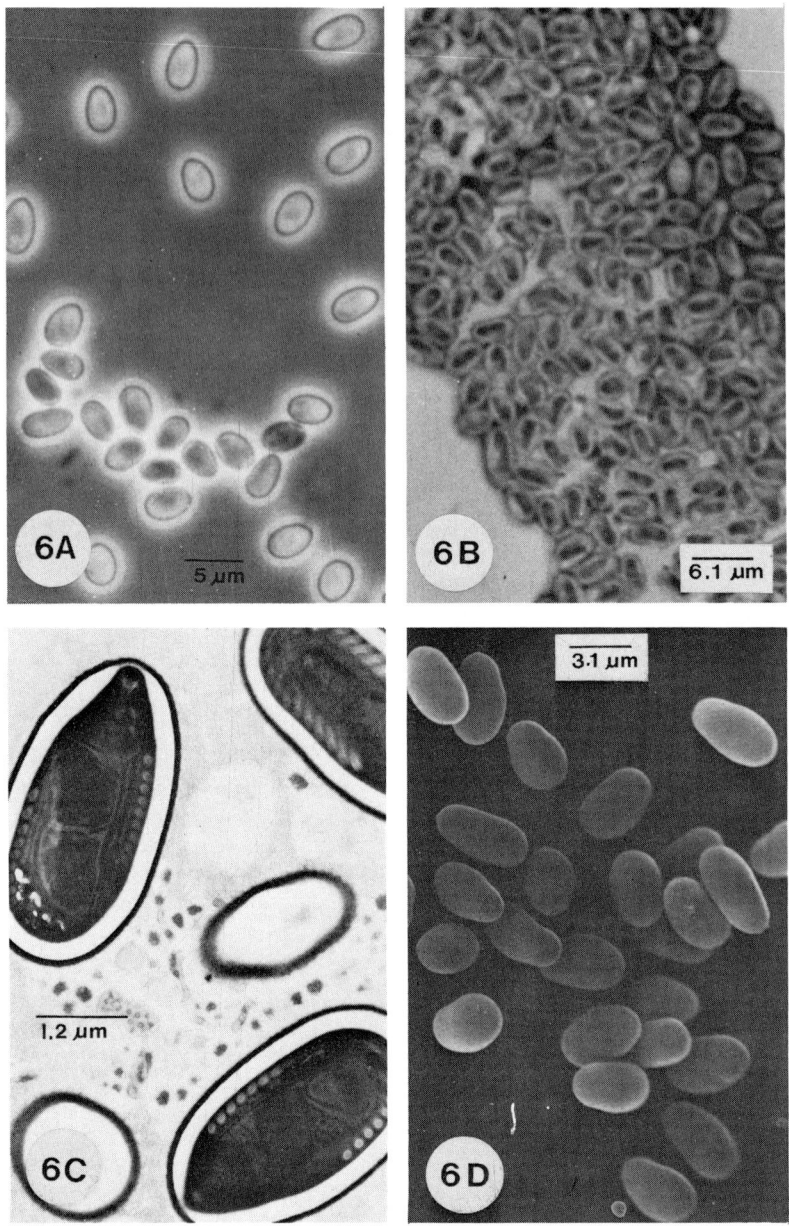

FIGURE 6. Spores of *Nosema algerae*. (A) Fresh, wet-mount preparation, phase (courtesy of J. V. Maddox); (B) stained with Giemsa (courtesy of J. V. Maddox); (C) transmission electron micrograph (courtesy of S. W. Avery, U.S. Department of Agriculture, Agricultural Research Service, Gainesville, Fla.); (D) scanning electron micrograph (courtesy of J. V. Maddox).

Description and Life Cycle[201]

Meronts — The initial meronts are small, about 1 μm in diameter, and in Giemsa-stained preparations, contain a deeply stained diplokaryon. These binucleate meronts typically produce tetranuclear forms which divide to produce two binucleate meronts. Occasionally nuclear division is independent of cytoplasmic cleavage and chains of diplokaryotic cells containing up to four cells are formed. Multinucleate forms containing up to six diplokarya

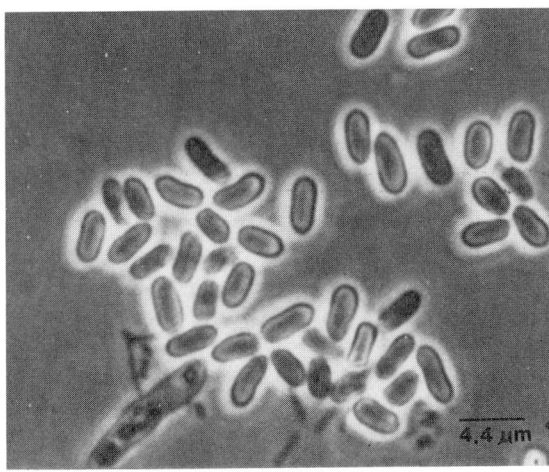

FIGURE 7. Spores of *Nosema fumiferanae*.

may also be formed which divide by budding or fragmentation to produce binucleate meronts. These forms may continue merogony or transform into sporonts. The terminal meronts are binucleate, rounded, about 5 μm in diameter, and in stained preparations, the nuclei are less deeply stained and their cytoplasm is usually vacuolated. Percy[347] found that the ultrastructure of meronts of *N. fumiferanae* is typical of that of other species of *Nosema* with each meront containing a diplokaryon along with smooth and rough endoplasmic reticulum and golgi complexes in the cytoplasm.

Sporonts — The sporogonic stages are usually larger and more elongate than the merogonic stages. In stained preparations, the cytoplasm of sporonts stains lightly and is generally streaked and mottled. Early sporonts are crescent-shaped, about 7 μm in length, and contain a diplokaryon. With nuclear division, a tetranucleate sporont is formed that divides to produce two binucleate sporoblasts.

Sporoblasts — The elongate sporoblasts are about 5 μm in length and each contain a diplokaryon. As with other species of microsporidia, the sporoblasts are also characterized by cytoplasmic and nuclear condensation.

Spores — Fixed spores measure 2 × 3 to 4 μm while fresh spores (Figure 7) are 2 × 3 to 5 μm. They are variable in shape ranging from ovoid to ovocylindrical in nature. A large vacuole is usually present at one end of the spore. Polar tubes, extruded under the stimulus of hydrogen peroxide, ranged from 65 to 105 μm in length, averaging about 80 μm. Although Thomson[201] indicated that the nuclei of spores apparently fuse in the mature spore, it is highly unlikely that this is the case based on more detailed investigations of the nuclear relationships of spores of similar species in the genus *Nosema*.

Natural Occurrence

According to Wilson,[320] *Nosema fumiferanae* was probably first reported by Graham[349] from three widely separated districts in Ontario. Thompson[201] indicated that the microsporidium was present in varying degrees in most budworm populations throughout Ontario, and later reported percentage infections of 36.5, 43.6, and 56.1 respectively, from 1955 to 1957 in overwintering larvae of the spruce budworm from the Uxbridge Forest in Ontario.[350] In an additional report, Thomson[351] also included similar data obtained for the years 1958 and 1959, reporting an infection rate of 69.1 and 81.3%, respectively for each year. From a survey of budworms in Parkinson Township, Ontario, Wilson[352] presented data for the years 1971 to 1973 with infection rates ranging from 22.4 to 50.6% for 1971, 33.3 to 41.6%

Table 11
PREVALENCE OF INFECTION BY *NOSEMA FUMIFERANAE* IN *CHORISTONEURA FUMIFERANA*, UXBRIDGE FOREST, ONTARIO, CANADA[353]

Year	Collection date	Predominant instars	No. examined	Infection (%)
1973	May 29	III-V	216	12.9
	June 29	IV-P	354	35.9
1974	June 4	V-VI	231	18.6
1975	June 5	VI-P	467	43.0
1976	June 9	VI-P	341	56.0
1977	June 1	VI-P	434	56.2
1978	June 14	VI-P	258	69.0

for 1972, and 66.8 to 82.0% for 1973. More extensive data obtained by Wilson[353] from Uxbridge Forest populations of the spruce budworm for the years 1973 to 1978 are shown in Table 11. Thus, *N. fumiferenae* is a common microsporidium associated with natural populations of the spruce budworm, often at epizootic levels.

Nosema gasti **(McLaughlin, 1969)**

Although McLaughlin[29,354] initially referred to this microsporidium from the cotton boll weevil, *Anthonomus grandis*, as a *Nosema* sp., he[203] subsequently described it as a member of the genus *Glugea* based on the disporous nature of the sporont. The species had been cited by Ignoffo and Garcia[355a] as a species of *Pleistophora*, probably because host cells packed with spores were considered as pansporoblasts.[203] Streett et al.[204] transferred it to the genus *Nosema* after the disporous nature of this genus was definitely established.[197,356]

Description and Life Cycle[203]

Meronts — Two merogonic phases can be distinguished in the life cycle of *Nosema gasti*. The meronts of the first phase are characterized, in stained material, by eccentrically located vesicular nuclei and cytoplasm with a thicker or darker region at the periphery of the meront. Binucleate sporoplasms (the stage released from a spore) divide into mononucleate bodies whose nucleus apparently divides mitotically to produce binucleate meronts. The second merogonic phase is initiated by these binucleate meronts characterized by a condensation of nuclear material into an irregular mass and well stained cytoplasm. Nuclear division results in the formation of tetranucleate meronts and cytokinesis produces two binucleate meronts. Occasionally octonucleate meronts are also formed. Second stage meronts are typically larger than first stage meronts and the cytoplasm and nuclei stain more homogeneously and intensely. Nuclei typically occur as diplokarya, especially in the terminal stage binucleate meronts.

Sporonts — Although not confirmed ultrastructurally, McLaughlin[203] cited evidence in Giemsa-stained preparations strongly suggestive of autogamy. The nuclei of early stage sporonts enlarge and apparently fuse. In some sporonts nuclear material is stranded with thickened, granular portions on the strands and no definitive nuclear membrane or boundary is seen. Nuclear configurations suggestive of meiosis are also seen. Thus, through nuclear division of the uninucleate sporont, tetranucleate sporonts are formed which divide to produce two sporoblasts. Nuclei, again, occur as diplokarya and the sporont cytoplasm is diffuse and stains lightly. The sporont also typically elongates prior to cytokinesis when two sporoblasts are formed.

Sporoblasts and spores — Sporoblasts are characterized by their gradual assumption of the typical shape of spores and the formation of the spore membrane. Apparently nuclear

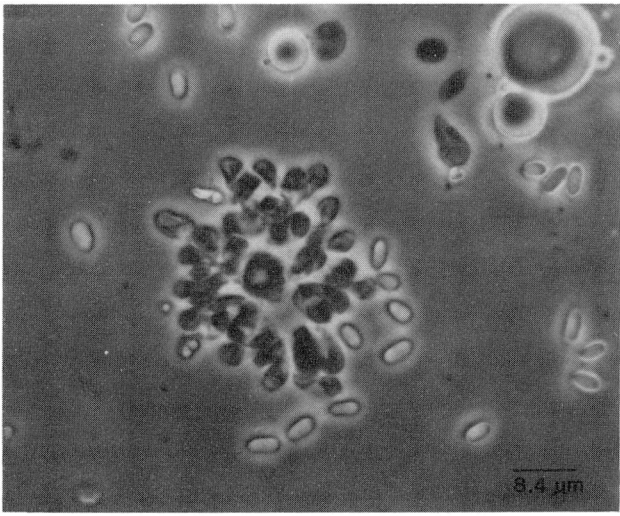

FIGURE 8. Spores of *Nosema gasti*. (Courtesy of R. E. McLaughlin, U.S. Department of Agriculture, Agricultural Research Service, State College, Miss.)

fusion again occurs and the uninucleate sporoblasts are characterized by a large nuclear mass that stains deeply and a polar granule that occurs in the polaroplast area. Spores with two nuclei are eventually formed that are ovoid with bluntly rounded ends and slightly reniform in shape. Polar tubes range in length from 49 to 97 μm, averaging 76 μm. Fresh spores (Figure 8) from adult boll weevils measure $4.34 \pm 0.3 \times 2.3 \pm 0.2$ μm and from larvae measure $4.16 \pm 0.3 \times 2.3 \pm 0.2$ μm. The life cycle from spore to spore is apparently completed within 24 hr in the midgut of larvae or adults, although both merogonic and sporogonic stages are common in the entire alimentary track and most other host tissue for at least 4 to 6 days postinfection.

Natural Occurrence[203]

Nosema gasti was originally isolated in 1962 from colonies of the cotton boll weevil maintained in laboratories in Mississippi, Texas, and South Carolina. However, cultures had been exchanged previously between these laboratories and had been started from weevils collected in the field from the U.S. and Mexico. Thus, the exact origin of the microsporidium is unknown and no data have been obtained on its natural occurrence.

Nosema locustae Canning, 1953

This microsporidium was first referred to as "probably a new species" of the genus *Nosema* by Steinhaus[357] and was described subsequently by Canning[205] as *N. locustae*. According to Canning,[206] it is also similar to or identical to a protozoan mentioned in several earlier reports by Goodwin[358-360] and Goodwin and Srisukh.[361]

Description and Life Cycle[205,206]

Meronts — Early meronts are uninucleate and spherical in shape, measuring 2.5×3.5 μm in diameter. As nuclear division occurs the cytoplasm of the meronts stains less homogeneously and frequently is vacuolated in appearance. Apparently cytokinesis of the tetranucleate meronts results in the production of uninucleate stages which may continue merogony or they may transform directly into sporonts.

Sporonts — Although somewhat unusual when compared to the life cycle stages of most other species of *Nosema*, the uninucleate products of merogony apparently develop directly

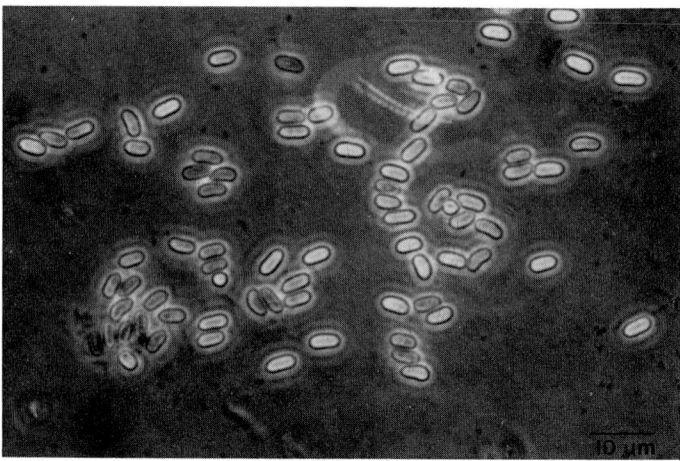

FIGURE 9. Spores of *Nosema locustae*.

into sporonts which become oval in shape. The nucleus divides to produce a binucleate sporoblast. More recently Henry and Oma[321] claim to have observed binucleate sporonts in the life cycle of *N. locustae* which undergo karyokinesis followed by cytokinesis — thus indicating the disporous nature of species.

Sporoblasts and spores — The wall of the sporoblast becomes thickened gradually to give rise to a mature spore. Spores are generally oval or ellipsoidal in shape, binucleate, and measure 4 to 6.5 × 2.5 to 3.5 µm in size in the fresh state (Figure 9). A few are very small (3 × 1.5 µm) while others are quite large (7 × 3.5 µm). Fixed spores are smaller, with the majority measuring between 3 to 5 µm in length by 2 to 2.5 µm in width. Polar tubes, extruded by mechanical pressure, vary between 75 to 156 µm in length.

In a classic study on the ultrastructure of a microsporidian spore, Huger[362] examined spores of *N. locustae* from the fat body of the locust, *Locusta migratoria migratorioides*. While agreeing incorrectly with others that the polar tube was a solid structure, Huger demonstrated the multilayered nature of the spore wall, described and named the polaroplast and suggested its function in the extrusion of the polar tube and sporoplasm, and described the other major architectural features of the spore internally including the polar tube and the sporoplasm. Up to 18 coils of the polar tube were counted in some sections but the average number has not been determined.

Natural Occurrence

Nosema locustae was originally described from a laboratory colony of *Locusta migratoria migratorioides* in England,[205] and it has also been identified from laboratory colonies of several species of grasshoppers from Montana.[357,363] It is also a common pathogen of a large number of various species of field-collected grasshoppers, having been found in natural populations from Montana, North Dakota, Minnesota, Oregon, Wyoming, Colorado, Arizona and Idaho in the U.S.,[321] and in Ontario[364] and Saskatchewan[365] in Canada. Henry[363] listed 47 species of grasshoppers which were found to be naturally infected from various areas in the western U.S. Prevalence of infection was relatively low (Tables 12 and 13) in most species with the highest levels of infection occurring in certain locations where three species of grasshoppers were consistently present.[366] In addition to its natural occurrence in North America and in laboratory colonies in England, Issi and Lipa[367] identified *N. locustae* in a grasshopper from the Soviet Union; undoubtedly, it also occurs in other regions where grasshoppers are common.[321]

Table 12
PREVALENCE OF INFECTION AMONG SUBFAMILIES OF GRASSHOPPERS COLLECTED IN CAMAS COUNTY, IDAHO[366]

	Subfamily					
	Cyrtacanthacridinae		Oedipodinae		Acridinae	
Year	No. examined	Infected (%)	No. examined	Infected (%)	No. examined	Infected (%)
1963	5,644	3.5	82	6.1	67	1.5
1964	10,833	3.9	251	2.0	164	2.4
1965	8,635	6.5	288	1.4	426	0.7
1966	3,683	8.7	233	6.0	365	3.3
1967	491	2.0	148	2.0	234	0.4

Table 13
PREVALENCE OF INFECTION OF *NOSEMA LOCUSTAE* IN THE MOST ABUNDANT SPECIES OF GRASSHOPPERS FROM CAMAS COUNTY, IDAHO, 1963—1967[366]

Species	No. examined	Infected (%)
Amphitornus coloradus	117	3.4
Arphia pseudonietana pseudonietana	101	1.0
Aulocara elliotti	261	2.7
Camnula pellucida	324	3.7
Chorthippus curtipennis	370	1.6
Dissosteira carolina	333	0.9
Melanoplus bivattatus	3,324	4.1
M. femurrubrum femurrubrum	525	4.0
M. foedus foedus	397	5.3
M. sanguinipes sanguinipes	23,196	5.5
Oedaleonotus enigma	1,551	4.9
Pseudopomala brachyptera	399	1.0

Nosema pyrausta Paillot, 1927

In his initial, brief description of this species, Paillot[208] proposed the name *Perezia pyrausta* for the microsporidium from the European corn borer, *Ostrinia nubilalis,* but subsequently used the name *P. pyraustae* in his more detailed description published in 1928.[209] The life cycle of the species has been studied by several authors.[210-213] Weiser[212] transferred it to the genus *Nosema* but its correct specific name (*pyrausta*) was not used until 1974 by Lewis and Lynch.[368] Although *Nosema pyrausta* is now accepted as the proper name for this microsporidium,[152] it is referred to as *Perezia pyraustae* in most of the extensive literature on the species as a parasite of *O. nubilalis.*

Description and Life Cycle[208-211,213]

Meronts — While Hall[210] suggested that the earliest meronts are uninucleate, most other authors[208,209,211,213] found only small, spherical bodies with two nuclei, usually in diplokaryotic arrangement. Nuclear division produces a tetranucleate meront which usually undergoes cytokinesis to give rise to two daughter, binucleate meronts. Characteristically, nuclear

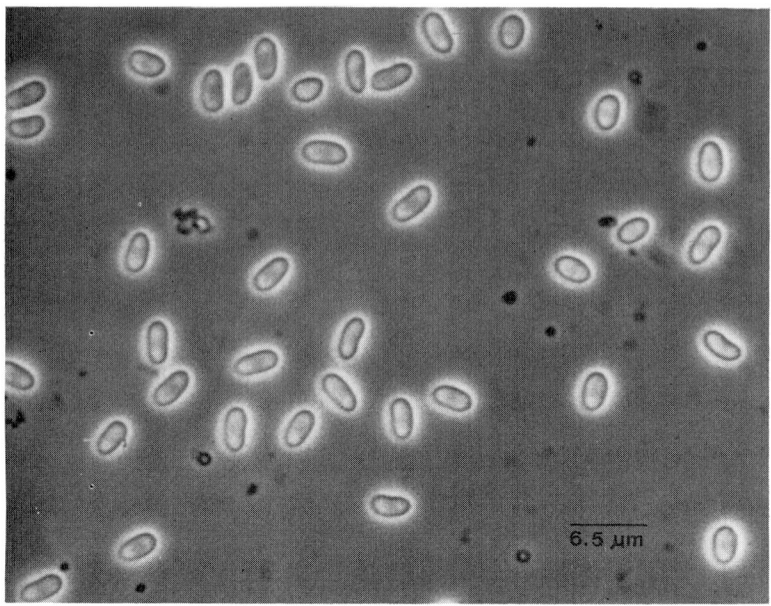

FIGURE 10. Spores of *Nosema pyrausta*. (Courtesy of T. C. Andreadis.)

division in some of the meronts of this species proceeds independently of cytokinesis to give rise to ovoid to spherical-shaped meronts with up to 16 nuclei;[211,213] however, 4 or 8 nuclei have been most commonly reported. These multinucleate meronts may give rise to chains of binucleate meronts or bud exogenously to form groups of binucleate meronts. Meronts range in size from 3 to 4 μm in diameter in the binucleate forms and up to 4 to 7 μm in diameter in the multinucleate forms. Multinucleate chains may range in length from 8 to 14 μm.[211]

Sporonts — Developing directly from binucleate meronts, the sporonts are initially binucleate and with nuclear division give rise to tetranucleate forms. In Giemsa-stained material, the cytoplasm of the sporonts is usually stained and vacuolated in appearance. They vary in size from 9 to 10 μm in length by 4 to 5 μm in width and are broadly elliptical in shape. Two sporoblasts are formed from each of the tetranucleate sporonts by cytoplasmic cleavage.

Sporoblasts — Although initially elliptical in outline, the binucleate sporoblasts become ovoid in form after cytokinesis of the sporont. They vary from 5 to 6 μm in length by 2.5 to 3.5 μm in width. The sporoblast cytoplasm is also vacuolated and with further condensation and membrane formation give rise to a mature spore.

Spores — The spores are typically ovoidal to ovocylindrical in shape (Figure 10) but many spores are also highly variable in shape. Some are reniform, didymiform, tubular, or highly irregular in form.[369] According to Kramer,[369] many of the latter types are probably anomalies and were also noted by Paillot[209] and Hall.[210] While Hall[210] concluded that nuclear fusion results in a uninucleate condition in the mature spore, Kramer[211] found only binucleate spores and also noted that the extruded sporoplasm was binucleate. The binucleate condition was recently confirmed by Andreadis[370] in electron micrographs which also illustrated typical nosematid-like ultrastructural features of spores of *Nosema pyrausta*. The polar tube has 10 to 12 coils. Various measurements on the length of the extruded polar tube are given in Table 14 as well as the dimensions of spores.

Table 14
MEASUREMENTS OF SPORES AND POLAR TUBES OF *NOSEMA PYRAUSTA*

Host	Stage	Length (μm) x̄	SE	Range	Width (μm) x̄	SE	Range	Ref.
Junonia coenia	Fixed spores	4.5	—	3.5—6.0	2.0	—	1.8—3.0	210
	Polar tubes			30—65				
Ostrinia nubilalis	Fixed spores	4.2	—	3.2—4.7	2.0	—	1.8—2.6	211
	Polar tubes	110.0		90—126				
O. nubilalis	Fixed spores							369
	Ovocylindrical	—	—	2.29—3.99	—	—	106—1.84	
	Oviodal	—	—	2.90—4.60	—	—	1.13—2.35	
O. nubilalis	Fixed spores	4.0	—	3.1—4.8	2.2	—	1.7—2.4	213
	Polar tubes	81.6	—	69.2—138.4				
O. nubilalis	Fresh spores	4.16	0.06	—	1.76	0.02	—	370
Macrocentrus grandii	Fresh spores	4.23	0.06	—	1.75	0.02	—	370

Natural Occurrence

Nosema pyrausta is a very common parasite of the European corn borer having been found in France,[208,209] China,[213] Canada,[213a] and from many localities within the U.S. The species was first identified in *Ostrinia nubilalis* from the U.S. by Steinhaus;[357] and, as suggested by Kramer,[371] is likely found throughout the distributional range of its host. Observations on the prevalence of infection in natural populations of *O. nubilalis* have been reported from at least 14 states[26,357,371-376a] and extensive surveys on the seasonal prevalence of infection by *N. pyrausta* were reported for Illinois,[371] Connecticut,[376a] and Nebraska.[375] In the latter study, Hill and Gary[375] reported the results of an extensive 16-year survey for two counties in Nebraska which indicated that infection in one county was mostly enzootic but reached epizootic proportions twice in the other county. Thus, they concluded that a high host density and a wide-spread spatial distribution are requisites for the epizootic development of *N. pyrausta* in field populations of the European corn borer.

Pleistophora schubergi schubergi (Zwölfer, 1927)

This species was originally described by Zwölfer[224] from larvae of the brown tail moth, *Euproctis chrysorrhoea,* and the gypsy moth, *Lymantria dispar.* Since its description, a number of similar species of *Pleistophora* have been described from other lepidopterous hosts. These were morphologically identical in most respects to Zwölfer's species and distinguished primarily on the basis of host specificity. In his monograph Weiser[212] treated *P. aporiae, P. pandemis,* and *P. hyphantriae* as subspecies or forms of *P. schubergi.* Two other species, *P. balbianii* and *P. noctuidae* are also presently recognized as subspecies;[152,377,378] and, a new subspecies, *P. schubergi neustriae* was recently described by Purrini.[379] Thus, there are seven subspecies presently recognized with Zwölfer's species being referred to as *P. schubergi schubergi.* Although there are a number of descriptions of this species,[224-227] they are rather brief in nature as are the results of studies on the ultrastructure of the species[224a] and on the subspecies *P. schubergi neustriae.*[379] Purrini[379] indicated that *P. schubergi neustriae* conformed in its characteristic features to those of the genus *Pleistophora* as redefined by Canning and Nicholas;[379a] however, as a result of a recent examination of the ultrastructure of *P. schubergi,* Maddox[379b] concluded that the species probably belongs to the genus *Vavraia* as redefined by Canning and Hazard.[250] Milner and Briese[224a] also concluded that this species fits more closely into the genus *Vavraia* but did not propose its transfer since they found no ultrastructural evidence that the spo-

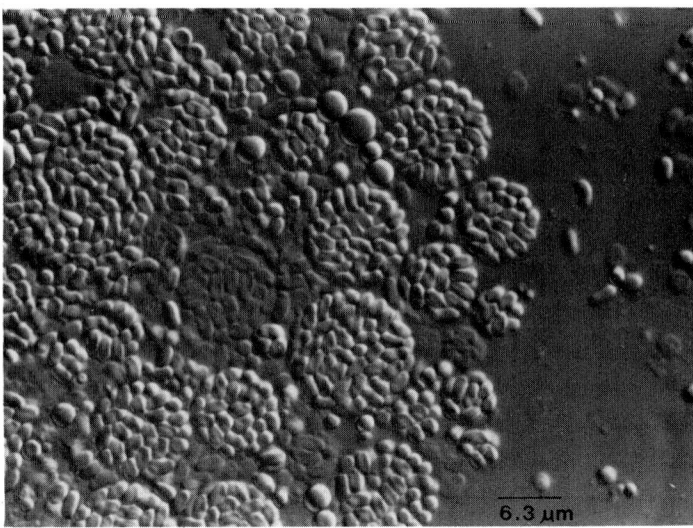

FIGURE 11. Sporophorous vesicles containing spores of *Pleistophora schubergi*, Nomarski interference contrast. (Courtesy of J. V. Maddox.)

rogonial plasmodia undergo multiple fission by rosette formation into uninucleate sporoblasts. Thus, at the present time it is still necessary to refer to this microsporidium as *P. schubergi*. This review will also deal primarily with reports where the species is referred to as *P. schubergi;* data on the other subspecies will not be included.

Description and Life Cycle[224,224a,226,227]

Meronts — The initial meronts are small, from 1 to 4 μm in diameter, and have a single nucleus. Older forms (4 to 7 μm in diameter) contain two or four nuclei and usually give rise to multinucleate forms which may be cylindrical or chain-like in appearance and up to 20 μm in length. Cytoplasmic cleavage gives rise to uninucleate or binucleate forms which develop into sporonts.

Sporonts — In Giemsa-stained preparations, the sporonts are lighter in color and usually vacuolated in appearance. Large, round multinucleate forms (plasmodia) are eventually formed which may contain up to 32, 64, or more nuclei. The plasmodia vary in size from 10 to 30 μm. Each nucleus serves as the focal point for cytoplasmic condensation and fragmentation, giving rise to the sporoblast.

Sporoblasts — With cytoplasmic condensation and spore-membrane formation, the oval sporoblasts become spores (Figure 11), surrounded at least initially by a sporophorous vesicle.

Spores — The spores are oval in shape and relatively small in size (Table 15). According to Lipa[226] spore size varies with the number of spores formed within each sporophorous vesicle. The fewer the number of spores per vesicle the greater the spore size. The sporophorous vesicle is not persistent,[212] and the spores are eventually released with its rupture. The polar tube is short in length (Table 15) and can be extruded with acetic acid, Lugol's fluid,[226] or upon drying and rewetting of spores.[224a]

At the electron microscopic level, Milner and Briese[224a] demonstrated that the sporoblasts and spores were formed within a sporophorous vesicle that they suggested was apparently part of the host cell's membrane system. Spores were uninucleate and the polar tube had about eight coils.

Natural Occurrence

This species has been reported as naturally occurring in populations of several species of

Table 15
MEASUREMENTS OF SPORES AND POLAR TUBES OF *PLEISTOPHORA SCHUBERGI*

Host	Stage	Length (μm) x̄	Length (μm) Range	Width (μm) x̄	Width (μm) Range	Ref.
Euproctis chrysorrhea and *Lymantria dispar*	Spores					224, 225
	Stained	2.5	—	1.5	—	
	Polar tubes		Longest = 35			
E. chrysorrhea	Spores	2.5	—	1.5	—	226
	Polar tubes	—	25—50			
Choristoneura fumiferana	Spores					227
	Fresh	2.5	1.9—2.9	1.4	1.2—1.8	
	Stained	2.4	1.0—2.7	1.4	1.2—1.7	
	Polar tubes	42	21.3—53.2			
Archips cerasivoranus	Spores					382
	Stained	2.2	1.7—2.9	1.6	1.2—1.7	
Anaitis efformata	Spores					224a
	Fresh(?)	1.7	—	1.0	—	
	Polar tubes	36.8	30—48			

lepidopterous pests of forest trees.[224,226,227,328,380,381] Zwölfer[224] reported a prevalence of infection of 70% of field-collected larvae of *Lymantria dispar* and about 84% in *Euproctis chrysorrhoea*. Lipa[226] recorded an infection rate of 3 to 8% in 1st and 2nd instar larvae of *E. chrysorrhoea* in Poland which increased to 45 to 50% in 3rd and 4th stage larvae collected the following spring. In 1978, 35.9% of the larvae of *Archips cerasivoranus* were infected with *Pleistophora schubergi* while only 3% were found infected in 1979.[382] Thus, this microsporidium appears to be endemic in natural populations of several lepidopterous hosts and occasionally increases in prevalence to epizootic proportions.

Vairimorpha necatrix Kramer, 1965

A microsporidiosis of the armyworm, *Pseudaletia unipuncta*, was first reported by Tanada and Chang[383] in Hawaii. Tanada[384] suggested that it appeared to be a mixed infection of a *Nosema* and *Thelohania* species. Kramer[244] described the microsporidia as *Nosema necatrix* and *Thelohania diazoma* from an Illinois population of *P. unipuncta*. Subsequent studies by Maddox,[385] Tanabe,[386] Fowler and Reeves,[387] Pilley,[245] and Maddox and Sprenkel[247] demonstrated that only a single, dimorphic species was involved. Although Fowler and Reeves[387] proposed retaining the name *N. necatrix* for the species, its dimorphic development sharply distinguished it from most other species of *Nosema*. In 1976, Pilley[245] established it as the type species of the genus *Vairimorpha*.

Description and Life Cycle

Characteristic of its status as a dimorphic species, the developmental stages of *V. necatrix* have been variously referred to as being on the one hand as *Nosema*-like or disporoblastic and on the other hand as *Thelohania*-like or octosporoblastic.[245] Although sporogenic development is disporoblastic initially and later also octosporoblastic, sporogony is also temperature dependent with disporoblastic development only at 32°C and both di- and octosporoblastic development occurring at lower temperatures (15 to 25°C).[245,385,387]

Meronts — The meronts of forms destined to undergo disporoblastic development are typically binucleate, generally spherical in shape, and variable in size. Nuclear division gives rise to tetranucleate forms that divide to give rise to binucleate daughter meronts. According to Pilley,[245] the octonucleate forms described by Kramer[244] are actually part of

Table 16
MEASUREMENTS OF GIEMSA-STAINED STAGES OF *VAIRIMORPHA NECATRIX*[245]

Stage	Measurements (no.)	Length (or diameter) (μm)			Width (μm)		
		Mean	SD	Range	Mean	SD	Range
Nosema type							
Early binucleate meront	25	2.3	0.3	1.6—3.0	—	—	—
Uninucleate meront	10	3.0	0.6	2.3—4.5	—	—	—
Presporont	25	4.1	0.3	3.5—4.7	—	—	—
Early sporont	25	6.4	0.8	5.0—7.9	2.8	0.3	2.4—3.7
Late sporont	25	9.9	1.3	7.9—12.1	2.8	0.7	1.7—4.9
Sporoblast	25	5.8	0.5	4.5—6.9	2.3	0.3	1.8—2.7
Spore	50	4.3	0.2	3.9—5.0	2.3	0.2	2.0—2.7
Thelohania type							
Giant binucleate meront	25	7.1	0.8	5.2—8.8	—	—	—
Giant tetranucleate meront	5	7.7	0.6	6.8—8.2	—	—	—
Giant octonucleate meront	5	8.5	1.2	6.6—10.0	—	—	—
Binucleate sporont	25	4.8	0.5	4.0—5.8	—	—	—
Tetranucleate sporont	25	5.6	0.7	4.6—6.5	—	—	—
Octonucleate sporont	25	5.6	0.6	4.7—7.0	—	—	—
Pansporoblast	25	6.4	0.5	5.6—7.4	—	—	—
Eight-spore group	25	6.9	0.3	6.0—8.5	—	—	—
Individual spore	50	1.9	0.1	1.5—2.0	1.1	0.1	1.0—1.3

the merogonic developmental cycle of the octosporoblastic form. Since both developmental patterns are exhibited at low temperatures, differentiation of the merogonic stages is difficult and somewhat open to individual interpretation. However, Pilley[245] found that the meronts of forms undergoing octosporoblastic development are frequently large in size and contain up to eight nuclei. In both developmental forms, binucleate daughter meronts apparently transform directly into sporonts although Kramer[244] considered that nuclear fusion results in the formation of uninucleate sporonts which divide mitotically to produce binuculeate sporonts. Measurements of meronts and other stages are presented in Table 16.

Sporonts — Early stage sporonts are characterized by their spherical shape and diplokaryotic nuclei.[244,245] In forms undergoing disporoblastic development these sporonts elongate and with nuclear division give rise to tetranucleate forms, generally spindle shaped and greatly elongated, which undergo cytokinesis to produce two sporoblasts. Nuclear division in forms undergoing octosporoblastic development gives rise to spherical sporonts with four and then eight nuclei. With cytoplasmic cleavage, eight sporoblasts are formed.

Sporoblasts — Disporoblastic development in the *Nosema*-like forms results in the formation of two sporoblasts, each of which develops independently into a mature spore. In Giemsa-stained preparations, the two nuclei are deeply stained in contrast to the palely stained and vacuolated cytoplasm.[245] In the *Thelohania*-like forms, each of the eight sporoblasts contains a single nucleus. The sporoblasts develop into mature spores held together by the sporont membrane. Thus, the sporogonial plasmodium gives rise to eight spores enclosed with a persistent, sporophorous vesicle.[244,247]

Spores — The spores (Figure 12A) produced as a result of disporoblastic development were originally recorded by Kramer[244] from fixed and stained preparations as varying in size from 5.06 to 6.05 × 2.82 to 3.98 μm. Spore measurements of similar material from *Spodoptera exempta* by Pilley[245] (Table 16) indicate a slightly smaller spore size. However, Maddox[385] and Maddox and Sprenkel[247] have shown that spore length of the *Nosema*-like

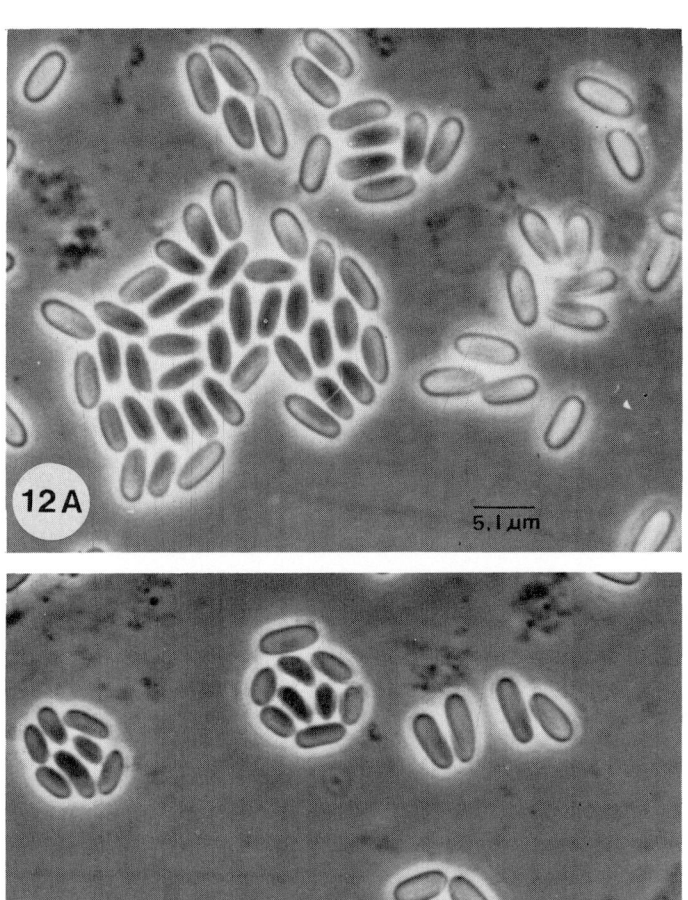

FIGURE 12. Spores of *Vairimorpha necatrix*. (A) Spores produced as a result of disporoblastic development at 26.7°C; (B) spores (within sporophorous vesicles produced as a result of octosporoblastic development at 21.1°C.

form is highly dependent upon temperature. Spores produced in hosts held at lower temperature are longer than spores produced at high temperatures (Table 17). The *Thelohania*-like spores (Figure 12B) are smaller and less variable in size. In *Pseudaletia unipuncta*, Kramer[244] recorded the spores as ranging from 2.82 to 3.98 μm in length by 1.74 to 2.40 μm in width. Pilley's[245] measurements (Table 16) indicate a slightly smaller spore size in *S. exempta* but it is not known if this variation is due to either host species or temperature.

The ultrastructural features of this species have not been examined although Fowler and Reeves[388] showed the smooth, nonornamented nature of the spore wall in scanning electron micrographs. The fine structure of this species is probably quite similar to that recently described by Malone and Canning[248] for *Vairimorpha plodiae*, a dimorphic species also infectious for lepidopterous larvae. While factors affecting polar tube extrusion have been examined for *V. necatrix*,[389] few measurements on tube lengths have been presented. Kramer[244]

Table 17
AVERAGE SIZE OF SPORES OF *VAIRIMORPHA NECATRIX* PRODUCED BY DISPOROBLASTIC DEVELOPMENT IN LARVAE OF *PSEUDALETIA UNIPUNCTA* REARED AT DIFFERENT TEMPERATURES[247]

Temp. (°C)	Stained spores[a]			
	Length (μm)			Width (μm)
	Mean	SE	Range	
32	3.3	0.7	1.7—4.5	1.9
27	3.9	0.3	3.0—4.6	1.9
21	4.4	0.5	3.5—6.1	1.7
15	4.7	0.7	2.8—5.9	1.8
10	5.2	0.7	3.5—8.0	2.0
	Fresh spores[a]			
32	4.1	0.8	2.4—5.2	2.2
27	4.9	0.4	4.0—5.7	2.4
21	5.2	0.4	4.2—6.0	2.4
15	5.7	0.5	4.2—6.9	2.4

[a] n = 100.

indicated that the polar tube of the *Nosema*-type spores ranged up to 90 μm in length while the partially extruded tubes of the *Thelohania*-type spores were less than 20 μm in length.

Natural Occurrence

Despite the absence of microsporidia in over 2000 larvae of the armyworm, *Pseudaletia unipuncta*, collected during 1955 and 1958 to 1960, Tanada and Chang[383] reported that 54.1% of 110 specimens collected during 1961 from rangeland in Hawaii were infected. In samples taken in 1962, Tanada[390] reported that 37.6% of 109 larvae were infected from the same locale where the pathogen had been isolated the previous year. However, in an adjacent site, only 1 of 55 larvae was infected. More extensive sampling in 1965 revealed that only very low rates of infection were apparent in a low level population of *P. unipuncta* from the initial site of discovery.[391] In various populations of the fall webworm, *Hyphantria cunea*, in Illinois, Nordin et al.[392] also found relatively high rates of infection by *V. necatrix* in colonies and larvae (Table 18). However, despite its natural occurrence in larvae of 14 species of Lepidoptera,[247,326] infections in most species from field crops are usually rare.[326] Thus, *V. necatrix* does not appear to be an important natural regulating factor of field crop pests and may only cause high levels of infections in situations conducive to rapid dispersal (e.g., within webworm colonies).

Vavraia culicis Weiser, 1947

Although originally referred to as *Plistophora kudoi* by Weiser in 1946,[167] this parasite of mosquitoes was subsequently renamed *P. culicis*[249] after he realized that priority for *P. kudoi* had already been established for another microsporidium. The initial descriptions by Weiser[167,249] were supplemented by Canning.[327] Another species, *Plistophora culisetae*, was later described from mosquitoes by Weiser and Coluzzi,[393] but was subsequently placed

Table 18
**PERCENTAGE OF *HYPHANTRIA CUNEA*
COLONIES AND LARVAE INFECTED
WITH *VAIRIMORPHA NECATRIX* FROM
ILLINOIS IN 1970[392]**

Region	No. colonies	Infected (%)	No. larvae reared[a]	Infected (%)
Northern	9	89	225	34
Central	17	41	425	12
Southern	29	21	735	7

[a] 25 larvae individually reared from each colony.

in synonymy with *P. culicis*.[394] In his new classification for the microsporidia, Weiser[149] proposed *P. culicis* as the type species for the new genus *Vavraia* defined as possessing a sporogonial plasmodium which divides five times to produce 32 spores within a persistent or subpersistent sporophorous vesicle. More recently, Canning and Hazard[250] confirmed the generally accepted view that this species produces a variable rather than fixed number of spores from the sporogonial plasmodium and redefined the genus based primarily on the unpaired nature of the nuclei during merogony and sporogony and the formation of a cyst around the sporogonial plasmodium during sporogony.

Description and Life Cycle
Meronts[250,327] — The early stages of infection are represented by meronts with two or more unpaired nuclei. Nuclear division ultimately produces plasmodia with up to 16 nuclei. Ultrastructurally, the meronts are coated externally by a thin, irregular layer of amorphous material which also divides during cytoplasmic cleavage. Thus, an amorphous coat also surrounds the daughter meronts, strands of which may persist for a time as connections between the coats of the uninucleate products of division.

Sporonts — In Giemsa-stained preparations the nuclei of sporonts are similar in size to those of the meronts but are more numerous and may be closely packed in individual sporogonial plasmodia. The cytoplasm does not become vacuolated although the formation of cleavage lines at maturity may result in a vacuolated appearance.[327] In electron micrographs the surface membrane of the sporogonial plasmodium can be observed to retract from the amorphous coat which is retained as a cyst wall surrounding the plasmodium during sporogony. The plasmalemma becomes the surface covering of the sporogonial plasmodium and undergoes involution with division of the plasmodium into sporoblasts.[250]

Sporoblasts — The uninucleate sporoblasts are formed by multiple fission of the sporogonial plasmodium by rosette formation and subsequent condensation of cytoplasm around each nucleus. As the sporoblasts mature into spores, the cyst wall undergoes reorganization into a double-layered structure. No evidence of meiotic division has been seen during sporulation.[250]

Spores — With formation of the spore wall, the sporoblasts develop into mature spores and occur as groups (Figure 13) of 10 to 60 or more within the cyst wall.[167,327] The cysts vary considerably in size, ranging from 10.3×11.8 μm to 26.5×17.6 μm.[327] Canning and Hazard[250] established that 8, 16, and 32 spores were the commonest number present within cysts but cysts with 64 spores were also seen. Thus, as indicated earlier, the attempt by Weiser[149] to establish the genus *Vavraia* based on the five synchronous nuclear divisions to produce 32 spores with *V. culicis* as the type species has required modification.[250] Weiser[167]

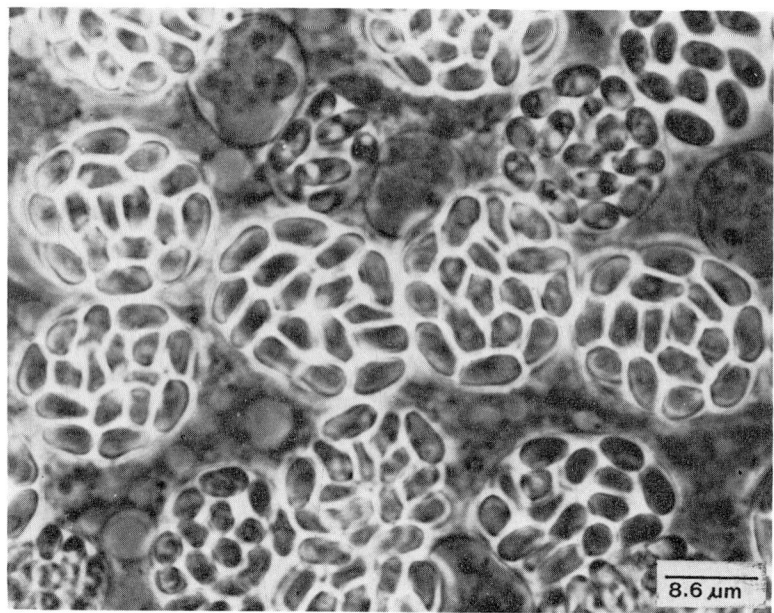

FIGURE 13. Cysts containing sporoblasts and spores of *Vavraia culicis*. (Courtesy of E. I. Hazard, U.S. Department of Agriculture, Agricultural Research Service, Lake Charles, La.)

indicated that spores measured 4 × 2.5 µm while Canning[327] found evidence of spore dimorphism. Macrospores measured 5.1 × 3.7 µm in fresh preparations and 5.9 × 3.7 µm in smears. Microspores measured 3.7 × 2.2 µm when fresh and 4.0 × 2.6 µm in smears. The spores are uninucleate and the polar tube has about 15 coils.[250] The amorphous coat forming the cyst wall becomes double layered with spore maturity.

Natural Occurrence

Vavraia culicis was originally found in natural populations of *Culex pipiens* in Czechoslovakia[167] and has also been reported from *Culiseta longiareolata* from Italy,[393-395] *Anopheles dureni* from West Africa,[396,397] *Aedes polynesiensis* from the Tokelau Islands,[318] and *Culex territans, Cu. salinarus,* and *Ae. triseriatus* from Louisiana.[398,399] While no data on the prevalence of infection was provided in most of these reports, Laird[318] indicated that 39.6% *Ae. polynesiensis* larvae were infected with *Vavraia culicis* collected in 1968 and 1.9% in 1980 from the Tokelau Islands. It also has been found as a common parasite of various laboratory colonies of mosquitoes around the world.[327,399-404a]

HOST SUSCEPTIBILITY AND ACTIVITY

The discussion in this section will be centered on those factors affecting the susceptibility and resistance of insects as hosts to entomogenous protozoa as well as a consideration of the activities and properties of protozoa as etiological agents of disease. Susceptibility and its counterpart resistance may be considered, from a classical standpoint, as involving innate or natural properties of a host which influence the preinvasion events as well as the internal defense mechanisms of the host (humoral and cellular immunity) which influence the post-invasion events of the host-pathogen relationship. Both the preinvasion and postinvasion aspects may also be influenced by various environmental factors. Pathogen activity and properties to be discussed include methods of transmission, infective processes, mode-of-action, and virulency.

Susceptibility and Resistance

The innate or natural immunity of insects to most microorganisms is well established even though the biochemical basis or those factors involved in determining the susceptibility or resistance of a host insect to a microorganism is largely unknown. Thus, insects are not susceptible to pathogens of other animals or plants and even among the entomopathogens with a relatively wide host range, the hosts are generally related either taxonomically or ecologically. In many cases an insect may *escape* infection from a potentially pathogenic agent due to its habits or behavior or its ecological separation from the normal habitat of the pathogen.

The majority of the entomogenous protozoa, like most species of entomogenous bacteria and viruses, are infectious *per os*. Infection thereafter involves penetration of the intestinal epithelium of the host by an invasive stage of the pathogen, usually contained within the ingested spore. The chitinous exoskeleton and chitinous lined portions of the fore and hindgut, impervious to penetration by most protozoa, are the primary barriers to external invasion. In some insects the midgut, while lacking a chitinous cuticle, is also protected by the chitin-containing peritrophic membrane. Although this membrane is quite effective in protecting epithelial cells of the midgut from entry by most microorganisms, the entomogenous protozoa have devloped invasive mechanisms to bypass the peritrophic membrane barrier. Thus, the midgut serves as the primary site for establishment of an infection. Physiological conditions within the digestive tract of the host may also constitute effective barriers to invasion. The gut pH, oxidation-reduction potentials, presence of various cations and digestive enzymes, or the presence of various antimicrobial substances significantly affect not only the number and kind of microorganisms that survive in the insect gut but also act to stimulate or inhibit germination of ingested spores.

Other preinvasion factors involved in determining the natural susceptibility or resistance of an insect to a protozoan include the age, stage, physiological state, or genetic strain of the host. The differential susceptibility of various genetic strains of honey bees and silkworms to microsporidia is also well documented.[405] In most cases, insects are susceptible to protozoa only in the larval stage of development and often exhibit "maturation immunity" or an increase in resistance to infection as the insect matures. Similarly, insects under stress (crowding, disease, unnatural food, high or low temperature, etc.) are usually less resistant to infection. As Weiser[405] has recently reviewed factors affecting the natural resistance and susceptibility of insects to entomogenous protozoa, further discussion will focus on postinvasion events that contribute to the internal defense mechanisms of insects and the influence of environmental factors on these postinvasion events. Defense mechanisms of insects are traditionally divided into cellular and humoral immunity.

Humoral Immunity

The presence of antibody-like substances in insects is well accepted,[406-409] although it is generally agreed that antibodies, as traditionally known in mammals (modified gamma-globulins or immunoglobulins), are not present in insects. Most of the literature on humoral immunity in insects deals with innate or induced bacteriolytic or bactericidal factors[409,410] and essentially no attempts have been made to examine for the possible existence of humoral responses to pathogenic protozoa in insects.[405,409] Weiser[405] cites several examples of atretisation or black degeneration of spores of the microsporidia *Nosema baetis*, *N. stegomyiae* (= *N. algerae*) and *Vavraia culicis* in various aquatic insects. Spores were typically compressed into clumps and darkened by deposition of melanin or melanin-like pigmentation, sometimes referred to as coagulated polyphenols. The apparent lack of the participation of hemocytes in the immune process is suggestive of possible humoral immunity. Similar observations of atretisation have also been reported by Hostounský[411] in larvae of three species of hymenopterous parasitoids of *Pieris brassicae* infected with *Nosema mesnili*;

Weiser and Coluzzi[394,395] in mosquitoes infected by *V. culicis*; and Vávra and Undeen,[192] Canning and Hulls,[193] and Hazard and Lofgren[336] in mosquitoes infected with *N. algerae*. However, the incidental nature of such observations and the apparent participation by hemocytes in depositing melanin on encapsulated spores of various entomogenous protozoa (see following section) suggest that the role of hemocytes in such cases of atretisation or black degeneration needs further study.

Another example of a possible humoral-related response to protozoa is provided by Weiser[412] who found that simultaneous exposure of the brown-tail moth, *E. chrysorrhoea*, to spores of the microsporidia *Nosema lymantriae* and *Thelohania similis* resulted in a mixed infection. As the host is resistant to *N. lymantriae* alone, the presence of *T. similis* apparently altered the resistant host to allow the simultaneous development of *N. lymantriae*.

Cellular Immunity

While the cellular responses of insects to various microorganisms, parasitoids, and other antigens have been extensively studied (see recent reviews of Salt,[413,414] Stephens,[406] Briggs,[408] Shapiro,[415] and Jones[410]) the responses of insects to protozoan pathogens generally have been neglected.[405,416] Observations on cellular responses, particularly phagocytosis, are generally incidental in nature and are scattered throughout the literature. Spores and vegetative stages of various microsporidia, coccidia, and neogregarines are readily phagocytized by hemocytes but often without any detrimental effects.[12,405] In his extensive review of the immune responses of insects to protozoa, Weiser[405] indicated that hemocytes do not normally aggregate in protozoan infected tissues. Only one example[417] was cited where hemocytes formed giant cells in the fat body of an ephemerid nymph *Ecdyonurus venosus* infected with the microsporidium *Nosema baetis*. The existence of inflammatory responses in insects to protozoa, however, has been well illustrated, particularly in the recent literature (Table 19). These reports demonstrate that insects often respond to protozoa in a manner similar to hosts reacting to other microorganisms or injury. Thus, the internal resistance of insects to protozoa appears to be primarily associated with cellular activities of hemocytes that act to phagocytose and encapsulate various protozoan stages to form nodular cysts that eventually melanize and are subsequently destroyed.

Environmental Factors

Few studies have been conducted on the influence of environmental factors on the postinvasion aspects of protozoan-insect associations. As in other host-pathogen associations, temperature may affect pathogenesis by altering the rate of pathogen development. In a few cases the differential susceptibility of protozoa to high temperature has been exploited as a means of partially eliminating protozoa from infected hosts (see review by Weiser[405]). In addition, such factors as irradiation may also significantly alter the host-pathogen association.[418] Whether or not such factors affect the host's defensive mechanisms or interfere with the development of the pathogen has not been studied.

Pathogen Activity

With the exception of a few ciliates, most protozoa are passively dispersed by the activities of the infected host, a nonhost carrier, or through the action of air or water. The host-protozoan association is usually initiated by chance encounter of the host with the protozoan.

Methods of Transmission

As previously mentioned, the majority of entomogenous protozoa are infectious *per os*, i.e., infection is initiated by ingestion of an infective stage of the protozoan. The infective stage is usually a spore or cyst but in certain cases as with the flagellatoses or ciliatoses, it may be vegetative or nonencysted reproductive forms of the protozoan. Horizontal trans-

Table 19
EXAMPLES OF CELLULAR OR INFLAMMATORY RESPONSES OF INSECTS TO PROTOZOA

Order	Protozoan	Host	Nature of observation	Ref.
Amoebida	*Malameba locustae*	*Lepisma saccharina*	Encapsulation and melanization of cysts in fat body and musculature	284
		Locusta migratoria, Locustana pardalina, Paracinema tricolor	Melanotic capsules around intact and damaged areas of Malpighian tubules, most extensively in *L. migratora*	277
		Melanoplus differentialis	Encapsulation of ruptured Malpighian tubules to produce globular, tumor-like swellings which later underwent melanization	24, 77
		Melanoplus spp.	Hemocytic clumping and encapsulation of cysts	84
		Schistocerca gregaria	Extensive encapsulation of infected cells in many host tissues accompanied by blackening	283
Eugregarinida	*Ascogregarina culicis*	*Aedes epactius, Ae. sollicitans, Ae. stimulans, Ae. vexans, Culiseta inornata, Psorophora columbiae*	Encapsulation and melanization of extracellular stages	311, 419
	Ascogregarina sp.	*Ae. polynesiensis*	Encapsulated and melanized spores	318
	Diplocystis major and *D. minor*	*Gryllus domesticus*	Encapsulated gametocysts	420
	D. major	*G. morio*	Melanization of sporocysts and nodule formation	421
Neogregarinida	*Lymphotropha tribolii*	*Tribolium castaneum*	Brown capsules composed of several cell layers surrounding degenerative trophozoites or gametocysts	102
	Mattesia dispora	*Anagasta kuehniella*	Cyst formation by phagocytes in fat body	422
Eucoccidiida	*Adelina cryptocerci*	*Cryptocercus punctulatus*	Brown oocysts walled off by closely packed host tissue	139
	Klossia aphodii	*Aphodius fimetarius* and *A. contaminatus*	Encapsulation of gamonts followed by deposition of a black pigment	139a
Microsporida	*Nosema algerae*	*Anopheles stephensi*	Brownish spots consisting of spores covered by thick layer of dark substance, probably melanin	192
		Mamestra brassicae	Encapsultation by aggregating hemocytes with melanized centers	191a
	N. acridophagus	*M. sanguinipes* and *Schistocerca americana*	Formation of inflammatory nodules or giant cells, later characterized as tumorous lesions	423, 424
	N. baetis	*Ecdyonurus venosus*	Megalocyte or giant-cell formation	417
	N. epilachnae	*Epilachna varivestis* and *Heliothis zea*	Hemocytic infiltration, nodule formation, and melanization	417a

Table 19 (continued)
EXAMPLES OF CELLULAR OR INFLAMMATORY RESPONSES OF INSECTS TO PROTOZOA

Order	Protozoan	Host	Nature of observation	Ref.
	N. heliothidis	Heliothis virescens, H. zea, Trichoplusia ni	Nodule formation, hemocytic infiltration, and occasionally melanization	416
		H. virescens and H. zea	Localized inflammatory response in fat body characterized by "whorl-like" or "swirling" appearance of hemocytes around infected cells	425, 426
	N. invadens	Cadra cautella and C. figulilella	Encapsulation, nodular masses, and hemocyte infiltration	427
	N. longifilium	Othiorhynchus fuscipes	Cyst surrounded by connective tissue-like capsule	428
	N. nepae	Nepa cinera	Spores encased in brown cyst of connective tissue and hemocytes	429
	N. sphingidis	Manduca sexta	Formation of inflammatory nodules with intense hemocytic infiltration and melanization	430, 431
		H. virescens, H. zea, T. ni	Nodule formation, hemocytic infiltration, and occasionally melanization	416
	N. transitellae	Paramyelois transitella	Encapsulation of infected fat body lobes and melanization	432
	N. trichoplusiae	H. virescens, H. zea, T. ni	Nodule formation, hemocytic infiltration, and occasionally melanization	416
	Pleistophora milesi	Maorigoeldia argyropus	Encapsulation and melanization of spores	433
	Vairimorpha heterosporum	Plodia interpunctella	Nodule formation and hemocytic infiltration	434
	V. necatrix	H. virescens and H. zea	Clumping and hypertrophy of hemocytes	426
	V. necatrix and N. heliothidis	H. virescens and H. zea	Intense hemocytic activity and nodule formulation	426

mission of all of the protozoan candidates listed in Table 4 is by the peroral route. This includes the contamination of food, placement of spores in the environment of the host, and cannibalism via feeding on weak or moribund infected individuals. In some species (e.g., *Nosema locustae*) cannibalism is the principal means of transmission within natural populations of certain grasshoppers.[366]

Some protozoa can also be transmitted to their offspring by either contamination of the surface of eggs (transovum) or via an ovarian infection (transovarian) in the female host. Transovum or transovarian transmission is usually in addition to the normal *per os* route and its significance, as a means of vertical transmission, varies greatly among species. Transovum transmission probably is not very important in maintaining natural levels of protozoan infection in host populations. Transovarian transmission, however, is often a highly significant means of transmission by microsporidia such as *N. pyrausta*,[26,322,435,435a] *N. fumifernae*,[350] and *N. heliothidis*.[350b] The extent or efficiency of transovarian transmission varies considerably among various protozoan-host associations. For example, Thomson[350] found that nearly all of progeny of females of the spruce budworm infected with *N. fumiferanae* were also infected. On the other hand, the percentage of transovarian transmission may be very low (e.g., an average of 12.5% for *Plodia interpunctella* infected with *V. plodiae*[350c]) or it may range up to 50% or more as in the case of *O. nubilalis* infected with *N. pyrausta*.[322] In some microsporidian genera such as *Amblyospora*, transovarian transmission was considered until recently to be the only known type of transmission. These unique microsporidia are polymorphic. Lethal infections are produced in male mosquito larvae, yielding haploid spores that are not infectious for other mosquito larvae. However, infections in females are nonlethal and yield diploid spores which give rise to direct infection of oocytes in the female adult and subsequent transmission to the next host generation.[436-438] While the existence of the perorally noninfectious spores from larval mosquitoes has been known for years, their haploid nature was only recently determined by Andreadis and Hall[436,437] and Hazard et al.[438] Subsequent studies have demonstrated the occurrence of karyogamy,[438a] gametogenesis, and plasmogamy[438b] in the life cycles of various other polymorphic microsporidia of mosquitoes. And, the haploid spores (meiospores) which could be germinated in vitro[438c] but are not infectious to mosquito larvae, may be infectious to various alternate or intermediate hosts such as copepods as recently shown for several species of *Amblyospora*.[438d,438e] Some of these polymorphic species are known to produce a third type of spore which is also uninucleate and is infectious for mosquito larvae.[438b]

A less common but interesting type of transmission involves the ovipositional activities of various hymenopteran parasitoids. In his review of host-parasite-pathogen interrelationships of protozoa, Brooks[439] indicated that many workers have suggested that some protozoa are mechanically transmitted via contaminated ovipositors. However, only a few studies have actually confirmed this mode of transmission.[440-442b,442e] Most evidence for mechanical transmission is based on circumstantial field observations of a close association between incidence of protozoan infection and incidence of parasitization (see Brooks[439]) or on laboratory studies where transmission was effected by the intrahemocoelic injection of spores, thus mimicking the possible action of a parasitoid's ovipositor.[442c,442d] In a few instances, female parasitoids, which developed in protozoan-infected hosts, also became infected and were capable of transovarially transmitting the protozoan to their own progeny.[442e,443-446]

A few protozoa can also penetrate the cuticular barrier of the host integument. Studies by Clark and Brandl[258] and Corliss and Coats[259] confirmed the suggestions by Muspratt[447,448] and others[257,316] that ciliates of the genus *Lambornella* invade mosquito larvae by means of cuticular cysts. Other ciliates may penetrate natural breaks or abrasions of the integument[257] but cannot penetrate the epithelial barrier of the host. The precise mode of entry of three recently described entomogenous species of *Tetrahymena* has not been determined[260a,260b,260i] but the presence of melanized cuticular wounds in larvae of *Sialis lutaria* with early stage infections of *T. sialidos* suggests that at least this species possesses the ability to breach the integument of its host.[260i]

Infective Processes

The infective stage of most entomogenous protozoa is represented by a spore and to a lesser degree, in some species, by a cyst. Once the spore is ingested by a susceptible host, usually the larval or nymphal stage of the host, it must germinate in order to initiate the infective process. In many cases the exact stimuli for spore germination or the effects of various stimuli are unknown. The mechanisms through which various species release the infective stage contained within the spore or cyst are species specific. As examples, the spores of various gregarines and neogregarines release sporozoites by rupture of the spore wall or through dissolution of polar plugs of the spores. In contrast spores of microsporidia release a sporoplasm by ejection through an everted polar tube. Detailed discussion of the infective processes will be limited to the neogregarines or microsporidia, since these are the groups which show the most promise as microbial control agents.

The mature spores or oocysts of the Neogregarinida are navicular, oval, or spherical in shape and possess one or two polar pores. The pores are covered or plugged with material that is dissolved, apparently by the action of the host's digestive fluids.[12] The motile, vermiform sporozoites emerge through the pores and, depending upon the species, migrate into the Malpighian tubules or directly penetrate the midgut epithelium and invade fat-body cells or other tissues. In some species the spores are thin-walled and simply rupture in the gut to liberate sporozoites. Few studies, if any, seem to have dealt with the influence of specific stimuli on germination of neogregarine spores. What is known indicates that the sporozoites emerge from spores under the influence of the host's digestive fluids or, more simply, upon ingestion by the host.

The microsporidian spore, in contrast to the general lack of study of neogregarine spores, has long been an object of fascination to parasitologists. It is among the smallest and most complicated of eukaryotic cells. Recent studies of ultrastructure and spore germination have resulted in a generally accepted concept of spore morphogenesis and function.[449,450] Despite its small size (averaging 3 to 6 μm × 1 to 3 μm), the spore contains an extrusion apparatus which is uniquely adapted to place the infective unit (the nucleated sporoplasm) into a host cell. The extrusion apparatus consists of two principal parts: the polaroplast and the polar tube. In mature spores the polar tube extends from the polar cap at the anterior end of the spore posteriorly through the polaroplast and is generally coiled spirally in the peripheral region of the posterior portion of the spore. At its anterior base the polar tube is attached to the polar cap which is an integral part of the polar sac-polaroplast complex that anchors the filament upon eversion. The polaroplast, an extremely laminated structure, serves as a swelling organelle. Once ingested by a susceptible host, the spore is stimulated to evert its polar tube which serves as a needle through which the sporoplasm is injected (Figure 14). Various stimuli permit water to enter the spore osmotically. The polaroplast lamellae swell to supply the intrasporal pressure essential to eversion of the hollow polar tube and subsequent extrusion of the sporoplasm. In some species the posterior vacuole also may provide pressure necessary for eversion.[449] Under pressure, the rigid tube penetrates the peritrophic membrane and epithelial gut cells and deposits the sporoplasm directly into various host cells.

In various in vitro studies of spore germination,[192,389,451-455b] spores of several species of microsporidia responded to various hydrogen-ion concentrations, temperature, and various concentrations of cations and anions. Undeen[389] demonstrated that a two-step germination procedure (activation and germination) was involved for several species of microsporidia. Spores germinated in response to a pH reduction only after activation in a highly basic solution. In addition, spores of *Nosema algerae*, from mosquitoes, required pretreatment in water and then germination occurred only after exposure to various alkali metal-ions and temperature. Dall[389a] recently proposed an interesting theory that attempts to explain the infuence of pH, and concentration and species of cations on polar tube extrusion through their effects on the activity of carboxylic ionophore molecules in spore membranes. The possibility that polar tube extrusion is under ionophoric control deserves further consideration.

In a recent series of papers Weidner,[456,457] Weidner and Byrd,[458] and Weidner et al.[458a] provided a physiochemical explanation of the process of polar tube activation, extrusion, and assembly for several nonentomogenous species of microsporidia. The internal displacement of calcium was an important initial step in the swelling response of polarplast lamellae that precedes discharge of the polar tube in spores of *Glugea hertwigi*.[458] Internally the polar tube consisted of a homogenous coil of polar tube protein which emerged from the growing tip, by an eversion process upon discharge, and was subsequently incorporated into the wall of the discharge tube.[456,457] Apparently, a buildup in osmotic pressure resulted in the eversion

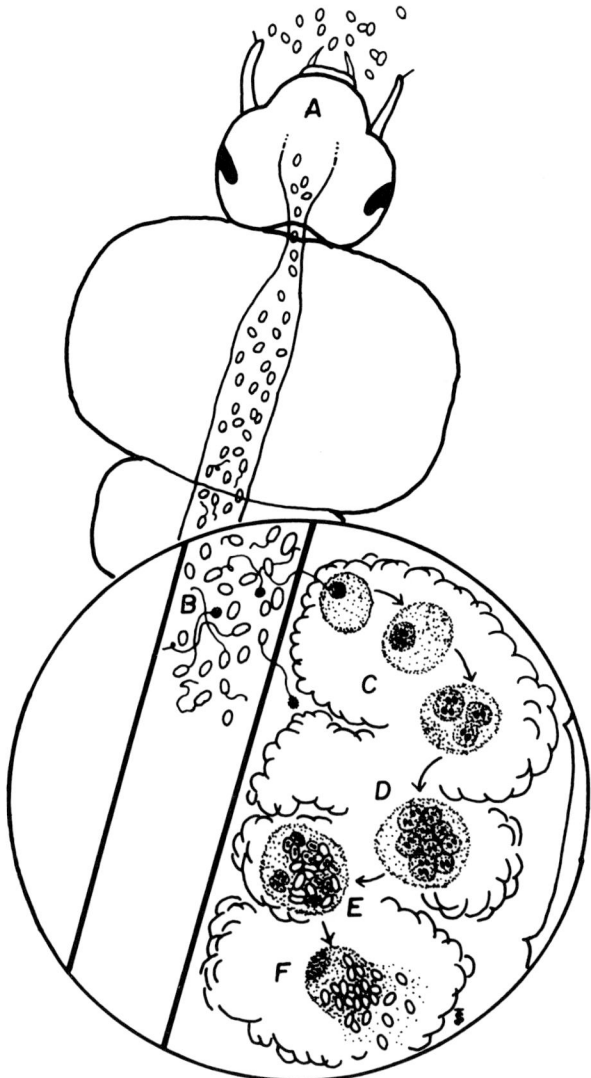

FIGURE 14. Schematic diagram of the infective process and life cycle of *Nosema algerae*. (A) Spores ingested by mosquito larva; (B) spores are stimulated to germinate within midgut; polar tubes are everted and act as inoculation needles through which the sporoplasm within the spore is ejected and deposited at the tip of tubes directly into host cells (fat body); (C) intracellular growth of sporoplasm occurs and by the process of merogony, many bi- and tetranucleate meronts are formed; (D) sporogony is initiated and binucleate sporonts undergo nuclear division and cytokinesis; (E) sporoblasts are formed which give rise to spores; (F) spores formed within host cells are liberated into host's environment with the rupture and disintegration of infected cells and host death.

of the polar tube protein to form the discharge tube through which the sporoplasm was subsequently ejected. The polaroplast organelle membrane is also apparently discharged to form the plasma membrane of the discharged sporoplasm.[458a] Although the stimuli required for spore germination of microsporidia vary (related to host type and its ecological habitat), the process of discharge and assembly of the polar tube, as proposed by Weidner,[456,457] may prove to be similar for many microsporidia.

Once the sporoplasm is deposited within a host cell it begins to develop and ultimately undergoes merogony and sporogony to produce new spores (Figure 14). Although the process is not clearly understood, some meronts escape or are released from intact or ruptured cells, invade adjacent cells, and eventually invade all susceptible cells and tissues.

In species exhibiting transovarial transmission, infection may be effected by development of the microsporidium within various cells and tissues of the female or male reproductive systems. Thus, eggs deposited by infected females contain various stages of the protozoan (meronts, sporonts, and/or spores) within nurse cells or in developing cells of the embryo. Protozoan development in the reproductive systems is usually indicative of an extensive systemic invasion of host tissue. In mosquitoes infected with certain species of *Amblyospora*, however, transovarian transmission is effected in a unique manner.[436] The protozoan is restricted to oenocytes in the female host which circulate or migrate with some eventually lodging near adult ovaries. Sporulation usually occurs only after a blood meal by the female but in some mosquitoes it may be under hormonal control.[436a,436b] The short-lived spores, formed within the oenocytes, are stimulated to germinate within a few hours. Adjacent ovaries are infected either by direct penetration by a polar tube, passive absorption, or active invasion of oocytes by sporoplasms extruded into the hemolymph.[436c] The ovipositor of the female acts as an injection needle in species transmitted by hymenopterous parasitoids. Various vegetative stages, which contaminate the ovipositor following oviposition in an infected host, are then mechanically transmitted to the hemocoel of a noninfected host.

Mode-of-Action

Hosts with a protozoan infection often die prematurely or their reproductive activities may be seriously imparied; however, most infections in insects are debilitory. The cause of death or reproductive impairment is usually difficult to determine. Competition for nutrients may be involved when the extracellular phase of protozoan development is extensive, but few reports substantiate this mechanism. As examples, the studies of Harry[459] and Dunkel and Boush[460] on eugregarines of beetles revealed only adverse effects in hosts subjected to the stress of suboptimal diets or starvation. In addition, no obvious pathological effects of the flagellate, *Herpetomonas muscarum*, were detected in eye gnats when extensive masses of flagellates were solely confined to the alimentary tract.[461,462] Thus, most insects appear to tolerate extracellularly developing protozoa relatively well.

Few studies have focused on the cytopathological effects induced by entomogenous protozoa. In some microsporidioses infected cells and their nuclei may be considerably hypertrophied and filled with developing stages of the protozoan.[212,405,463] In other hosts, infected cells are essentially normal in size and appearance; after extensive protozoan development cellular cytoplasm is gradually replaced by vegetative and spore stages of the protozoan and only cell membranes and nuclei remain intact. The formation of complex, neoplastic xenomas by some microsporidia is confined to certain species parasitic in fish; none are known among the entomogenous protozoa.[463,464] A few entomogenous microsporidia, however, do produce a type of xenoma termed a syncytium.[464] These syncytial xenomas are characterized by infected cells that form a multinuclear plasmodium, as individual cellular membranes are lost and nuclei become hypertrophied. In addition to nuclear and cellular hypertrophy, infected cells of insects may exhibit extensive alterations of cytoplasmic organelles such as chromosomes,[465-471] endoplasmic reticulum,[472-474] mitochondria,[471-474] lysosome bodies,[473] ribosome bodies,[472] protein granules,[474] and vacuoles.[472,473] Despite these extensive alterations induced by the developing intracellular stages of microsporidia, there is little evidence to suggest how such changes are mediated. Weiser[405] indicated that entomogenous protozoa apparently do not produce toxins. Since adjacent, noninfected cells appear normal, any protozoan-produced toxin is confined to the infected cell. Martins and Perondini[475] found that the nuclear hypertrophy caused by microsporidian development in the striated, parietal muscle of a sciarid fly was directly proportional to the proximity of the nucleus to the pathogen. They suggested that nuclear hypertrophy resulted from an increase in cellular metabolism induced by juvenile hormone-like substance in an effort to repair damage caused by the microsporidium. Cellular alterations may also result from competition between developing parasites and normal cellular constituents for nutrients within the cell, thus inducing atrophic changes in the cytoplasm.[463] Certain stages of the protozoan may also produce proteolytic enzymes which result in cellular lysis but this has never been confirmed.

In most insects, signs and symptoms of protozoan infections are generally related to the infected tissues and organs. The functional capacity of the organ is progressively impaired with gradual replacement of the cells of a tissue or organ with developing stages of the protozoan. In fat body infections, the lack of nutritional reserves may result in pupal or adult deformities or adults with shortened longevity. Development in reproductive tissues may result in sterile adults or adults with reduced fecundity. Protozoa, causing systemic or midgut infections, are usually more virulent; death may occur due to general debilitation or to the disruption of the digestive activities of the gut and/or through the disrupted gut serving as a portal of entry for the development of secondary septicemia-producing organisms in the hemocoel. Thus disease, due to most protozoa, is generally the result of functional lesions resulting from cellular destruction with death occurring after irreversibility is reached.

Virulence

Probably the most limiting characteristic of the entomogenous protozoa when considering their use as microbial control agents is their low virulence. While some species are highly virulent, resulting in immediate death, the majority produce chronic and often nonlethal infections in their hosts. In addition, the existence of few, if any, strains of protozoa which differ in virulence in contrast to strains of differing virulence with many bacteria, fungi, and viruses also serve to limit their selection as potential microbial control agents.

Little experimental effort has been made to select for highly virulent strains of entomogenous protozoa using classical methods.[8] Hazard and Lofgren[336] discovered that the virulence of *Nosema algerae* was greatly increased after 20 passages through the mosquito, *Anopheles quadrimaculatus*. The greater virulence was associated with more rapid development of the microsporidium in late-stage larvae but records documenting the unanticipated increase in virulence were not kept. An additional 40 passages of the pathogen through the mosquito did not result in an enhancement of virulence.

Efforts to mass produce certain microsporidia in alternate hosts can result in either a decrease or increase in virulence. Henry et al.[476] found that spores of the microsporidia *Nosema acridophagus* and *N. cuneatum* were more virulent for grasshoppers when mass produced in *Heliothis zea* than in the normal grasshopper host. Such spores not only caused higher mortality among grasshoppers but the LT_{50} for grasshoppers was also shortened. Undeen and Alger[477] reported that spores of *N. algerae* produced in larvae of *H. zea* were as infective as spores produced in *Anopheles stephensi*. Spores produced in the dipteran *Phormia regina*, however, were considerably less infective than those produced in either *H. zea* or *An. stephensi*. On the other hand, Hall[25] found that the microsporidium *N. infesta* gradually became less virulent when cultured for several generations in the potato tuberworm, *Phthorimaea operculella*, an alternate more easily reared host than the original host, *Tehama (Crambus) bonifatella*. Although many protozoa may have wide host ranges or have been produced in different hosts, few additional attempts have been made to assess for changes in pathogen virulence. Similarly, there appears to be almost no effort to search for natural strains of various protozoa which might differ in virulence. As pointed out by Henry and Oma,[478] the effectiveness of some protozoa as microbial control agents might be enhanced by the selection of strains with shorter developmental periods to enhance virulence by selecting for increased spore production at the expense of some virulence and host specificity.

The virulence and pathogenicity of protozoa, like other entomopathogens, are also related to dose, the stage and host species involved, and various environmental factors. High doses generally result in acute infections, and, as indicated previously, host susceptibility is often related to age and environmental factors such as temperature. The pathogen is usually most virulent for the combination of young larvae and high temperature. Virulence is also related to the site of pathogen development within the host. Chronic infections usually are the result of pathogen development in the fat body while acute pathogenesis results from systemic or midgut and musculative infections. Reduced infectivity or virulence also may be associated with prolonged storage of protozoan spores under artificial conditions. In most cases, however, this is more indicative of loss in spore viability. Since most of the entomogenous protozoa are obligate pathogens, there are few, if any, reports of the influence of repeated culture on artificial media on virulence. Similarly, little is known of the virulence of the few species of microsporidia that have been cultured for several passages in tissue culture.

STABILITY, SENSITIVITY, AND PERSISTENCE

Data on factors affecting the survival of protozoa (mostly the spore stage) are primarily limited to a few microsporidia pathogenic for economically important terrestrial insects.[4,39,479-482] Previous studies focused on effects of moisture and temperature on spore stability or influence of solar radiation on spore persistence. In his review on microsporidia, Kramer[479] indicated that spore survival varies from species to species. In dried feces or in host cadavers at room temperature, spore survival may range from a few weeks to over a year. In cold, clean aqueous solutions spores may survive 7 to 10 years. Maddox[481] concluded that (1) spores of most species of microsporidia do not persist in the general environment for more than 1 year; (2) temperatures above 35°C greatly reduce the viability of most spores; (3) drying is harmful; and (4) naked spores cannot withstand exposure to direct sunlight for more than a few days. It is beyond my scope to review all of existing data on survival of the entomogenous protozoa. Thus, comments will be confined to those factors affecting stability, sensitivity, and persistence of protozoa in general. The results of various studies on those agents discussed in "Potential Candidate Microbial Insecticides" will be summarized in more detail.

Stability

Spores of most microsporidia and neogregarines have been stored, generally for only a few weeks, in water at temperatures slightly above freezing (0 to 6°C); spore viability is usually excellent under these conditions and infectivity is similar to that of fresh suspensions of spores. Many protozoa may be stored dried simply by placing infected cadavers at low temperature. The presence of contaminating microorganisms, however, is usually detrimental to survival and antibiotics may be needed to control these contaminants. Attempts to determine viability of spores, under various conditions of prolonged storage, will be discussed as they relate to storage at above-freezing temperatures, subfreezing temperatures, and after lyophilization.

Survival at Above-Freezing Temperatures

Good short-term survival of spores either in aqueous suspension or in various dried preparations of spores can be obtained at temperatures ranging from 0 to 6°C. Long-term survival of spores at low or high temperatures has only been determined for a relatively few species, mostly microsporidia. Dry spores of a few microsporidia generally do not survive temperatures above 35°C for more than a few days: maximal survival time is measured in minutes when exposed to temperatures above 50°C.[481] Kaya[483] reported a significant loss in the infectivity of spores of *Vairimorpha necatrix* after exposure to 35°C for 6 days while

Table 20
LONGEVITY RECORDS OF PROTOZOAN SPORES HELD DRY IN VARIOUS FORMULATIONS AT 0 TO 6°C

Order	Species	Formulation	Longevity	Ref.
Microsporida	Microsporidium hyphantriae	Surface glass slide	>3 months	417, 488
		Host cadaver	2 months	
	Nosema apis	Host cadaver	3.5 months	489
		Surface glass slide	7 months	
	N. lymantriae	Surface glass slide	>3 months	417, 488
		Host cadaver	2 months	
	N. muscularis	Surface glass slide	>3 months	417, 488
		Host cadaver	2 months	
	N. oryzaephili	Flour	>9 months	490
	N. whitei	Flour	>15 months	491
	Octosporea muscaedomesticae	Fecal deposits on glass slide	>15 months	479
		Vacuum dried in water	<7 days	492
		Vacuum dried in sucrose	>2 years	
		Vacuum dried cadavers	>2 years	
	Thelohania similis	Surface glass slide	>3 months	417, 488
		Host cadaver	2 months	
	Vairimorpha necatrix	Corn meal	>10 months	493
		Host cadaver	>12 but <18 months	
	V. plodiae	Fecal material	>3 but <4 months	246
Neogregarinida	Mattesia trogodermae	Cellulose powder	>6 months	484

Maddox[482] indicated spores remained viable for 30 min at 60 to 70°C, 90 min at 55°C, 5 hr at 50°C, and 3 weeks at 40°C. Similarly, Nara et al.[484] found that dry spores of the neogregarine *Mattesia trogodermae* did not survive 30 min at 73°C or higher temperatures.

Spore survival at more moderate temperatures (20 to 30°C) is apparently related to the species, the environment, and the nature of the spore material. In infected cadavers spore survival at or near 25°C ranged from 18 days for *Nosema destructor*[485] to over 1 year for *N. apis*[486] and *N. bombycis*.[487] Spores in cadavers infected with *N. muscularis*, *N. lymantriae*, *Microsporidium* (= *Thelohania*) *hyphantriae*, and *Thelohania similis* remained viable for over 13 months when held at high humidity but were not viable after 2 months as dried cadavers at room temperature.[417,488] Spores of *N. pyrausta*, in vacuum-dried cadavers of *Ostrinia nubilalis*, were still moderately infective after storage for 6 months (under vacuum) at 22 to 24°C.[368] Dry, naked spores or spores in fecal deposits or tissue smears on glass surfaces at or near room temperatures have survived from 1 month to over 1 year. Dried spores of most species of microsporidia, however, do not survive more than 2 or 3 months at room temperatures.[479-482] Drying may be instantly lethal for spores of parasites of aquatic insects. Spores of *Vavraia culicis* did not germinate after dehydration when stored as a squashed tissue preparation on glass slides for 24 hr.[404a] White[489] found that spores of *Nosema apis* survived for 1 to 2 weeks in the presence of fermentative bacteria. In clean suspensions the survival time of other microsporidian species ranged from 9 to 12 months.[246,417]

Most investigators have conducted spore survival studies at temperatures ranging from 0 to 6°C. Survival of dry formulations of spores varied from a few months for *Vairimorpha plodiae* (in fecal material) to over 2 years for *Octosporea muscaedomesticae* (vacuum dried in sucrose or in vacuum dried cadavers) (Table 20). Since spores of protozoa infecting stored product insects may germinate prematurely when wet, spore survival up to 15 months at low temperatures would significantly enhance their potential for use as microbial control agents. Spores of protozoa infecting terrestrial insects generally are stored in aqueous sus-

Table 21
LONGEVITY RECORDS OF MICROSPORIDIAN SPORES IN AQUEOUS MEDIA AT 0 TO 6°C

Species	Longevity	Ref.
Microsporidum hyphantriae	>13 months[a]	417
Nosema apis	7—9days[b]	489
	>3 months[b]	
	Nearly 7 years	494
N. bombycis	10 years	495
N. destructor	>6 months	221
N. fumiferanae	1.5—2.5 months	350
N. infesta	6—9 months	25
N. lymantriae	>13 months[a]	417
N. muscularis	>13 months[a]	417
Nosema species	6 years	496
Octosporea muscaedomesticae	>2 years	479
Thelohania similis	>13 months	417
Vairimorpha necatrix	2—2½ years	385, 481
	30—36 months[c]	497
	>23 months[c]	493
	>6 but <12 months[c,d]	
	>21 months	498
V. plodiae	>9 months	246
Vavraia culicis	>4 months	404a

[a] Host cadavers at high relative humidity.
[b] In presence of fermentative microorganisms.
[c] In presence of antibiotics.
[d] Triturated host cadavers in water.

pensions at low temperatures. Survival of spores of some species in clean suspensions has been detected after 6 to 10 years (Table 21). The spores of most species will survive periods up to 1 year, particularly if the growth of adventitious microorganisms is prevented through purification techniques or use of antibiotics.[497]

Survival at Subfreezing Temperatures

Spore survival only has been studied for relatively few protozoa at subfreezing temperatures (Table 22). Spores of *O. muscaedomesticae* survived 2 to 3 weeks in frozen water or in cadavers at −20°C[492] but survived over 7 years[499] with little loss in infectivity when frozen in liquid nitrogen. The survival time of frozen spores of other species generally varies from nearly 1 year to up to 2 years (Table 22). In liquid nitrogen spores of at least 16 species of microsporidia from terrestrial insects survived for 1 year,[500] although some species such as *Nosema algerae* from aquatic insects cannot be preserved in liquid nitrogen.[499] The capacity of some species to survive freezing may be improved by the use of infected cadavers or by the addition of various cryoprotectants. For example, Maddox[481,482] found that spores of *Vairimorpha necatrix* survived for only 2 months when frozen in water at −34°C while Fuxa and Brooks[493] determined that spores of the same species frozen in a 1:1 water glycerine suspension survived 23 months at −15°C. Spores of *V. necatrix*, stored in liquid nitrogen for up to 8 years,[500] were as infectious as freshly prepared spores.

Survival after Lyophilization

Relatively few attempts (Table 23) have been made to assess the effects of lyophilization on the viability of protozoan spores or the potential for long-term storage of freeze-dried

Table 22
LONGEVITY RECORDS OF PROTOZOAN SPORES AT SUBFREEZING TEMPERATURES

Order	Species	Preparation	Temperature (°C)	Longevity	Ref.
Microsporida	*Nosema apis*	Midguts in saline or as dried cadavers	−23	2 years	501
	N. fumiferanae	Cadavers	−5	8 months	350
		Water	−5	4—6 months	
	N. infesta	Water	−30	12 months	25
	N. locustae	Cadavers	−10	1 year	36
		Water	−10	5 years	
	N. pyrausta	Vacuum-dried cadavers	−12	12 months	368
	N. trichoplusiae	Water	Liquid N$_2$	7 years	499
	Octosporea muscaedomesticae	Water	−20	ca. 2 weeks	492
		Cadavers	−20	ca. 3 weeks	
		Water	Liquid N$_2$	7 years	499
	Vairimorpha necatrix	Cadavers	−15	23 months	493
		Cadavers + glycerine	−15	23 months	
		Corn meal	−15	9 months	
		Water + glycerine	−15	23 months	
		Water	−34	ca. 2 months	481, 482
		Water	Liquid N$_2$	8 years	482
Neogregarinida	*Mattesia trogodermae*	Cellulose powder	−19	15 months	484
		Cellulose powder	−30	9 months	

Table 23
LYOPHILIZATION STUDIES WITH MICROSPORIDIAN SPORES

Species	Formulation	Storage conditions	Affect on longevity	Ref.
Nosema algerae	Spores in 50% sucrose	—	Slight loss in infectivity after lyophilization	492
N. apis	Host abdomens	18°C over P_2O_5	Slight decrease in infectivity after 1 month	503
N. pyrausta	Cadavers	22—24°C under pressure	About 20% loss in infectivity initially and about 60% loss after 6 months	368
N. whitei	Cadavers	—	No loss in infectivity immediately after lyophilization	492
Octosporea muscaedomesticae	Spore suspension in 20% sucrose	21°C under vacuum	Complete loss of infectivity	499
	Spore suspension in water; 12% skim milk; 10% glycerol; equine serum; 5% mesoinositol; 5% ascorbic acid + 5% thiourea	5°C with N_2 gas	Complete loss of infectivity immediately after lyophilization	492
	Spores in 50% sucrose or in host tissues	5°C with N_2 gas	No apparent loss in infectivity for up to 2 years	
Vairimorpha necatrix	Cadavers	Room temp. under vacuum	20% loss in infectivity initially and about 30% after 2 years	502
	Cadavers	−15°C under vacuum	Only slight loss in infectivity after 23 months	493
	Spore suspension in water	−15°C under vacuum	Obvious high loss in infectivity at low dosages initially and slightly greater loss after 23 months	
	Spore suspension in 20% sucrose	21°C under vacuum	After 2½ years, loss in infectivity comparable to that in water at 4°C	499

spores. The viability of lyophilized spores, although based only on observations on six species of microsporidia, appears related to the nature of the lyophilized spore formulation. Spores suspended in water or in materials commonly used as cryoprotectants (see Table 23) exhibit extensive reductions in viability or are completely inactivated after lyophilization.[492,493] Spores lyophilized in 20 or 50% sucrose or in host tissues, however, may be highly infectious immediately after lyophilization and only show slight losses in infectivity when held at various temperatures for up to 2 years.[492,493,499,502] The importance of the quantity of a cryoprotectant as well as its nature and concentration is illustrated by lyophilization experiments with spores of *Octosporea muscaedomesticae*. Vávra and Maddox[499] found that spores of this species did not survive lyophilization when suspended in 20% sucrose while Teetor-Barsch and Kramer[492] obtained excellent survival of lyophilized spores for up to 2 years when suspended in 50% sucrose. Although lyophilization appears to be a promising alternative for preservation of protozoan spores, additional efforts will be needed for other species of protozoa, an evaluation of survival over extended periods of time, and the type of cryoprotectant. Only two species, *O. muscaedomesticae* and *Vairimorpha necatrix*, of lyophilized spores have been examined after periods up to 2 years.[492,499] Infectivity of lyophilized spores was not much better than spores kept in water at low temperatures or at subfreezing temperatures.

Table 24
SENSITIVITY OF MICROSPORIDIAN SPORES TO SUNLIGHT[506]

Pathogen	Bioassay host	Medium	Decline in incidence of infection	Ref.
Nosema algerae	Anopheles albimanus	Aqueous suspension	None in 4 hr but significant decline in intensity	506
N. apis	Apis mellifera	Crude aqueous suspension	100% in 37 hr	489
		Tissue smear on glass	100% in 15 hr	
N. fumiferanae	Malacosoma pluviale	Spores on cherry leaves	21% in 4 hr, 90% in 5 hr	507
N. heliothidis	Heliothis zea	Spores on cotton leaves	90% in 4 hr	508
N. trichoplusiae	H. zea	Spores on artificial diet	100% in 18 hr	482
		Spores on glass slides	100% in 6 hr	
Octosporea muscaedomesticae	Phormia regina	Aqueous suspension	100% in 3 hr	509
Vairimorpha necatrix	Pseudaletia unipuncta	Spores on corn leaves	100% in 3—4.5 hr	385
		Spores on corn leaves	100% in 5 hr	482
		Spores on glass slides	100% in 6 hr	
	H. zea	Spores on artificial diet	100% in 20 hr	
		Spores mixed with soil	20—30% in 28 hr	
	Estigmene acrea	Spores on bean leaves	100% in 78 hr	483
Vavraia culicis	Anopheles albimanus	Aqueous suspension	LT_{50} = 3.1 hr, LT_{90} = 6.1 hr	328

Sensitivity

This section will focus on the effects of environmental factors (other than temperature and moisture), i.e., sunlight — ultraviolet (UV) radiation, various chemicals, insecticides, and therapeutic agents. However, this is not to underemphasize the influence of temperature or humidity on spores sensitivity. The principal threat to spore survival is undoubtedly dehydration.[480]

Sunlight or UV Radiation

The adverse effects of sunlight or UV radiation on several species of microsporidia is well documented (Tables 24 and 25). The rate of inactivation by sunlight is affected by the degree of purity of the spores and the carrier. Purified spores in aqueous suspension or dry spores on the surface of various types of leaves or glass are rapidly inactivated by sunlight (Table 24) or UV light (Table 25). Complete inactivation in bright sunlight usually occurs within minutes or hours. Spores exposed on artificial diet survive up to 20 hr. Spores mixed with various substrates such as soil or fecal material may withstand inactivation for a few hours when exposed to artificial UV or up to 28 hr when exposed to natural sunlight. While inactivation rates may vary, dependent upon prevailing atmospheric conditions, substrate, and purity, spores of most microsporidia are inactivated relatively rapidly; thus spore dispersal and persistence is highly dependent on being shielded from the harmful sunlight-UV. Despite the well demonstrated harmful effect of UV radiation on entomopathogens,[504,505] little is known of the mechanism of inactivation. Ignoffo et al.[505] speculated that some substance (perhaps peroxide radicals) is produced from photooxidation of one or more amino acids and that this radical(s) reduces both virulence and viability of entomopathogens exposed to sunlight-UV.

Insecticides

Relatively few observations have been made of the sensitivity or compatibility of protozoa with insecticides. Although Hinks and Ewen[512a] found that males of *Melanophys sangfui-*

Table 25
SENSITIVITY OF MICROSPORIDIAN SPORES TO UV RADIATION FROM GERMICIDAL LAMPS[506]

Pathogen	Bioassay host	Light source	Measured radiation (W/cm²)	Medium	Decline in incidence of infection	Ref.
Nosema agrotidis	Agrotis segetum	Unreported	13.7—1026	Aqueous suspension	LD_{50} = 1000 W/cm²	510
N. algerae	Anopheles albimanus	GE 30T8 germicidal lamp	121	Aqueous suspension	100% in 8 min	506
N. apis	Apis mellifera	GE 15T germicidal lamp	Unreported	Spores in sugar syrup	100% in 5 hr	511
N. fumiferanae	Choristoneura fumiferana	WL 30 germicidal lamp	Unreported	Spores on artificial diet	100% in 5 hr	507
	Malacosoma pluviale	WL 30 germicidal lamp	Unreported	Spores on cherry leaves	100% in 4 hr	507
Octosporea muscaedomesticae	Phormia regina	GE 30-W germicidal lamp	5500	Aqueous suspension	100% in 30 min	509
				Dry spores on glass	100% in 15 min	
				Spores in feces	No effect in 3 hr	
				Dry spores with uric acid on glass	No effect in 30 min	
Pleistophora schubergi	Agrotis segetum	Unreported	13.7—1026	Aqueous suspension	LD_{50} = 600 W/cm²	510
Vairimorpha necatrix	Estigmene acrea	Chromato-Vac CC-20	Unreported 254 nm	Aqueous suspension	100% in 4—10 min	483
			Unreported 366 nm	Spores on bean leaves	50% in 6 hr	
				Aqueous suspension	No change in 6 hr	
	Heliothis zea	GE F-30-T8 BL lamps with peak radiation at 254 nm	140 (254 nm) and 1800 (365 nm)	Spores on H. zea eggs	96% in 4 hr, half life = 2.1 hr	505

Trichoplusia ni	Lighted growth chamber with 16-hr photoperiod	15	Dry spores on glass	100% in 1 day	498
			Aqueous suspension	100% in 10 days	
			Spores on collard plants	100% in 10 days at lower dosages to 100% in 29 days at highest dosage	

nip[es injected with *Malameba locystae* were more resistant to cypermethrin than healthy males, most workers[308,512,513a,513b] have demonstrated increased susceptibility of protozoan-infected hosts to insecticides. Maddox[482] presented direct observations on the effect of mixing spores of *Vairimorpha necatrix* with four insecticides for various time intervals. Spore viability was largely unaffected by tetramethylpyrophosphate (TMPP), tetraethylpyrophosphate (tepp), pyrenone, and a 0.5% concentration of malathion. However, spores were inactivated when exposed to 5% malathion for 30 min. McCray et al.[296] found that mixing spores of *A. culicis* with DDT did not affect the normal developmental cycle of the eugregarine in larvae, pupae, and adults of a DDT-resistant strain of *Aedes aegypti*. Although Stapp and Casten[308] found that larvae of a DDT-resistant strain of *Ae. aegypti* heavily infected with *A. culicis* were more susceptible to DDT, McCray et al.[296] did not find any significantly consistent differences in the mortality rates of larvae of four strains of *Ae. aegypti* infected and noninfected with *A. culicis* when exposed to DDT, dieldrin, malathion, carbaryl, and temephos. In an extensive study, Mussgnug and Henry[514] found that malathion and spores of *Nosema locustae* acted independently with additive results when simultaneously administered to grasshoppers. At low spore doses the LD_{50} of malathion was only slightly affected; however, at a high spore dose (10^7 spores), the LD_{50} was significantly lowered. They concluded that the percentage of infection by *N. locustae* was not affected by the addition of malathion and suggested use of wheat-bran formulations containing both materials. Mussgnug[515] established the compatibility of spores of *Nosema locustae* with low concentrations of carbaryl, dimethoate, and carbofuran; similar studies with carbaryl and dimethoate were also conducted by Morris.[515a] Subsequent work has focused on determining the potential of bait formulations of *N. locustae* spores and carbaryl for grasshopper control.[286a,516,517] Similarly, compatibility of spores of *N. pyrausta* and carbaryl and carbofuran was demonstrated in field studies by Lublinkhof and Lewis[518] and Lublinkhof et al.[519] Since the intensity of infection by *N. pyrausta* was reduced under laboratory conditions and treatments were not simultaneously adminstered in field tests, additional research will be needed to clarify whether spores are inactivated by contact with insecticides.

At least two other studies on combinations of microsporidian spores and insecticides have been conducted.[385a,385c] Richter and Fuxa[385a] evaluated the interaction of *V. necatrix* with carbaryl, methyl parathion, and permethrin in assays with the velvetbean caterpillar, *Anticarsia gemmatalis,* by adding spores to the surface of artificial diet-insecticide mixtures. Combinations of spores and methyl parathion were antagonistic while the other combinations were additive. However, the periods of lethal mortality were greater for the various combinations than with each agent alone. As *V. necatrix* is relatively nonvirulent for *A. gemmatalis*[385a,385b] and spores were not mixed directly with the insecticides, additional work is also needed with this species to assess its compatibility with insecticides. Khan and Selman[385c] found that a combination of pirimphos methyl and spores of *N. whitei* increased the larval period of *Tribolium castaneum* as well as their weight and length at maturity compared to control larvae. As larvae treated with spores alone were even larger, they suggested that reductions obtained in the combination treatment could have been due to the debilitative effects of the insecticide on the host or to an effect of the insecticide on the microsporidium, reducing spore yields.

Therapeutic Agents

Unlike other chemicals in the environment whose effects are usually related to spore viability, therapeutic agents are evaluated based on their effect on developmental stages in the living host. Most research on chemotherapeutics has been aimed at prevention of disease (nosemosis of the honey bee, *Apis mellifera*), or elimination of microsporidia in laboratory colonies of insects. Fumidil B, a commercial preparation of the bicyclohexyl amine salt of fumagillin acid ($C_{26}H_{34}O_7$), has been extensively used to control infections by the microsporidium *Nosema apis* in honey bees (see reviews by Goetze and Zeutzschel,[520] Gochnauer et al.,[521] and Furgala and Mussen[522]). The effectiveness of fumagillin against *N. apis* was first demonstrated by Katznelson and Jamieson,[523] who suggested that the compound was active at the time of spore germination. Bailey[524] indicated that infection was arrested during the intracellular phase of development through destruction of the vegetative forms of the protozoan. In more extensive studies on the mode of action of fumagillin, Przelecka and Hartwig[525] and Hartwig and Przelecka,[526] a conclusion also supported by Liu,[527] indicated that the primary effect was inhibition of DNA replication in cells of the microsporidium. In contrast, Jaronski[528] found that the primary effect of fumagillin was on RNA synthesis and that loss of DNA and cessation in cell replication were secondary effects. Although many other compounds have been tested, Fumidil B is the only effective drug for control of *N. apis* in honey bees.[529,530]

Fumidil B also has been successfully used to eliminate or suppress infection by various microsporidia in *Anopheles gambiae*,[338] *Ostrinia nubilalis*,[531,532] *Anthonomus grandis*,[533] *Choristoneura fumiferana*,[534] *Tribolium castaneum*,[535] *Phormia regina*,[536] and *Drosophila willistoni*.[537] However, the relatively high dosage rates necessary to eliminate certain microsporidioses may also be associated with adverse side effects on the host, i.e., increased larval period and mortality and reduced pupal weight, fecundity, and egg hatch.[531,533,534] Fumidil B also has been used (continuous exposure for 35 days) to eliminate *Nosema disstriae* from hemocyte cell cultures of the forest tent caterpillar, *Malacosoma disstria*.[538] In contrast Bayne et al.[539] and Kurtti and Brooks[540] found that fumagillin was ineffective in eliminating microsporidian infection from their cell lines. Briese and Milner[548a] also found that fumagillin had no effect on the level of infection by *Pleistophora schubergi* in the geometrid *Anaitis efformata*.

Although fumagillin was originally shown to have amoebicidal activity,[541] little effort has been made to evaluate its activity against entomogenous protozoa other than microsporidia. Bailey[542] reported that under field conditions fumagillin was ineffective against the honey bee amoeba *Malpighamoeba mellificae*. In addition, Fumidil B appeared to be of no value as an anticoccidian agent against *Adelina tribolii* when used at dosages effective against *N. whitei* in *Tribolium castaneum*.[535] Shinholster[535] suggested, however, that high dosages may be effective since deformed sporocysts were found in infected larvae. Dunkel and Boush[543] also were unable to eliminate the eugregarine *Pyxinia frenzeli*, at concentrations up to 10% of the diet with fumagillin, from cultures of the black carpet beetle *Attagenus megatoma*.

Only one other compound, benomyl, has been investigated extensively as an antimicrosporidian agent. A systemic fungicide (methyl 1-(butylcarbamoyl)-2-benzimidazole-carbamate), benomyl is used extensively to control a variety of fungal phytopathogens of field crops. Its antimicrosporidian activity was discovered by Hsiao and Hsiao.[544] Several other researchers have also demonstrated the suppressive action of benomyl against microsporidia.[535,537,545,546,548a] However, as in some cases with Fumidil B,[532] the suppressive action of benomyl may be only temporary. Harvey and Gaudet,[545] Brooks et al.,[546] and Briese and Milner[548a] reported strong resurgences of microsporidioses during the pupal and adult stages of insects reared as larvae on benomyl-treated diet. In addition benomyl, at high dosages,[537,545,547] also may be toxic to the host insect and is ineffective against some microsporidioses.[530,548] At the same time, Sohi and Wilson[538] found that *Nosema disstriae* was completely eliminated in hemocyte cell cultures of *Malacosoma disstria* exposed for a period of 35 days to benomyl or fumagillin. They suggested that the optimal activity of benomyl is dependent on continuous long-term exposure since only the vegetative stages of the pathogen are affected by the drug. Thus, since spores do not germinate in the cell culture medium, autoinfection is not possible and the pathogen may have been eventually eliminated through serial subculturing. In another cell line study, Bayne et al.[539] reported that microsporidian development was inhibited; however, elimination did not occur even after 8 weeks exposure to benomyl or 15 weeks exposure to Fumidil B. Since their cell line was not mitotically active, the microsporidian remained in an arrested stage and could not be eliminated through serial subculturing.

Hsiao and Hsiao[544] suggested that the effect of benomyl was on sporogenesis, an explanation supported by the observations of Shinholster[535] who noted that dark, translucent, and nonrefractile spores were produced when *N. whitei* was exposed to benomyl. Shinholster[535] also suggested that the actively growing vegetative stages were affected. Brooks et al.[546] found that benomyl produced aberrant meronts and sporonts of *Nosema heliothidis* and resulted in the rapid disappearance of vegetative stages from host tissues. The production of abnormal spores was also noted and they suggested that benomyl affected both merogony and sporogenesis.

Prior to the more recent studies on fumagillin and benomyl, few other antibiotics or chemical compounds have received extensive examination as antimicrosporidian agents. Goetze and Zeutschel[520] and Bailey[549] reviewed studies on various chemotherapeutic agents used to control Nosema disease of honey bees. Nosemack or merthiolate, a complex organic mercury compound, was successfully used against *Nosema apis* in the early 1950s. However, these compounds were associated with some deleterious effects on bees and have been replaced by the more effective and safer Fumidil B. Other compounds potentially effective and those found ineffective against *N. apis* are summarized by Bailey.[549] Aside from work with *N. apis*, only buquinolate (ethyl 4-hydroxy-6,7-diisobutoxy-3-quinoline-carboxylate) has shown any promise as a chemotherapeutic agent for other invertebrate microsporidioses. Overstreet[548] and Overstreet and Whatley[547] found that of seven drugs tested, only buquinolate produced a reduction in the number of blue crabs infected with *Nosema michaelis*.

Several antibiotics have also been examined for the control of amoebic disease of grasshoppers.[84,281,550] The prophylactic use of Thipyrimeth, a triple sulfa drug containing 3% sulfamethazine sodium, 6% sulfathiozole sodium, and 4% sulfapyridine sodium, has been recommended as the most effective means for control of *M. locustae* in laboratory colonies of grasshoppers.[84,550] Thipyrimeth was also the only one of five amoebicidal drugs tested by Donaldson[281] which gave control of *Malameba* in grasshoppers in South Africa. In a later study Henry and Oma[550] did not obtain adverse effects of Thipyrimeth on either grasshopper survival or fecundity and indicated that a 4-sulfa drug, TPM_2, also was effective in controlling *M. locustae*. Apparently, Thipyrimeth arrests trophozoite development but the mode of action of the drug has not been determined. Cyst viability is evidently not affected as the Thipyrimeth will not cure established infections and must be used prophylactically.[84,550] A few drugs have also been tested against *M. mellificae* in the honey bee but without success.[542,551]

Most other antibiotics (those used to control bacterial or fungal contaminants in artificial diet, tissue culture medium, or aqueous suspensions of spores during storage) have not been shown to adversely affect vegetative development or spore viability of various microsporidia.[339,493,497,499,538,543,552] However, Higby et al.[552] found that certain concentrations of nystatin reduced the infectivity of spores of *Nosema eurythremae* when used in combination with other antibiotics to control various bacterial and fungal contaminants of spores harvested from infected rediae of the trematode *Fasciola hepatica*.

Other Chemicals and Antimicrobial Agents

The literature on the toxic effect of chemicals on spores of the entomogenous protozoa other than *N. apis* in honey bee combs is greatly scattered, and few systematic studies have been aimed at determining their adverse effects. Formalin and acetic acid have been used to disinfect combs contaminated with spores of *Nosema apis*,[549] but more recent efforts have concentrated on the fumigant ethylene oxide (see reviews by Gochnauer et al.[521] and Furgala and Mussen[522]). Carbolic acid also has been shown to inactivate spores of *N. apis*[489] and formalin and acetic acid were suitable as general disinfectants of microsporidian spores.[212]

The sensitivity of spores of various microsporidia is dependent on the germicide, its concentration, and duration of exposure. Maddox[482] found that spores of *Vairimorpha necatrix* were inactivated by concentrations of ethyl alcohol higher than 50% but spore viability was little affected by concentrations lower then 25%. Fuxa and Brooks[493] reported that of several germicides tested at various concentrations, only mercuric chloride was harmless to spores of *V. necatrix*. Spores of *Nosema algerae* were inactivated by 95% ethanol for 20 min or 0.1% formalin overnight.[553] Sodium hypochlorite (the active ingredient in commercial bleach) is detrimental to microsporidian spores[482,493,547,554] but did not inactivate spores of the neogregarine *Mattesia grandis* suspended in 0.05% solution for 15 min.[554] Quaternary ammonium compounds (Zephiran Chloride and Hyamine 10-X) also have varying effectiveness as disinfectants or germicides depending upon the species, dosage, and duration of exposure.[493,554]

Various microbial inhibitors such as sorbic acid and methyl-*p*-hydroxybenzoate frequently have been used in artificial diets for rearing insects infected with microsporidia. While few attempts have been made to determine their specific effect on protozoan infections of insects, Hsiao and Hsiao[544] found that neither sorbic acid nor methyl-*p*-hydroxybenzoate inhibited microsporidian development in the alfalfa weevil, *Hypera postica*. However, Dunkel and Boush[543] found that the eugregarine *Pyxinia frenzeli* was eliminated from *Attagenus megatoma* at a concentration of 8% sorbic acid in the diet.

In recent studies with an insect growth regulator, methoprene did not appear to have any significant affects on the infectivity of spores of the microsporidium *V. necatrix* for larvae of *Heliothis virescens*[555] or sporocysts of the eugregarine *Ascogregarina culicis* for larvae of *Aedes aegypti* or *Ae. epactius*.[419] And, although infected larvae of *Ae. aegypti* were not more susceptible than noninfected larvae, the mortality rates of *Ae. epactius* larvae infected with *A. culicis* were significantly greater than those of noninfected larvae exposed to methoprene.[419]

Persistence

The field persistence of a protozoan is a function of both biotic and abiotic environmental factors. It also includes the dispersal and transmission of a protozoan by primary and secondary hosts, and various parasites, predators, and nonhost carriers. Relatively few studies have addressed the effects of biotic factors and the limited efforts on abiotic factors have concentrated on temperature, moisture, and solar radiation. Since abiotic factors were discussed in previous sections, the following discussion will center on attempts to determine persistence of various protozoa under field conditions.

Emphasis on the use of protozoa as microbial control agents has been placed on its efficacy as a short-term control agent. These tests involved the use of repeated applications, thus precluding an accurate determination of persistence. Weiser and Veber[27,28] were among the first to attempt to control insect pests with protozoa, and concluded that poor persistence and dispersal of *Microsporidium hyphantriae* precluded its use as a long-term control agent of the fall webworm, *Hyphantria cunea*. Infected larvae ceased feeding shortly after infection, spores were not disseminated in the feces, and infected larvae were selectively preyed upon by birds. Weiser[556] also indicated that an attempt to control *E. chrysorrhoea* with *Pleistophora schubergi* also was unsuccessful due to removal of spores by rain and/or their inactivation by sunlight. The first attempt to determine the persistence of a microsporidium under field conditions was carried out by Kaya.[557] Spores of *P. schubergi* were applied on northern red oak trees in aqueous suspension with and without the UV protectant "Shade" and then bioassayed in the laboratory with larvae of *Anisota senatoria*. Protected spores were infective 6 days post-treatment resulting in 31% infection while naked spores were noninfective at 6 days and infected only 10% of the larvae after the 4th day. In similar experiments with *Vairimorpha necatrix*, Gardner et al.[558] found that naked spores sprayed on soybeans produced 16% infection in *Heliothis zea* larvae 10 days after application. Along with UV radiation, heavy precipitation during the test period was suggested as being at least partially responsible for the decrease in pathogen activity. A substantial reduction in infectivity also was found with unprotected spores of *V. necatrix* within 5 days after spraying on cabbage plants.[498]

More extensive observations of the field persistence of *V. necatrix* were made by Fuxa and Brooks.[559] Spores with and without the UV protectants SAN 285 and Shade were sprayed onto foliage of the row crops tobacco, cotton, and soybeans. Laboratory bioassay tests conducted up to 14 days postapplication with *H. zea* larvae revealed that the UV protectants generally increased spore half-life on cotton and soybeans by 1.7- to 2.1-fold. No significant differences in spore survival, as related to crops, were apparent until the 4th day after spraying. At that time spore survival was generally better on tobacco than on cotton, perhaps due to the effect of shading or to the sticky surface of the tobacco leaves that prevented the loss of spores. On day 7 (and even more so on days 10 and 14) spore survival was consistently better on soybeans than on tobacco or cotton, a difference attributed primarily to the growth characteristics of the crops. Sprayed soybean leaves were often overgrown and shaded by new leaves within 7 days whereas new tobacco growth was removed by suckering and the cotton grew in a flat canopy with the older leaves still broadly exposed to sunlight.

In a similar study with *V. necatrix* and *Nosema pyrausta*, Lewis[560] measured spore persistence by infesting corn during the whorl and pollen-shedding stages of plant development with egg masses of *Ostrinia nubilalis* at various time intervals postapplication. Spore persistence of *N. pyrausta* was not significantly affected by the UV protectant during either the whorl or pollen-shedding stages of corn development. Significantly lower rates of infection, however, were obtained from larvae placed on plants at 5 or 10 days postapplication during the whorl stage of development as opposed to no significant difference in the infection rate during the pollen-shedding stage. A similar decline in spore viability, as measured by intensity of infection, was also noted during the whorl stage of corn development. Another field test, comparing the persistence of *N. pyrausta* with *Vairimorpha necatrix*, indicated that spores of both species remained viable for up to 12 days postapplication. Spore longevity was attributed to protection from sunlight provided by the corn plant by placement of spores in the whorl or behind the sheath collar during the pollen-shedding stage.

Almost no attempts have been made to examine the persistence of spores in the soil or in aquatic habitats. Chu and Jaques[498] determined that spores of *V. necatrix* were rapidly inactivated when applied to soil in field plots during the summer. Less than 50% of the original activity was retained 13 days after application and no activity was detected after 3 months. However, spores in refrigerated soil (held at 4°C) retained about 50% activity after 12 months; spores mixed with soil in pots and embedded in the field during the winter months were still highly infective after 4 months (from late January to late May) but were completely inactivated during the summer months of June through September. More recently, Germida[561] carried out studies on the field persistence of spores of *Nosema locustae* in soil using light microscopy. Although phase contrast and bright-field microscopy (with six different stains) were not suitable for visualizing spores in soil, fluorescence microscopy with the stain fluorescein isothiocyanate allowed qualitative and quantitative determination of spores in soil debris. Spores, under laboratory conditions, persisted at a high level for over 8 weeks when incubated at 5°C but exhibited a 1000-fold decrease after incubation for 1 week at 27°C. Spore recovery increased with increasing sand content in the soils, and no decrease was detected in sterile sand at 5 or 27°C for 8 weeks. Persistence apparently was related to the temperature-dependent activity of the indigenous soil microflora. In soil samples from a field treated with a spore-bait mixture for grasshopper control, spores were detected at low levels for over a 2-month period. Spores, presumably identical with *N. locustae*, were also detected from nontreated fields with high grasshopper populations. Germida et al.[561a] also monitored the survival and persistence of spores of *N. locustae* in field soils in a 3-year study. Low numbers of spores were found to persist in soil but not on vegetation for 3 to 4 months after each field application. Spores appeared to disappear from the soil as a result of leaching and their interaction with soil microorganisms.

In field tests conducted with *N. algerae* in an aquatic habitat, Anthony et al.[562] found that infection rates in *Anopheles albimanus* were dosage dependent and required repeated applications. Other studies by Alger and Maddox[563] showed that spores of *N. algerae* settled out of the feeding zone of mosquitoes within 24 hr and long-term control was not obtainable. Wang[404a] also found that spores of *Vavraia culicis* settled rapidly in water. Thus, environmental factors, such as radiation or moisture, may be less important in an aquatic habitat where control depends on developing more suitable formulations of the protozoan that keep spores suspended in the feeding zone of the target plots.

SPECIFICITY

While specificity in its strictest sense is limited to the species taxon, it is usually discussed in terms of various taxa such as the species, genus, family, or higher taxonomic categories.[564,565] Specificity is also generally discussed as in vivo specificity (in plants, invertebrates, and vertebrates) and in vitro specificity (in tissue explants or cell lines). With increased consideration of entomopathogens as potential microbial control agents, emphasis has been placed on selecting candidates with wide host ranges for target pests and minimal safety risk to nontarget organisms, i.e., beneficial insects, plants, and vertebrates. In contrast to the fairly extensive published literature on entomopathogenic viruses, fungi, and bacteria, much of the studies of in vivo specificity of protozoa to invertebrates and vertebrates is in unpublished reports. Although specificity of the microsporidia has been most extensively studied, data on host range exists for almost all of the described species of entomogenous protozoa. Detailed discussion on specificity here will be primarily limited to candidates listed in Table 4: a more general coverage on other groups or species of protozoa also will be presented. The discussion will be devoted exclusively to the entomogenous protozoa even though many closely rated species are known pathogens of plants, invertebrates, and vertebrates. Despite the close taxonomic affinity of entomogenous species to vertebrate pathogens, there are no data to suggest that vertebrate pathogens can be transmitted to insects or other invertebrates;[566] and, as will be seen in the following discussion, there is little evidence to indicate that the entomogenous species are infectious for vertebrates.

In Vitro Specificity

As discussed more thoroughly by Ignoffo,[564] the susceptibility of tissue explants or cell lines to various pathogens should not be interpreted as duplicating in vivo specificity. However, a high or low degree of in vitro specificity may be an important consideration in assessing the potential hazards involved in developing a microbial control agent. Relatively few attempts have been made to evaluate the in vitro susceptibility of entomogenous protozoa. All studies to date relate to growth and development of various microsporidia in tissue explants or established cell lines.[567] There are no known reports of in vitro plant specificity with entomogenous protozoa and only a few attempts have been made to examine the specificity of invertebrate microsporidia in mammalian cell lines.

Only 11 invertebrate microsporidia have been studied in in vitro systems, with three of these species occurring naturally in hosts other than insects (Table 26). Most microsporidia have been studied in tissue explants or primary cell cultures of their natural hosts but a few cell lines also were susceptible to alien microsporidia. *Nosema disstriae*, a natural pathogen of the forest tent caterpillar, *Malacosoma disstria*, was successfully grown in primary cell cultures or cell lines of *M. americanum*, *Galleria mellonella*, and *Heliothis zea*.[539,568-571] An

Table 26
IN VITRO CULTIVATION OF INVERTEBRATE MICROSPORIDIA

Pathogen	Inoculum	In vitro system	Response	Ref.
Nosema algerae	Spores	Pig kidney cells	Vegetative growth and spore maturation at 26 and 35°C; vegetative growth but no spores at 37°C; no infection at 38°C	339
	Spores	Trichoplusia ni cell line TN-368, Heliothis zea cell line 1PLB-1075, and Mamestra brassicae cell line 1ZD-Mb-0503	Vegetative growth and spore maturation; cessation of spore production by 6th subculture	341
	Spores	Dipteran cell lines ATC15 (Aedes albopictus), MOS43 (Anopheles stephensi), MOS55 (An. gambiae), MOS20 (Ae. aegypti), ARM (Armigeres subalbatus), and DM1 (Drosophila melanogaster)	Vegetative growth and spore maturation but limited in comparison with lepidopteran cells	574
	Spores	Embryonic rat brain cells, frog cell line XTC-6 (Xenopus laevis), and Chang liver cells	Vegetative growth and spore maturation at 27 and 34°C but not at 38°C (also no growth in embryonic chick brain cells at 38°C)	574a
N. apis	Explants taken from adults exposed per os to pathogen	Midgut explants from Apis mellifera	Vegetative growth and spores	575
N. bombycis	Hemolymph from infected larvae	Ovarian tissues of Bombyx mori	Vegetative growth and spore maturation	576
	Spores	B. mori cell from ovarian explants	Vegetative growth and spore maturation	577
	Spores	B. mori primary cell culture	Vegetative growth, spore maturation, and cell-to-cell spread	199

Organism	Inoculum	Cell culture	Results	Ref.
	Spores	Primary cell cultures of rat, mouse, rabbit, and chick embryos	Limited growth in all cultures, best in rat and chick embryo cells; spores formed only in rat embryo cells; no growth at 37°C	572
	Spores	Manduca sexta cell line MRRL-CH1	Vegetative stages and spores	572a
N. disstriae	Spores	Antheraea eucalypti cell line	Vegetative stages and spores	572b
	Explanted hemocytes from infected larva	Malacosoma disstria hemocyte cell line	Vegetative growth, spore maturation, and cell-to-cell spread	578
	Subculture of infected hemocyte or ovarian cell line	M. disstria hemocyte cell line (Md 66) and ovarian cell line (Md 63)	Vegetative growth, spore maturation, and cell-to-cell spread	579
	Subculture from persistently infected hemocyte cell line	M. disstria hemocyte cell line IPR1-MD-66	Vegetative growth, spore maturation and cell-to-cell spread; elimination of pathogen in time at 35°C	538, 580
	Explanted tissues from infected larvae and pupae	Explants of M. disstria Hemocytes Immaginal cells Pupal gonadal cells Silk gland cells Silk gland cells	Vegetative stages and spores Spores only Vegetative stages and spores Spores and few vegetative stages Spores and few vegetative stages	568
	Spores	M. americanum primary hemocytes cultures	Vegetative cells and spores	568
	Spores	Galleria mellonella primary hemocyte culture	Transient infection with vegetative stages	568
	Spores	Heliothis zea ovarian cells line IPLB-1075	Vegetative growth and spore maturation	540, 569, 570
	Persistently infected hemocyte cell line IPR1-MD-66	M. disstria hemocyte cell line UM-MDH-1	Vegetative growth, spore maturation, and cell-to-cell spread	571
	Persistently infected hemocyte cell line IPR1-MD-66	M. disstria hemocyte cell line UMN-MDH-1, H. zea ovarian cell line IPLB-1075, and Tria-	Vegetative growth, spore maturation, and cell-to-cell spread	571a

Table 26 (continued)
IN VITRO CULTIVATION OF INVERTEBRATE MICROSPORIDIA

Pathogen	Inoculum	In vitro system	Response	Ref.
	Persistently infected hemocyte cell time IPR1-MD-66	*toma infestans* embryo cell line BTC-32	No growth	571, 571a
		Blattella germanica embryo cell lines		
N. eurytremae	Spores	Frog cell line Xen (*Xenopus laevis*) and mosquito cell line 61 (*Aedes pseudoscutellaris*)	Vegetative growth, spore maturation, and cell-to-cell spread	552
		Cell lines 60 (*Aedes malayensis*), alb (*Ae. albopictus*), and **frog** (*Spodoptera frugiperda*)	No growth	
	Spores	Embryonic rat brain cells and frog cell line XTC-6	Vegetative growth and spore maturation at 27 and 34°C but not at 38°C (also no growth in embryonic chick brain cells at 38°C)	574a
N. heliothidis	Spores	*Heliothis zea* ovarian cell line IPLB-1075	Vegetative growth and spore maturation	570
N. mesnili	Explants taken from larvae exposed *per os* to pathogen	Tissue (gut and fat body) explants of *Pieris brassicae*	Vegetative growth and spore maturation	581
N. michaelis	Spores	Crab hemocytes and epithelial cells, human erythrocytes, neuroblastoma C13OD, ascites leukemia EL4, and mouse macrophages	Initial stage (sporoplasm) in all cell lines; lysis of erythrocytes, fibrous corona round pathogen in C13OD, EL4, and mouse macrophages; no observations on ultimate fate of sporoplasm within cells	573
Pleistophora sp.	Explants from infected snails	Gonadal and heart explants of snail *Biomphalaria glabrata*	Vegetative growth and spore maturation	539

Vairimorpha necatrix	Explanted tissue from infected larvae	Explants of adipose tissue from Pseudaletia unipuncta	Spore maturation	499
Vavraia culicis	Spores	Mamestra brassicae cell line IZD-Mb-0503	Vegetative growth and spore maturation	574
	Spores	Ovarian cell lines from Ae. albopictus, Ae. pseudoscutellaris (AP-61), Culex pipiens quinquefasciatus, and Culiseta incidens	Vegetative growth and spore maturation	404a

even greater ability to grow in cells of diverse origin is shown by *N. algerae*. This species was successfully cultured in three lepidopteran cell lines and six dipteran cell lines (Table 26). Both *N. algerae* and *N. bombycis* were also cultured in mammalian cells but only at temperatures less than 37°C.[339,572] A low degree of in vitro specificity for microsporidia is also indicated by the successful culture of *N. eurytremae*, a parasite of trematode larvae, in a dipteran and an amphibian cell line[552] as well as the successful invasion of various mammalian cells by the crab microsporidium *Nosema michaelis*.[573] Thus, on the basis of limited observations, the in vitro specificity of the microsporidia for invertebrates is low. Although mammalian cells also may be susceptible, the inability of microsporidia to produce infections in cells at or above 37°C (the homeothermic vertebrate body temperature) limits any concern for hazards of microsporidia primarily to invertebrates and to the poikilothermic vertebrates, especially fish.

There have not been any studies on specific factors affecting in vitro specificity. Since cell lines of diverse origin are susceptible to microsporidia, specificity may be more related to spore germination or the infection process rather than nutrition. Spores usually do not germinate in most tissue culture media but have to be "primed" by pH manipulation or the use of alkali metal ions. However, since some cell lines were not susceptible or only transient infections were produced by the use of primed spores,[552,568,573] host cell nutrition may also be a factor in determining in vitro specificity of microsporidia. Infection in some cell lines also may elicit a cellular response;[573] whether or not such a response prevents subsequent development of the parasite is not known.

In Vivo Specificity

Most of what is known of in vivo specificity of the entomogenous protozoa pertains to the microsporidia, and most studies have been concerned only with invertebrate specificity. More recent studies have assessed the host range of protozoa among species and genera closely related to the habitual host. Until recently the entomogenous protozoa, especially the microsporidia, were considered host specific and new species were frequently described because of their presence in a new host. Greater emphasis was placed on host specificity when it was realized that some species are not only infectious for species in closely related genera but also for species in other families and even orders of hosts. Most species, however, still tend to be relatively specific, with natural hosts being closely related and usually occurring in similar habitats. Host specificity is also related in part to the method of administration of the pathogen to suspect hosts. As most protozoa are perorally infectious, bioassays are usually conducted by adding spores to the food or the environment. In some cases in vivo susceptibility has also been determined by intrahemocoelic injection of spores into suspect hosts despite the fact that this is not the natural route of infection. Further discussion of in vivo specificity will follow taxonomic groups with emphasis on the host range of the potential microbial agents previously listed (Table 4).*

Flagellates

There is little in vivo specificity among the genera of entomogenous trypanosomatids.[16] Flagellates of mosquitoes and calliphorid or muscoid flies are cross transmissible *per os* or by intrahemocoelic injection to various members of the host family as well as between hosts of different families and even orders of insects. For example, Wallace,[16] in his extensive

---

* In Tables 27 to 41 that list host ranges of these microbial agents, emphasis has been placed on documenting the first reported natural occurrence of a pathogen in either a field population and/or laboratory colony of the host even though some hosts may have been recorded earlier as susceptible by experimentally induced infections. In some instances where this initial report does not refer to the specific protozoan under discussion by its correct name, the reference is cited as a consequence of a subsequent reference which indicated the distinct likelihood that the protozoan referred to in the earlier report is actually the pathogen under discussion.

review of the trypanosomatid parasites of insects, lists 26 species of diptera as natural hosts for *H. muscarum* and also provides lists of other insects which were susceptible *per os* or by intrahemocoelic injection. Similarly, *Crithidia fasciculata* was detected in 21 different species of mosquitoes and was cross transmissible (*per os* or by intrahemocoelic injection) to some mosquitoes and a few other insects.

The extensive early literature on attempts to infect vertebrate hosts with insect trypanosomatids is critically covered and analyzed by Wenyon.[66] Most attempts were either negative or have been discredited as a result of more careful studies. There has not been any recent attempt to investigate the vertebrate specificity of other entomogenous flagellates, probably because of their low virulence and other unfavorable attributes as microbial control agents.

Amoebae

Remarks on host specificity of the amoebae will be limited because so few strictly entomogenous amoebae are known and only one, *Malameba locustae*, has any obvious potential as a microbial control agent. *M. locustae* has a wide host range among grasshoppers in various genera and subfamilies of the Acrididae (Table 27) but has been found naturally, usually at low levels of infection, in only a few species. The amoeba has been more commonly encountered in laboratory colonies of various grasshoppers. In fact, its presence in a culture of *Schistocerca gregaria* served as the apparent source of infection in a free-living population of the silverfish, *Lepisma saccharina*.[284] Although Corbel[584] indicated that a few species of crickets were susceptible to the amoeba, the only confirmed record is that of Ernst and Baker[287] in an unspecified species of the genus *Pteronemobius*. Certainly its low specificity among grasshoppers and its infectiousness for a host in another order of insects indicates the need to determine its host range, at least among other invertebrates, prior to its use as a microbial control agent.

A more recently described species of this same genus, *Malameba scolyti*, is known from natural populations of bark beetles.[80,582] The honey bee pathogen *Malpighamoeba mellificae* appears to be restricted to adult bees.[583] *Malpighiella refringens* was described from the northern rat flea, *Ceratophyllus fasciatus*, but, according to Lipa,[10] may also infect *Ctenocephalides canis*. The few other amoebae associated with insects are so poorly known as to preclude discussion of host specificity.

Eugregarines

There are many described eugregarines (about 1300 species); however, since most of the literature is taxonomic, only a precursory discussion of host specificity is possible. Many are entomogenous while still others are found in hosts from almost all invertebrate phyla. As a whole the eugregarines occur only in a single host species or in closely related species.[12,91,97] Various families and genera of eugregarines typically have specific hosts but a few species infect hosts outside of closely related species of the same genus. Since most species are considered harmless parasites or commensals, few have been studied as potential biological control agents or examined extensively as to host specificity. However, a few of the acephaline gregarines (species that occur in the intestinal tract or hemocoel of insects) may be associated with host pathology,[12] and at least one species, *Ascogregarina culicis*, has been considered as a microbial control agent for mosquitoes. *A. culicis* is a common and cosmopolitan parasite in natural populations of the yellow fever mosquito, *Aedes aegypti*, and also has been recorded from several other species of *Aedes* as well as other genera of mosquitoes (Table 28). Reports of *A. culicis* occurring in *Ae. geniculatus* by Ganapati and Tate[293] and Kramar[294] are not included in this table since Vávra[110] concluded that the gregarine they described was probably *A. barretti*. Although the data (Table 28) suggest a wide host range for *A. culicis*, the species appears to be fairly host specific under natural conditions as are the other species in this genus (Table 2). Several species of mosquitoes tested under

Table 27
HOSTS SUSCEPTIBLE TO *MALAMEBA LOCUSTAE*

Order	Family	Host	Infection[a]	Ref.
Orthoptera	Acrididae	*Ageneotettix deorum*	E	24
		Arphia pseudonietana	E	24
		A. sulphurea	E	24
		A. xanthoptera	E	24
		Austroicetes cruciata	L	282
		Campylacantha olivacea olivacea	E	24
		Chloealtis conspersa	E	24
		Chorthippus longicornis	E	24
		Chortoicetes terminifera	L	282
		Chortophaga viridifasciata	E	24
		Diabolocatantops axillaris	N	286a
		Dichromorpha viridis	E	24
		Dissosteira carolina	E	24
		Encoptolophus sordidus sordidus	E	24
		Hadrotettix trifasciatus	N	24
		Hippiscus rugosus	E	24
		Locusta migratoria migratorioides	E	85
		Locustana pardalina	L, N	279, 280
		Melanoplus angustipennis	E	24
		M. bivattatus	L	357
		M. confusus	E	24
		M. dawsoni	L	357
		M. differentialis	L	77
		M. femurrubrum	L	77
		M. flavidus flavidus	E	24
		M. foedus	E	24
		M. gladstoni	E	24
		M. gracilis	E	24
		M. keeleri luridus	E	24
		M. packardii	E	24
		M. sanguinipes	L	36
		M. scudderi	E	24
		M. viridipes viridipes	E	24
		M. walshii	E	24
		Oedaleus senegalensis	N	286a
		Orphulella speciosa	E	24
		Paracinema tricolor	E	277
		Pardalophora apiculata	E	24
		P. haldemanii	E	24
		Pardillana limbata	N	287
		Phoetaliotes nebrascensis	E	24
		Pseudosphingonotus savignyi	N	286a
		Schistocerca gregaria	L	284
		S. vaga vaga	N	84
		Spharagemon bolli	E	24
		S. collare	E	24
		Trachyrhachis kiowa kiowa	N	24
		Trimerotropis maritima maritima	E	24
	Gryllidae	*Pteronemobius* sp.	N	287
Thysanura	Lepismatidae	*Lepisma saccharina*	N	284

[a] E, experimental infection; N, natural infection in field populations; L, natural infection in laboratory colony.

Table 28
HOSTS SUSCEPTIBLE TO *ASCOGREGARINA CULICIS*

Order	Family	Host	Infection[a]	Ref.
Diptera	Culicidae	*Aedes aegypti*	N	113
		Ae. albopictus	N	313, 314
		Ae. alcasidi[b]	E	109
		Ae. argenteus	N	315
		Ae. desmotes	E	109
		Ae. epactius[b]	E	311
		Ae. excrucians	N	305a
		Ae. geniculatus	N	305a
		Ae. ingrami	N	585
		Ae. koreicus	N	312
		Ae. polynesiensis	N	317
		Ae. sierrensis[b]	E	112
		Ae. sollicitans[b]	E	311
		Ae. stimulans[b]	E	311
		Ae. vexans[b]	E	311
		Anopheles jeyporiensis	?	585
		An. maculipennis	N	305a
		An. sinensis	?	585
		Armigeres obturbans	N	313
		Ar. subalbatus	N	313a
		Culex pipiens pipiens	N	305a
		Cu. tritaeniorhynchus	?	586
		Culiseta inornata[b]	E	311
		Heizmannia taiwanensis[b]	E	109
		Psorophora columbiae[b]	E	311

[a] N, natural infection; E, experimental infection.
[b] Hosts in which development of the gregarine was abnormal or incomplete.

laboratory conditions or found associated with heavily infected natural populations of *Ae. aegypti* were not infected by *A. culicis*,[109,306,307,311-314] and development of the gregarine was abnormal or incomplete[109,112,311] in several of the species in Table 28. Few investigators have attempted to assess the suitability of experimental hosts to complete the life cycle of *A. culicis* or to produce infective oocysts. Additional research will be required to determine the host specificity of *A. culicis* as well as other gregarines.

Neogregarines

In contrast to the eugregarines, relatively few species of neogregarines have been described. However, most are pathogenic for their host and at least two species, *Mattesia grandis* and *M. trogodermae*, have been evaluated as microbial control agents. Most neogregarines have been reported from a single host or a few closely related species. Most effort at determining host ranges of neogregarines has been with *Mattesia*. For example, *Lymphotropha tribolii* is known only from the red flour beetle, *T. castaneum*,[102] but also is infectious experimentally for *Tribolium anaphae*. Another species, *Farinocystis tribolii*, which occurs most commonly in *T. castaneum*, was also infectious when fed in massive dosages for *Tribolium confusum*, *T. destructor*, and *Tenebrio molitor*.[12] Purrini[355] recently found *F. tribolii* in a number of stored grain pests including *T. castaneum*, *T. confusum*, *T. molitor*, *Tenebrionides mauretanicus*, *Calandra granaria*, *Laemophloeus ferrugineus*, *Gracilia albanica*, and two species in the genus *Attagenus*. A wide host range including

Table 29
HOST RANGE OF *MATTESIA GRANDIS*

Order	Family	Host	Infection[a]	Ref.
Coleoptera	Curculionidae	*Anthonomus grandis*	L	129
Hymenoptera	Braconidae	*Bracon mellitor*	E	444
Lepidoptera	Gelechiidae	*Pectinophora gossypiella*	E	355a
	Noctuidae	*Heliothis virescens*	E	355a
		H. zea	E	355a
		Trichoplusia ni	E	355a
	Pyralidae	*Anagasta kuehniella*	E	129
		Plodia interpunctella	E	129

[a] L, natural infection in laboratory colony; E, experimental infection.

Table 30
HOST RANGE OF *MATTESIA TROGODERMAE*

Order	Family	Host	Infection[a]	Ref.
Coleoptera	Dermestidae	*Trogoderma glabrum*	N, L	331
		T. granarium	L	331
		T. grassmani	N	330
		T. inclusum	N, L	330, 331
		T. ornatum	N	330
		T. simplex	N	330
		T. sternale	N	330
		T. teukton	E	331
		T. variabile	N	330

[a] N, natural infection in field population; L, natural infection in laboratory colony; E, experimental infection.

insects of several orders has also been reported for *M. dispora*, a species common to the Mediterranean flour moth, *A. kuehniella*.[355,355a] Similarly, *Mattesia grandis*, a neogregarine of the cotton boll weevil, *Anthonomus grandis*, has been experimentally transmitted to several lepidopterans and a hymenopteran (Table 29). On the other hand, *M. trogodermae* appears to be limited to closely related species of dermestid beetles in the genus *Trogoderma* (Table 30). Pounds,[329] in an extensive study (acute *per os* or acute inhalation tests) of the possible hazards and host range of *M. trogodermae*, found no evidence of infection or pathological effects in rats. Additional feeding or injection tests also were negative when tested against representative nontarget species from marine (brine shrimp, *Artemia salina*), aquatic (water flea, *Daphnia magna*; and emerald shiner, *Notropis atherinoides*), and terrestrial (greater wax moth, *Galleria mellonella*; black cutworm, *Agrotis ipsilon*; honey bee, *A. mellifera*; and red worm, *Eisenia rosea*) habitats. Spores germinated in the intestinal tract of several of these hosts but no evidence of infection or pathological effects were subsequently detected. Earlier, Canning[106] demonstrated that *M. trogodermae* was also not infective to *A. kuehniella*, *Plodia interpunctella*, and *Cryptolestes pusillus*.

Coccidia

Few exclusively entomogenous coccidia are known: most are parasites of vertebrates and few have been examined for host specificity. In addition, some species reported from various stored products insects and described as members of the genus *Adelina* are poorly differentiated taxonomically and may be identical.[12] Most species are known from a single species of host. Thus, *A. cryptocerci* was described from the wood cockroach *Cryptocerus punctulatus*,[139] *Legerella hydropori* from the dytiscid beetle *Hydroporus palustris*,[145] *Chagasella hartmanni* from the hemipteran *Dysdercus ruficollis*,[143a] and *Ithania wenrichi* from the crane fly *Tipula abdominalis*.[144] Other species as well as perhaps some of those described from single host species are less host specific. *Adelina sericesthis* is infectious for the scarabeid grubs *Melolontha hippocastani* and *Aphodius howitti* as well as its type host *Sericesthis pruinosa*.[12,140] *Adelina mesnili* infects the Mediterranean flour moth, *A. kuehniella*,[587a] the webbing clothes moth, *Tineola bisselliella*,[587b] the Indian meal moth, *Plodia interpunctella*,[587b] and the tobacco moth, *Ephestia elutella*.[355b] A similar if not identical species, *Adelina tribolii*, was described by Bhatia[588] from the red flour beetle, *T. castaneum*, but is also apparently infectious for the confused flour beetle, *T. confusum*, and the yellow mealworm, *T. molitor*.[12] More recently, Purrini[355b] recorded this species from *Attagenus pellio*, *A. piceus*, *A. fasciolatus*, *Laemophloeus ferrugineus*, *Trogoderma granaria* and *Gracilia albanica*. He also found it subsequently in *Ptinus pusillus*.[355c] Thus, few definitive remarks can be said about coccidian host specificity except that some species have a fairly wide host range and additional study is needed.

Microsporidia

The large number of microsporidia precludes an indepth discussion of host specificity of this group. Hosts are known from almost all invertebrate phyla and classes of the vertebrates; the majority of species are associated with arthropods and fishes. Until recently, most species were considered host specific, and many new species have been described based on their presence in an unreported host species. Most species are host specific under natural conditions where host range is limited to species living together and sharing a common food source. However, under laboratory conditions, many entomogenous species are infectious for many hosts of different families and even orders of insects. Some microsporidia appear to be fairly specific to certain groups of hosts and often various hosts groups are parasitized by host-specific species of microsporidia. For example, microsporidia of the genera *Pilosporella*, *Parathelohania*, *Hyalinocysta*, and *Amblyospora* are primarily (if not exclusively) parasites of mosquitoes and the species within the genera appear to be highly host specific. However, the possibility that some of these may have intermediate hosts as recently demonstrated for *Amblyospora*[438d,438e] and the noninfectious nature of the meiospores for mosquito larvae have hampered specific determinations of host specificity. Similarly, species of *Pegmatheca* occur only in blackflies while most species in the genera *Glugea* and *Pleistophora* are primarily host-specific parasites of fish. It appears likely that in future taxonomic revisions, most of the nonvertebrate species now included in the genus *Pleistophora* will be placed in other genera. Weiser[590] also suggests that species which develop in a wide range of host tissues are usually not very host specific. However, tissue-specific species are not necessarily host specific as exemplified by *Vairimorpha necatrix* and *Pleistophora schubergi* (see Tables 39 and 40). As of 1976,[589] over 700 species of microsporidia had been described. Subsequent discussion of host specificity will focus on species listed as potential microbial control agents (Table 4) with emphasis on their infectivity for nontarget organisms and (where applicable) for vertebrates.

Nosema algerae — Although this species is infectious for many culicids (Table 31), anopheline mosquitoes are probably the primary hosts.[591] Estimates of the comparative susceptibility of a number of species in different genera reveal that species of the genus *Anopheles* are generally highly susceptible to *N. algerae* while species in the genera *Aedes*, *Culex*, *Wyeomyia*, and *Armigeres* are less susceptible.[319,591] However, at least two anopheline mosquitoes, *Anopheles subpictus*[591] and *An. atroparvus*,[592] exhibited limited susceptibility and the mosquitoes *Psorophora ciliata* and *Toxorhynchites rutilus septentrionalis* were not susceptible to *N. algerae*.[319] The wide host range of this microsporidium (under experimental

Table 31
PER OS SUSCEPTIBILITY OF HOSTS TO NOSEMA ALGERAE

Phylum	Class	Order	Family	Host	Infection[a]	Ref.
Arthropoda	Insecta	Coleoptera	Chrysomelidae	Gastrophysa viridula	E	600b
				Leptinotarsa decemlineata	E	600c
				Oulema lichenis	E	600d
				O. melanopus	E	600d
		Diptera	Chironomidae	Chironomid sp., probably Goeldichironomus	E	319
			Culicidae	Aedes aegypti	E	192
				Ae. albopictus	E	597
				Ae. taeniorhynchus	E	319
				Ae. triseriatus	E	319
				Ae. unilineatus	E	597
				Anopheles albimanus	N, L	319, 337
				An. annularis	E	591
				An. atroparvus	E	592
				An. balabacensis	L	337
				An. crucians	E	598
				An. culicifacies	E	591
				An. gambiae	L, N	193, 338
				An. melas	L	338
				An. pulcherrimus	E	591
				An. quadrimaculatus	L	337
				An. stephensi	L	192
				An. subpictus	E	591
				An. triannulatus	E	319
				Armigeres subalbatus	E	343
				Culex nigripalpus	E	319
				Cx. pipiens pipiens	E	193
				Cx. pipiens quinquefasciatus	E	193
				Cx. restuans	E	319
				Cx. salinarius	E	336
				Cx. sitiens	N	599
				Cx. tarsalis	E	319
				Cx. territans	E	319
				Cx. tritaeniorhynchus	E	319
				Wyeomyia mitchellii	E	319
			Muscidae	Musca domestica	E	319
				Stomoxys calcitrans	E	319
		Hemiptera	Notonectidae	Notonecta undulata	E	600
		Lepidoptera	Noctuidae	Amathes c-nigrum	E	600a
				Heliothis zea	E	594
				Mamestra brassicae	E	191a
				M. dissimilis	E	600a
				Spodoptera exigua	E	563
			Pyralidae	Galleria mellonella	E	563a
Platyhelminthes	Trematoda	Digenea	Schistomatidae	Fasciola hepatica	E	601
				Schistosoma mansoni	E	602

[a] E, experimental infection; L, natural infection in laboratory colony; N, natural infection in field population.

Table 32
HOSTS SUSCEPTIBLE BY INJECTION WITH *NOSEMA ALGERAE*

Phylum	Class	Order	Family	Host	Ref.
Arthropoda	Crustacea	Decapoda	Astacidae	Crayfish	594
	Insecta	Diptera	Calliphoridae	*Phormia regina*	594
		Lepidoptera	Arctiidae	*Diacrisia virginica*	594
			Noctuidae	*Heliothis zea*	594
				Spodoptera ornithogalli	594
			Pieridae	*Pieris brassicae*	594
			Pyralidae	*Galleria mellonella*	594
		Hemiptera	Lygaeidae	*Oncopeltus fasciatus*	594
		Megaloptera	Corydalidae	*Corydalus cornutus*	594
		Odonata		Damselfly	594
		Orthoptera	Blaberidae	*Blaberus discoidalis*	594
			Blatellidae	*Blattella germanica*	319
Chordata	Mammalia	Rodentia	Muridae	*Mus musculus*[a]	595

[a] Localized and transient infections only at site of infection in cooler body parts of the tail, ears, and feet.

conditions) is also evident since it can be orally transmitted to two muscoid flies, four coleopterans, six lepidopterans, a hemipteran predator of mosquito larvae, and even two species of trematodes (Table 31). It also has been transmitted, by injection, to an even wider range of hosts, including representatives of six orders of insects, a crayfish, and white mice (Table 32). Despite the apparent lack of specificity of *N. algerae*, a wide variety of other insects as well as species from other phyla were not susceptible to this microsporidium administered orally or by injection (Table 33). In addition to the negative records of transmission (Table 33), several authors have reported other less specific data on *N. algerae*. Hazard[593] reported no evidence of infection in crayfishes, fresh-water shrimps, mosquito fishes, damselfly naiads, dragonfly naiads, dytiscid larvae, hellgrammites, chickens, and mice when fed larvae of *Heliothis zea* infected with *N. algerae*. Similarly, Savage[319] found no evidence of infection in representative species of various aquatic predators including Anisoptera, Zygoptera, Megaloptera, Coleoptera, and Hemiptera when fed mosquito larvae infected with *Nosema algerae*. Many nontarget mosquitoes of the genera *Culex* and *Aedeomyia* and various associated aquatic Hemiptera (Belostomatidae, Naucoridae, and Notonectidae), Coleoptera, Diptera, Odonata, Ephemeroptera, and Crustacea also were negative when areas in or near plots treated with spores of *N. algerae* in Panama[562] were examined. Thus, the natural host range of *N. algerae* appears limited to culicids; however, its wide in vitro and in vivo host range under experimental conditions indicates that this microsporidium will require more extensive examination prior to its widespread use in aquatic ecosystems as a microbial control agent.

The possible development of *N. algerae* in vertebrates has been primarily examined using mice.[594] Although intravenous injections of spores failed to cause infections, transient and localized infections were produced in mice injected subdermally with spores in the cooler body sites of the tail, ears, and feet.[595] Vegetative stages of *Nosema algerae* were observed in mice up to 10 days postinoculation and some presporal stages were found in some mice 27 days postinoculation. Temperature was the primary factor limiting development only to the extremities since the upper temperature limit for *N. algerae* is 35 to 37°C.[339] Observations also indicated infection, produced as a result of bites by infected adult mosquitoes, is unlikely.[563] In a more extensive study, Alger et al.[596] found vegetative stages of *N. algerae* in the tails of normal mice up to 8 days postinjection and up to 96 days in the tails of nude (immunodeficient) mice. They concluded that development of *N. algerae* is limited first by

Table 33
SPECIES NOT SUSCEPTIBLE TO *NOSEMA ALGERAE*[a]

Phylum	Class	Order	Family	Host	Per os	Injection	Ref.
Annelida	Hirudinea	Gnathobdellida	Hirudinidae	*Hirudo* sp.		x	594
Arthropoda	Crustacea	Decapoda	Astacidae	*Procambarus* sp.	x		600
	Insecta	Coleoptera	Dytiscidae	*Coptotomus interrogatus*	x		600
			Hydrophilidae	*Hydrophilus* sp.	x		600
		Diptera	Chironomidae	*Chironomus* sp.	x		594
			Culicidae	*Psorophora ciliata*	x		319
				Toxorhynchites rutilus septentrionalis	x		319
		Hemiptera	Belostomatidae	*Belostoma fluminea*	x		600
				Belostoma sp.	x		319
			Nepidae	*Ranatra australis*	x		600
		Lepidoptera	Arctiidae	*Diacrisia virginica*	x		594
			Noctuidae	*Agrotis ipsilon*	x		603
				Spodoptera ornithogalli	x		594
			Pyralidae	*Galleria mellonella*	x		594
		Megaloptera	Corydalidae	*Chauliodes rastricornis*	x		600
				Corydalus cornutus	x		594
		Odonata	Aeschnidae	*Anax junius*	x		600
		Orthoptera	Blaberidae	*Blaberus discoidalis*	x		594
				Nauphoeta cinera		x	319
				Panchlora nivea		x	319
			Blattidae	*Blatta orientalis*		x	319
				Eurycotis floridana		x	319
				Periplaneta americana		x	319
				Undetermined species of frog		x	594
Chordata	Amphibia	Salientia		*Bufo terrestris*	x		319
				Hyla cinerea	x		319
	Mammalia	Rodentia	Muridae	*Mus musculus*		x	594
	Pices	Cyprinodontiformes	Poeciliidae	*Gambusia affinis*	x		600
Platyhelminthes	Tubellaria	Tridadida		Undetermined species		x	594

[a] See text for additional discussion of other less specifically identified species found to be noninfected by *Nosema algerae*.

Table 34
HOST RANGE OF *NOSEMA FUMIFERANAE*

Order	Family	Host	Infection[a]	Ref.
Lepidoptera	Lasiocampidae	*Malacosoma americanum*	E	202
		M. californicum lutescens	E	202
		M. californicum pluviale	E	202
	Nymphalidae	*Nymphalis antiopa*	E	202
	Tortricidae	*Archips cerasivoranus*	E	604
		A. fervidanus	E	604
		Choristoneura fumiferana	N	605
		C. murinana	E	350
		C. pinus	E	350

[a] E, experimental infection; N, natural infection.

Table 35
HOST RANGE OF *NOSEMA GASTI*

Order	Family	Host	Infection[a]	Ref.
Coleoptera	Curculionidae	*Anthonomus grandis*	L	354
Lepidoptera	Gelechiidae	*Pectinophora gossypiella*	E	355a
	Noctuidae	*Heliothis virescens*	E	355a
		H. zea	E	355a

[a] E, experimental infection; L, natural infection in laboratory colony.

the mouse's body temperature and secondly by its immune system. Even in immunologically impaired nude mice, growth of the microsporidium was restricted to the injection site and limited by phagocytosis. Mice also were not infected when fed upon by adult mosquitoes heavily infected with *N. algerae* nor when injected *per os* with spores. Although a positive (indirect fluorescent antibody) test was obtained in a human accidentally injected with spores, negative results were obtained with a volunteer who allowed infected female mosquitoes to feed on her arm. In their summary, Alger et al.[596] concluded that *N. algerae* presents no threat to warm-blooded animals.

Nosema fumiferanae — An endemic pathogen of the spruce budworm, *Choristoneura fumiferana*, this microsporidium was originally thought to be restricted to the genus *Choristoneura*.[350] However, it has been subsequently shown to be infectious to a few other lepidopterans of other families as well as genera within the family Tortricidae (Table 34). While extensive efforts have not been made to determine the host range of *N. fumiferanae*, it was not infective for the lepidopterans *Acleris variana*, *Archips cerasivoranus*, and *Ma a americanum*,[350] or the hymenopterous parasites *Apanteles fumiferanae* and *Glypta fumiferanae*.[350,350a] Wilson[202] enlarged this list of nonsusceptible hosts to include the lepidopterans *Galleria mellonella*, *Bombyx mori*, and *Hyphantria cunea* as well as the cricket *Gryllus assimilis*. (*Archips cerasivoranus* is included as a susceptible host in Table 34 based on the citation by Thomson,[604] although in earlier work he[350] had indicated that this species was not susceptible to *Nosema fumiferanae*.) Thus, this species appears to be restricted primarily to Lepidoptera, occurring naturally only in populations of *C. fumiferana* and possibly in sympatric populations of *C. pinus*.[202,320]

Nosema gasti — This species is known only as a pathogen in laboratory cultures of the boll weevil, *Anthonomus grandis*, and its host range has not been extensively studied. Ignoffo and Garcia[355a] found it to be infectious for three species of Lepidoptera (Table 35) but not infectious for *Trichoplusia ni*. McLaughlin,[203] in laboratory tests, reported that *N. gasti* was not infectious for the plum curculio, *Conotrachelus nenuphar*. Several other coleopterans, known to have ingested a bait formulation of *N. gasti* under field conditions, also were not susceptible, i.e., the coccinellids *Coleomegilla maculata lengi*, *Coccinella novemnotata*, *Hippodamia convergens*, and *Chilocorus stigma* and the chrysomelids *Diabrotica undecimpunctata howardi* and *Trirhabda* sp., probably *virgata*.[33,203] Although the reported host ranges of *N. gasti* are limited, its infectivity for several lepidopterans indicates that its potential host range is probably much greater.

Nosema locustae — This microsporidium is known only as a pathogen of Orthoptera. Over 90 species of grasshoppers, in the family Acrididae, are susceptible to *N. locustae*, along with a single species from each of three other orthopteran families (Table 36). The only orthopteran tested and known to be nonsusceptible is the eastern house cricket, *Acheta domesticus*,[606] although two species of West African grasshoppers, from field plots treated with *N. locustae*[286b] were not infected. The lack of infection in the African grasshoppers may be because of the small sample size that was taken. The extensive host range among grasshoppers is likely due to the similarity in habitats and the sympatric nature of many grasshopper populations. It is also interesting that while the known host range of *N. locustae* is extensive, few attempts have apparently been made to test the susceptibility of nonorthopteran insects. There are no data on the potential susceptibility of various predators or parasites of grasshoppers to *N. locustae*.[321] The honey bee, *Apis mellifera*,[607] and the lepidopterans *Heliothis zea*[476] and *Agrotis ipsilon*[603] appear to be the only known nonorthopterans whose nonsusceptibility has been verified.

N. locustae has been extensively tested for potential hazards to vertebrates (Table 37): it has been exempted from the requirement of a tolerance by the U.S. Environmental Protection Agency.[276] No significant effects attributable to the microsporidium were detected when tested against several species of mammals, birds, and fish. These tests included maximum challenges where spores of *N. locustae* were injected intracerebrally, intraocularly, and intraperitoneally in rabbits and mice.[608] In addition to the data summarized in Table 37, a few other observations on the safety of *N. locustae* have been recorded. Henry[609] found that spores were not activated or did not germinate when exposed to the contents of the rumen of cows. Results were also negative in bioassay tests with grasshoppers using various tissues or the gastric contents of mice previously exposed *per os* to spores of *N. locustae*[610] In addition, no histological evidence of infection was obtained in either sparrows or various small mammals collected from a field previously treated (38 to 42 days) with spores of *N. locustae*.[611-613]

A recent study[613a] also addressed the possible impact of spores of *N. locustae* on microbial activity and nutrient cycling in soil. No adverse effect was found at the recommended field dosage rate but microbial activity was stimulated at a higher rate, probably due to spore degradation by indigenous soil microflora or their priming effect on degradation of soil organic matter.

Nosema pyrausta — This is another microsporidium whose host range has been little studied. It is an endemic pathogen of natural populations of the European corn borer, *Ostrinia nubilalis*. Most of its known hosts (Table 38) are dipterous or hymenopterous parasites associated with corn borer larvae or eggs infected with *N. pyrausta*. In fact, this may be the primary reason why populations of parasites such as *Lydella thompsoni*[623] and *Macrocentrus grandii*,[370,376,442e] have declined in areas where the natural prevalence of *N. pyrausta* is high. However, not all parasites of the corn borer are infected by *N. pyrausta*. Andreadis[624] found no evidence of infection in the tachinids *Aplomya caesar* and *Lixophaga* sp. or the ichneumonid *Eriborus terebrans* that had developed in larvae infected with *N. pyrausta*.

Pleistophora schubergi — This species is infectious for a wide range of lepidopterous larvae from many different families as well as a few species of hymenopterous sawflies

Table 36
ORTHOPTERAN HOSTS OF *NOSEMA LOCUSTAE*

Family	Host	Infection[a]	Ref.
Acrididae	*Acorphya glaucopsis*	E	286b
	Acrida bicolor	E	286b
	Acrotylus longipes incarnatus	E	286b
	Aeropedellus clavatus	N	363
	Ageneotettix deorum deorum	N	363
	Aiolopus simulatrix	E	286b
	Aleuas lineatus	E	614
	Allotruxalis strigata	E	614
	Amphitornus coloradus	E	34
	Arphia conspersa	N	363
	A. pseudonietana pseudonietana	E	363
	Aulocara elliotti	N, L	363
	Brachystola magna	N	363
	Bradynotes obesa	N	363
	Bruneria brunnea	N	363
	Camnula pellucida	L, N	206, 363
	Cataloipus cymbiferus	E	286b
	C. fuscocoerulipes	E	286b
	Chorthippus albomarginatus	N	367
	C. curtipennis	N, L	363
	Conozoa wallula	N	363
	Cordillacris occipitalis occipitalis	E	36
	Cratypedes neglectus	N	363
	Cryptocatantops haemorrhoidalis	E	286b
	Diabolocatantops axillaris	E	286b
	Dichroplus bergi	E	614
	D. conspersus	E	614
	D. elongatus	E	614
	D. maculipennis	E	614
	D. pratensis	E	614
	D. pseudopunctulatus	E	614
	D. punctulatus	E	614
	Dissosteira carolina	L, N	206, 363
	D. spurcata	N	363
	Drepanoptera femoratum	E	36
	Encoptolophus sordidus costalis	E	363
	Hadrotettix trifasciatus	N	363
	Hesperotettix viridis viridis	N, L	363
	Heteracris harterti	E	286b
	Hierglyphus daganesis	E	286b
	Homoxyrrhepes punctipennis	E	286b
	Hypochlora alba	E	363
	Kraussaria amabile	E	286b
	K. angulifera	E	286b
	Laplatacris dispar	E	614
	Locusta migratoria migratorioides	L	358
	Melanoplus alpinus	N	363
	M. bivittatus	L, N	357, 363
	M. borealis borealis	N	363
	M. bowditchi canus	E	363
	M. bruneri	N	363
	M. confusus	N	363
	M. cuneatus	N, L	363
	M. dawsoni	L, N	357, 363
	M. differentialis	N, L	363
	M. femurrubrum femurrubrum	N, L	363

Table 36 (continued)
ORTHOPTERAN HOSTS OF *NOSEMA LOCUSTAE*

Family	Host	Infection[a]	Ref.
	M. foedus foedus	N	363
	M. gladstoni	E	363
	M. infantilis	N, L	363
	M. keeleri luridus	E	36
	M. occidentalis occidentalis	N	363
	M. packardii	N, L	363
	M. sanguinipes sanguinipes	L, N	357, 363
	Mermiria maculipennis macclungi	N	363
	Metator nevadensis	N	363
	M. pardalinus	N	363
	Oedaleonotus enigma	N, L	363
	Oedaleus nigeriensis	E	286b
	O. senegalensis	E	286b
	Opeia obscura	N	363
	Phlibostroma quadrimaculatum	N	363
	Phoetaliotes nebrascensis	N	363
	Pseudopomala brachyptera	N	363
	Pseudosphingonotus savignyi	E	286b
	Psoloessa delicatula delicatula	N	363
	Pyrgomorpha cognata	E	286b
	Schistocerca americana	E	363
	S. cancellata	E	614
	S. gregaria	L	205
	S. vaga vaga	E	363
	Scotussa lemniscata	E	614
	Spharagemon equale	E	35
	Sphingonotus canariensis	E	286b
	Trachyrhachis kiowa kiowa	N	363
	Trimerotropis compestris	N	363
	T. fontana	N	363
	T. gracilis sordida	N	363
	T. inconspicua	N	363
	T. laticincta	N	363
	T. pallidipennis pallidipennis	N	363
	Xanthippus corallipes	N	363
	Zonocerus variegatus	E	286b
Gryllidae	*Gryllus* sp.	E	363
Tetrigidae	Undetermined sp.	E	363
Tettigoniidae	*Anabrus simplex*	N	286c

[a] E, experimental infection; L, natural infection in laboratory colony; N, natural infection in field population.

(Table 39). It naturally occurs in various lepidopterous pests of forest trees. Species known to be nonsusceptible include the lepidopterans *Bombyx mori*[224] and *Galleria mellonella*[227] as well as the coleopterans *Oryzaephilus surinamensis*, *Tenebrio molitor*, and *Tribolium castaneum*.[227] This wide host range, which includes several serious forest defoliators, is one reason why this species is considered to have good potential as a microbial control agent.[325] However, its possible effect on vertebrates and other nontarget organisms has not been tested.

Vairimorpha necatrix — This microsporidium is considered as an attractive candidate microbial control agent because of its high degree of virulence and its wide host range.[326] All known hosts are species of Lepidoptera (Table 40), and many are major pests of agri-

Table 37
SUMMARY OF IN VIVO TESTS CONDUCTED TO EVALUATE POSSIBLE HAZARDS OF *NOSEMA LOCUSTAE* TO VERTEBRATES

Type of test	Species	Number of animals	Length of study	Route	Dose (spores/animal)	Evaluation criteria	Results	Ref.
Primary skin irritation	Rabbits	6	3 days	Dermal to abraded and intact skin	11.5×10^7	Erythema and eschar formation, edema formation, and temperature	Negative	615
Acute dermal toxicity	Guinea pigs	10	14 days	Dermal to abraded and intact skin	9.2×10^7	Physical appearance, body weight, temperature, food consumption, survival, and gross and microscopic pathology	No lesions related to pathogen or other significant differences between test and control animals	616
Acute inhalation toxicity and pathogenicity	Rats	20	14 days	Inhalation	All animals exposed simultaneously to presumably 2.3×10^8 spores for hr[a]	Physical appearance, behavior, body weight, food consumption, survival, organ weights, and gross and microscopic pathology	One animal died apparently unrelated to pathogen, no other significant differences	617
Subacute oral	Rats	20	90 days	Dietary feeding	Presumably 1.2×10^7 spores over a 13-week period[a]	Physical appearance, behavior, growth, food consumption, survival, chemical lab data	Negative	618
Acute oral	Rats	40	21 days	Intubation	2.3×10^8	Physical appearance, behavior, growth, food consumption, body temperature, survival, clinical studies, organ	No significant differences from the controls, lesions observed unrelated to pathogen	619

Table 37 (continued)
SUMMARY OF IN VIVO TESTS CONDUCTED TO EVALUATE POSSIBLE HAZARDS OF *NOSEMA LOCUSTAE* TO VERTEBRATES

Type of test	Species	Number of animals	Length of study	Route	Dose (spores/animal)	Evaluation criteria	Results	Ref.
	Rainbow trout and bluegill sunfish	100 of each species	4 days	*Per os* topical	Five dose levels from 2.2×10^8 to 1×10^9 spores/ℓ	Survival, weights, and gross and microscopic pathology	No deaths, LC_{50} estimated to be greater than 1×10^9 spores/ℓ	620
	Mallards	10	204 days	Intubation	5×10^9 spores	Survival, behavior, body weight, blood parameters, gross and microscopic pathology	Death of one bird pathogen unrelated, no other significant differences; LC_{50} estimated to be greater than 5×10^9 spores/bird	621
	Ring-necked pheasant	10	51 days	Intubation	5×10^9 spores	Survival, body and organ weights, gross and microscopic pathology	Negative except for greater heart mass in treated birds	621
Acute-toxicity pathogenicity	Mice	20	14 days	Intraperitoneal	3×10^6 spores	Survival, body temperature, and weight gain	Negative	622
	Rabbits	24[b]	70 days	Intracerebral	2.9×10^7	Clinical monitoring including survival, weight gain, and body temperature; hematologic parameters including hemogram, total protein	No deaths or clinical signs of illness related to pathogen activity, few lesions relating only to traumatic injury, no evidence of	608
		24[b]	70 days	Intraocular	1.5×10^8			
		24[b]	70 days	Intraperitoneal	2.7×10^8			

Animal	No.	Duration	Route	Dose	Observations	Results	Ref.
Mice	36[c]	56 days	Intracerebral	1.5×10^8	and blood urea nitrogen; urine for spores; and gross and microscopic pathology	pathogen multiplication although spores found in some tissues shortly after injection	
	36[c]	56 days	Intraperitoneal	2.7×10^8	Clinical monitoring including survival and weight gain; gross and microscopic pathology	Same as above	608

[a] The exact spore dose involved was not specified in these original reports. Thus, the dose listed in the table is based on the dose used in several other studies conducted simultaneously by Hazleton Labs, i.e., the use of 1 mℓ of spore suspension containing 2.3×10^8 spores/mℓ.[615,616,619]

[b] Only 32 rabbits were used, 20 were simultaneously injected intracerebrally, intraocularly, and intraperitoneally while 4 additional rabbits were injected either intracerebrally, intraocularly, or intraperitoneally.

[c] Only 42 mice were used, 30 were simultaneously injected intracerebrally and intraperitoneally while 6 additional mice were injected either intracerebrally or intraperitoneally.

Table 38
HOST RANGE OF *NOSEMA PYRAUSTA*

Order	Family	Host	Infection[a]	Ref.
Diptera	Tachinidae	*Lydella thompsoni*	N	373
Hymenoptera	Braconidae	*Chelonus annulipes*	N	373
		Macrocentrus grandii	N	373
	Trichogrammatidae	*Trichogramma evanescens*	E	373a
Lepidoptera	Nymphalidae	*Junonia coenia*	E	210
	Pyralidae	*Ostrinia nubilalis*	N	208

[a] E, experimental infection; N, natural infection.

Table 39
HOST RANGE OF *PLEISTOPHORA SCHUBERGI*

Order	Family	Host	Infection[a]	Ref.
Hymenoptera	Diprionidae	*Gilpinia hercyniae*	E	227
		Neodiprion abietis	E	227
		N. sertifer	E	227
	Tenthredinidae	*Pristiphora erichsonii*	E	227
Lepidoptera	Arctiidae	*Estigmene acrea*	E	381
		Hyphantria cunea	E	381
	Geometridae	*Alsophila pometaria*	E	381
		Anaitis efformata	L	224a
		Erannis defoliaria	N	625
	Lasiocampidae	*Malacosoma americanum*	E	381
		M. disstria	L	626
		M. neustria	E	224
	Lymantriidae	*Euproctis chrysorrhoea*	N	224
		Leucoma salicis	N	378
		Lymantria dispar	N	224
		L. monacha	E	224
	Noctuidae	*Agrotis segetum*	E	510
		Mamestra brassicae	E	627
		M. dissimilis	E	627
		M. persicariae	E	627
		M. pisi	E	627
		Trichoplusia ni	E	325
	Notodontidae	*Phalera bucephala*	E	380
		Symmerista canicosta	E	381
	Olethreutidae	*Cydia funebrana*	E	628
		C. pomonella	E	628
		C. pyrivora	E	628
	Pieridae	*Aporia crataegi*	N	379
	Saturniidae	*Anisota senatoria*	N	381
		Hyalophora cecropia	E	381
	Sphingidae	*Deilephila nerii*	E	325
		Manduca sexta	E	325
	Thaumetopoeidae	*Thaumetopoea processionea*	N	378
	Tortricidae	*Archips cerasivoranus*	N	382
		Choristoneura fumiferana	N	227

[a] E, experimental infection; L, natural infection in laboratory colony; N, natural infection in field population.

Table 40
HOST RANGE OF *VAIRIMORPHA NECATRIX*

Order	Family	Host	Infection[a]	Ref.
Lepidoptera	Arctiidae	*Diacrisia virginica*	N	247
		Estigmene acrea	N	483
		Hyphantria cunea	N	392
		Isia isabella	E	385
	Bombycidae	*Bombyx mori*	E	385
	Citheroniidae	*Citheronia regalis*	N	247
		Eacles imperialis	N	247
	Glyphiptergyidae	*Homadaula anisocentra*	E	385
	Lasiocampidae	*Malacosoma disstria*	E	626a
	Noctuidae	*Acronicta oblinita*	N	247
		Agrotis ipsilon	E	385
		Anagrapha falcifera	N	247
		Anticarsia gemmatalis	E	385a
		Autographa californica	E	386
		Bellura gortynoides	N	247
		Caenurgina crassiuscula	N	247
		C. erechtea	N	247
		Feltia subterranea	E	632
		Heliothis virescens	E	633
		H. zea	N	634
		Peridroma saucia	E	385
		Plathypena scarbra	E	630
		Pseudaletia unipuncta	N	383
		Pseudoplusia includens	E	635
		Simyra henrici	N	247
		Spodoptera exempta	E	245
		S. exigua	E	636
		S. frugiperda	E	385
		S. ornithogalli	E	385
		Trichoplusia ni	N	247
	Notodontidae	*Datana integerrima*	E	385
		D. ministra	N	247
	Pieridae	*Colias eurytheme*	E	384
	Pyralidae	*Achroia grisella*	E	385
		Diatraea saccharalis	E	635
		Galleria mellonella	E	384
		Ostrinia nubilalis	E	385
		Pediasia trisecta	E	385
		Plodia interpunctella	E	247

[a] E, experimental infection; N, natural infection.

cultural crops. The natural occurrence of *V. necatrix* in 15 species and its wide experimental host range, including representatives of 9 families of Lepidoptera, indicate that it has an even greater potential host range. Insects that are not susceptible include the following lepidopterans: *Pieris rapae*;[629] *Phryganidia californica*;[386] *Depressaria pastinacella, Manduca sexta* and *M. quinquemaculata*;[385] the dipterans: *Musca domestica*[385] and *Bonnetia comta*;[385d] the hymenopterans: *Campoletis sonorensis* and *Cardiochiles nigriceps*;[630] and *Neurotoma fasciata*;[385] and, the coleopterans: *T. confusum*;[631] and *Hypera punctata, Acalymma vittata, Cerotoma trifurcata,* and *Diabrotica undecimpunctata howardi*.[385] Thus, *V. necatrix* appears restricted to Lepidoptera but not all leidopterans tested are susceptible.

Vavraia culicis — This species is infectious (*per os*) for a wide range of culicids (Table 41). Susceptibility ranged from very low for *Aedes aegypti* to very high for *Culex tarsalis*

Table 41
HOST RANGE OF *VAVRAIA CULICIS*

Order	Family	Host	Infection[a]	Ref.
Diptera	Culicidae	*Aedes aegypti*	E	404
		Ae. polynesiensis	N	637
		Ae. sollicitans	E	398
		Ae. taeniorhynchus	E	398
		Ae. triseriatus	N	399
		Anopheles albimanus	E	638
		An. atroparvus	E	404
		An. dureni	N	396
		An. franciscanus	E	638
		An. gambiae	L	400
		An. occidentalis	E	404a
		An. stephensi	L	400
		Culex pipiens pipiens	N	167
		Cx. pipiens quinquefasciatus	E	396
		Cx. salinarius	N	399
		Cx. tarsalis	E	398
		Cx. territans	N	398
		Culiseta incidens	L	404a
		Cu. inornata	E	398
		Cu. longiareolata	N	393
Lepidoptera	Bombycidae	*Bombyx mori*	E	404a
	Geometridae	*Sabulodes aegrotata*	E	404a
	Noctuidae	*Heliothis zea*	E	328
		Mamestra brassicae	E	639
		Spodoptera exigua	E	404a
		Trichoplusia ni	E	404a

[a] E, experimental infection; L, natural infection in laboratory colony; N, natural infection in field population.

and *Anopheles stephensi*. Efforts to produce *V. culicis* in alternate hosts have also established its infectivity for several species of lepidopterans. Wang[404a] found that the coleopteran *T. molitor* and the lepidopteran *G. mellonella* were essentially refractory to infection; only negligible to light infections were produced even when the hosts were intrahemocoelically injected with spores. Weiser[404b] also found that *G. mellonella* was refractory to infection by *V. culicis* as well as the other lepidopterans *Spodoptera litura* and *Feltia segetum*. However, it is interesting to note that there are no records of refractory species of culicids. Wang[404a] also found that *V. culicis* would not develop in mouse macrophage cells maintained at 37°C but the limits of the host range of this species and any potential hazards to vertebrates have yet to be determined.

Ciliates

As previously indicated, there are only a few species of entomogenous ciliates and little data exist on their host range. Most ciliates are parasites of mosquitoes[259,260h,260m] or black flies,[260a,260i,260n] and the absence of *Tetrahymena dimorpha* and *T. sialidos* in numerous other invertebrates from the habitats of their hosts suggests that these species may be host specific.[260b,260i] In addition, *Lambornella clarki* has only been found associated with *Aedes sierrensis*[258,259,260k,260l] and *T. rotunda* was found in only two of four *Simulium* species examined.[260a] The only experimental studies on host specificity of tetrahymenid ciliates, except for a few early reports,[260c,260d,260e] were carried out by LeBrun[260f,260g] who examined the virulence of 10 species for 12 species of mosquitoes in 4 genera. Corliss and Weiser[260h] discuss two cases of ciliatoses in vertebrates in which *T. pyriformis* may have been involved but the uncertain systematic status of the ciliates cited in these and other reports[259] preclude any further assessment of ciliate host specificity.

PRODUCTION

Relatively few attempts have been made to mass propagate species of protozoa pathogenic for insects.[4,640-642] Most entomogenous protozoa are obligate pathogens and must be produced in living cells. A few species have been cultured in liquid media or in cell-tissue systems. Thus, the microsporidia and neogregarines that have been mass produced for field trials have been propagated in laboratory-reared or field-collected insects using procedures similar to that used for the entomopathogenic viruses.[641]

Whole Organism Technology

The habitual host is usually used to mass produce entomogenous protozoa. Occasionally other hosts are substituted if the habitual host is small or is difficult to rear. Generally, neonatal or early stage larvae or nymphs are exposed to a low dose of spores placed on artificial diet or a suitable food substrate. The insects are then placed at a favorable temperature (generally 21 to 29°C) to permit maximum growth of both the host and the pathogen. Mature spores harvested from insects that are in an advanced stage of infection (larvae, pupae, or adults), are extracted by homogenization of the entire insects in a tissue grinder or blender. Large particles of cellular debris are removed by straining the homogenate through cheesecloth or a 80-200 mesh screen. The resulting filtrate, in sterile distilled water, is centrifuged to form a pellet of spores. Relatively pure suspensions of spores can be obtained with two or three repeated washings and centrifugations. Spores prepared in this manner are usually sufficient for field use. Purer suspensions of spores can be obtained by the triangulation method of Cole[643] or by density gradient centrifugation techniques using Ludox[644-645b] or Percoll.[645c] Spore concentrations are determined using a bacterial counting chamber. Representative yields of protozoan spores for various species produced for either field tests or laboratory studies are shown in Table 42. As previously stated, purified spore suspensions in water may be stored for a few weeks at 4°C prior to field use without any significant loss of viability. However, some investigators have accumulated and maintained their stock inoculum by freezing the infected hosts or purified spores in water at −12 to −10°C for up to 1 year prior to field use. Spores stored this way are generally much less active than spores stored in water at 4°C or at −10°C for only 1 to 4 months.

Economical, in vivo production of a protozoan is largely dependent upon development of suitable low cost techniques for mass producing the insect. This includes the result of determinations of optimal spore dose, host age, rearing temperature, and incubation period. The processes involved are best illustrated by the techniques used to produce the boll weevil pathogens *Mattesia grandis* and *Nosema gasti* and the grasshopper pathogen *N. locustae*. These are the only protozoan species that have been mass produced on a scale sufficiently large to permit extensive field tests and adequate cost analysis studies.

McLaughlin and Bell[554] further modified the techniques of Gast and Davich[646] in order to produce large amounts of spores of *M. grandis* and *N. gasti* in living weevils. Petri dishes, containing a semisynthetic agar-based diet, were implanted with eggs of the boll weevil and incubated at 30°C and 50 to 60% relative humidity. Spores of *M. grandis*, obtained from stock material stored at 4°C in distilled water, were purified by repeated washings and centrifugations, suspended in 0.05% sodium hypochlorite for 15 min at room temperature, and again centrifuged and resuspended in 0.005% sodium hypchlorite. Since spores of *N. gasti* were inactivated by sodium hypochlorite, these spores were treated with a concentrated solution of Zephiran chloride (1:500 dilution) for 15 min prior to centrifugation and resus-

Table 42
YIELDS OF ENTOMOGENOUS PROTOZOA PRODUCED IN VIVO

Order	Species	Host species	Per os	Intra-hemocoelic injection	Yields[a]	Ref.
Amoebida	Malameba locustae	Melanoplus differentialis	x		2.5×10^6/host/day	24
Eugregarinida	Ascogregarina culicis	Aedes aegypti	x		2×10^6	296
Microsporida	Microsporidium hyphantriae	Lymantria dispar	x		1×10^7	642
	Nosema acridophagus	Heliothis zea	x		2.2×10^8	476
		M. bivittatus	x		1.3×10^8/g host	478
		M. sanguinipes	x		8×10^8/g host	478a
	N. algerae	Anopheles stephensi	x		8.9×10^5	594
		H. zea		x	1.8×10^9	594
		Pieris brassicae		x	2.5×10^9	649
	N. cuneatum	H. zea	x		6.5×10^7	476
		M. bivittatus	x		1.2×10^9/g host	478
		M. sanguinipes	x		3×10^9/g host	478a
	N. equestris	Gastrophysa viridula	x		$5\text{---}13 \times 10^7$	649a
		Leptinotarsa decemlineata	x		$10\text{---}20 \times 10^7$	649a
	N. eurytremae	P. brassicae		x	6×10^8	648
	N. fumiferanae	Choristoneura fumiferana	x		1.4×10^8	654
	N. gasti	Anthonomus grandis	x		6.8×10^7	554
	N. gastroideae	G. polygoni	x		$1.7\text{---}3.1 \times 10^7$	649a, 657
		G. viridula	x		$2\text{---}6.3 \times 10^7$	649a
		L. decemlineata	x		$6\text{---}9 \times 10^7$	649a
	N. heliothidis	H. zea	x		2×10^9	643
	N. locustae	M. bivattatus	x		3.6×10^9	517a
		M. differentialis	x		7.1×10^9	517a
	N. lymantriae	L. dispar	x		3.5×10^7	642
	N. pyrausta	Ostrinia nubilalis	(probably transovarian)		8.6×10^7	372
	Pleistophora schubergi	Euxoa segetum	x		1.7×10^8	658
	Vairimorpha ephestiae	Galleria mellonella		x	4.6×10^8	639a
		Mamestra brassicae	x		5×10^8	639a
	V. heterosporum	G. mellonella		x	5×10^8	639a
		M. brassicae	x		5×10^8	639a, 659
	V. necatrix	H. zea	x		1.7×10^{10}	493
		Trichoplusia ni	x		1.6×10^9/g host	651
	V. plodiae	G. mellonella		x	8×10^8	639a
		M. brassicae	x		9×10^8	650
			x		2.3×10^9	639a
	V. sp. 696	H. virescens	x		4.9×10^9	639b
	Vavraia culicis	Ae. aegypti	x		1.4×10^4	404a
		An. occidentalis	x		8.7×10^3	404a

Table 42 (continued)
YIELDS OF ENTOMOGENOUS PROTOZOA PRODUCED *IN VIVO*

Order	Species	Host species	Inoculation technique Per os	Inoculation technique Intra-hemocoelic injection	Yields[a]	Ref.
		Bombyx mori	x		1.7×10^8	404a
		Culex pipiens	x		1.0×10^5	404a
		Culiseta incidens	x		2.6×10^6	404a
		H. zea	x		5.4×10^8	328
		M. brassicae	x		3×10^8	639
		Sabulodes aegrotata		x	2×10^8	404a
		Spodoptera exigua		x	1.4×10^8	404a
		T. ni	x		5.6×10^8	404a
Neogregarinida	Mattesia dispora	Plodia interpunctella	x		3.1×10^6	660
	M. grandis	Anthonomus grandis	x		1.7×10^6	554
	M. trogodermae	Trogoderma glabrum	x		1.8×10^6	661
		T. glabrum	x		1.7×10^7	132

[a] Expressed as number of spores/host unless otherwise indicated.

pension in a 1:1000 dilution of the germicide. The plates containing the larvae were inoculated by using a chromatographic sprayer calibrated to deliver 1 mℓ of spore suspension to each dish. Inocula of both species were adjusted to contain 1×10^6 spores per mℓ. Third instar larvae (8 to 9 days after hatching) were used for *M. grandis* and second instar larvae (2 to 4 days after hatching) were used for *N. gasti*. Treated larvae were held at 30°C until adult emergence. Adults infected with *M. grandis* were harvested 5 to 6 days after emergence while those infected with *N. gasti* were harvested 2 to 4 days after emergence. Infected adults were blended in water and filtered through cheesecloth to remove large tissue debris prior to centrifugation. Resuspended spores were passed through a 1/2-in. thick cotton pad before further purification by repeated washings and centrifugations. Spores were stored at 4°C in distilled water. Average yields (over a period of several months) were 1.7×10^6 spores per adult for *M. grandis* and $6.8 \times$ per 10^7 spores per adult for *N. gasti*. Estimated costs, based on a yield of 100 adults per plate, for spore production per hectare (including formulation into a bait for field use) ranged from ca. $3.06 for *N. gasti* to $125.00 for *M. grandis*. The cost differential in producing the two species was a result of the difference in yield. *M. grandis* infects only the adipose tissue and yielded 1.7×10^9 spores per 1000 adults while *N. gasti* produced a systemic infection and yielded 6.8×10^{10} spores per 1000 adults.

In contrast to rearing the host on an artificial diet, live plants and artificial diet were used to mass produce grasshoppers for production of *Nosema locustae*. Spores were produced in the grasshopper *Melanoplus bivittatus*, a species which is relatively large in size and tolerates extremely high spore concentrations,[606] to obtain sufficient spores for the extensive field-test program which preceded registration of *N. locustae* by the U.S. Environmental Protection Agency.[34-37] As described by Henry and Oma,[321] 1st-stage nymphs were reared (200 to 300 per tube) through the first molt. The grasshoppers were then transferred to large screened cages (Figure 15A) and reared to 5th-instar nymphs on a diet of lettuce, seedlings of balbo

FIGURE 15. Mass production of spores of *Nosema locustae*. (A) Grasshopper rearing facilities; (B) infected cadavers, previously frozen at $-10°C$ are thawed and crushed by grinding with a wheat mill; (C,D) initial phases in the filtering of crushed cadavers to extract and concentrate spores; (E) spores suspended in distilled water are subjected to repeated centrifugations for 60 min at 10,000 g under vacuum and refrigeration by the placement of 2,000 mℓ of spore suspension in each of 4 centrifuge tubes shown in the head of an L-3-40 Beckman centrifuge; (F) appearance of spore pellet as a finished product; bottle contains about 10^{12} spores or sufficient spores to treat 400 ha. (Courtesy of J. E. Henry, U.S. Department of Agriculture, Agricultural Research Service, Bozeman, Mont.)

 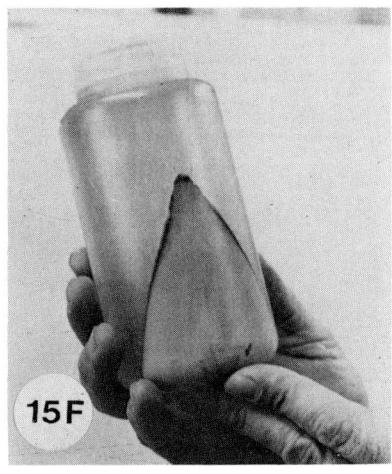

FIGURE 15. Continued.

rye, and wheat bran. These late-stage nymphs were inoculated by feeding lettuce sprayed with spores (ca. 10^6 spores per cage) for 2 consecutive days and again on the 4th day. This inoculation produced an infection rate of > 99% per cage for 300 to 500 adults per cage. The grasshoppers were transferred to small vials (2 adults per vial) 2 to 3 weeks postinoculation to reduce cannibalism and then fed until death on agar-based diet containing either wheat bran or crushed dog food. Infected cadavers were stored at $-10°C$. After thawing, the cadavers were crushed in a wheat mill (Figure 15B), suspended in distilled water, and filtered several times through cheesecloth (Figures 15C and 15D). The spores were concentrated and purified by differential centrifugation (Figures 15E and 15F) prior to being stockpiled in distilled water at $-10°C$. Continued improvements in technique resulted in a threefold increase in spore production (from 1×10^9 to 3×10^9 spores per grasshopper). Each infected grasshopper contained approximately 3×10^9 spores (enough to treat 1.2 ha). Thus, the estimated cost of spore production was only $0.21 to 0.25 per ha.[37,321] The cost of formulation into a bait and aerial application would extrapolate to a total cost of $1.86 per ha.[37] This cost is economical for a grasshopper-control program on rangelands where the average annual return is less than $12.00 per ha per year.[647]

More recently, Henry[517a] examined a number of factors that affect the efficiency of spore production of *Nosema locustae*. Average spore yields were found to be significantly higher in *Melanoplus differentialis* (7.1×10^9 spore per grasshopper) than in the standard host species *M. bivittatus* (3.6×10^9 spores per grasshopper), because of the higher survival potential of *M. differentialis* in laboratory culture. Analyses of other factors indicated that optimal cage density was about 250 grasshoppers per cage, a 60-W bulb in each cage provided maximum useful light intensity, and the technique of feeding treated lettuce was probably most appropriate for small production facilities. He concluded that feeding lettuce treated with a dose of 2.2 to 2.4×10^9 spores per cage to 5th instar *M. differentialis* (250 per cage) should result in the production of about 1×10^{10} spores per grasshopper and 1×10^{12} spores per cage. Thus, sufficient spores would be produced in a single cage (at the standard field application rate of 2.5×10^9 spores per ha[35]) to treat about 400 ha.

The production potential for other entomogenous protozoa also has been studied (Table 42). Yields generally range from 8.7×10^3 per host to 1.7×10^{10} per host. Most species of protozoa were produced in a manner similar to that used to produce the boll weevil and grasshopper pathogens. However, in some cases surrogate hosts or alternate inoculation techniques were utilized to increase efficiency. Undeen and Maddox[594] investigated the possibilities of producing spores of *Nosema algerae* in nonmosquito hosts, because of difficulties in rearing and infecting mosquitoes and the relatively small yield of spores per mosquito. Larvae of the corn earworm, *H. zea*, were susceptible by intrahemocoelic injection; each corn earworm larvae yielded as many spores as would be obtained from 2000 mosquito larvae. Similarly, an injection method was used to mass produce spores of *Nosema eurytremae*, a parasite of trematodes,[648] and *N. algerae*[649] in larvae of the cabbage white butterfly, *Pieris brassicae*. In some cases intrahemocoelic injections in alternate hosts have been bypassed by feeding spores to early stage larvae starved previously for 12 to 24 hr. Thus, *N. algerae* is routinely produced now by feeding spores on artificial diet to 4- or 5-day-old larvae of *H. zea* starved previously for 24 hr and then harvesting the spores from mature larvae[562] or adults.[506] Spores of *N. algerae* have also been obtained from pupae of *T. ni* by feeding starved, 3rd-stage larvae on treated diet.[341] Another mosquito parasite, *Vavraia culicis*, also has been produced by feeding spores to larvae of several species of lepidopterans.[328,404a,639] Hostounský and Weiser,[650] Weiser,[639] and Hostounský[649a] administered spores of several species of microsporidia to larvae by allowing larvae to drink spore-droplets placed on paraffin-coated slides or on a bacteriological inoculation loop. In a similar manner, Fowler and Reeves[651] inoculated larvae of *Heliothis zea* and *Trichoplusia ni* with spores of *Vairimorpha necatrix* by placing a single drop of suspension on the mouthparts or by dispensing the inoculum on the head capsule. Such inoculation techniques in most cases are too laborious for mass production and species such as *V. necatrix* can be more easily mass produced by feeding spores on artificial diet to larvae.[493] Similar procedures using artificial diet were used to produce spores of *Nosema fumiferanae* in larvae of *Choristoneura fumiferana*[652-654] and *Pleistophora schubergi* in larvae of *Malacosoma disstria*.[653,655] Spores of *N. pyrausta* have also been harvested from larvae of *Ostrinia nubilalis* obtained from infected colonies in which the microsporidium was maintained by transovarian transmission.[368,518,519,656] Nordin[555] incorporated methoprene, an insect growth regulator, into the artificial diet of *H. virescens* larvae and obtained significantly increased larval weights and yields of *V. necatrix* spores. These yields, however, were less than those obtained by Fuxa and Brooks[493] in *H. zea*. Thus, the producer must balance the added costs of using methoprene against the availability and potential difficulties of producing the pathogen in a surrogate host. For example, the decline in yields of *N. infesta* produced in the surrogate host *Phthorimaca operculella* may have resulted from a reduction in virulence of the microsporidium due to passage through its unusual host.[25] As with other obligate pathogens, the production of protozoa is ultimately limited by the amount of host tissue convertible to protozoan material. Host size may affect yield potential but growth rate and tissue specificity of the pathogens are also significant factors in spore production.[554,650]

Tissue Culture Systems

As discussed previously in the section on Specificity only a few microsporidia have been grown in tissue culture systems (Table 26). Most species were grown in tissue explants, primary cell cultures, or specific cell lines of their native hosts and to a lesser extent in cells of foreign hosts. Most studies have dealt with aspects of demonstrating growth of various microsporidia in tissue cultured cells or with techniques used to obtain infection, growth, and replication in various types of cells. While the potential of using cell cultures in mass producing spores is frequently mentioned, few investigators have specifically addressed its potential.

The possibility of producing spores, of at least some species of microsporidia, in tissue culture systems has been demonstrated (Table 26). Primed spores are usually used to inoculate cell cultures but the process can be complicated by the lack of techniques to induce spore germination in the culture media or by the introduction of microbial contaminants.[571] In their studies, Tsang et al.[571] and Kurtti et al.[571a] circumvented the use of spores and possible problems with contamination by transferring hemocytes from the *Malacosoma disstria* line persistently infected with *Nosema disstriae* to uninfected cell cultures of several species of insects. Under these conditions and even where spores were used,[341] high rates of cross infection were dependent upon the density of the host cell population. Streett et al.[341] found that infection rates, in three cell lines inoculated with spores of *N. algerae*, ranged from 23 to 32% within 72 hr postinfection. The cultures also continued to produce spores for several transfers but fewer spores were produced with each successive transfer until spore production ceased after the sixth subculture. In contrast, Tsang et al.[571] and Kurtti et al.[571a] achieved an average cell infection level of 97% in a cell line from *H. zea* by inoculating with a high proportion of infected to uninfected cells. A range of 1 to 4×10^7 spores per mℓ was obtained and no diminution in parasite yield was obtained after 20 subcultures. Spores of species produced in vitro were also infectious in vivo.[339,341,404a,552,579]

Despite the progress already made in the use of tissue culture to produce spores of some microsporidia,[567] the high cost of culture media, the low yields in spores, and the relatively infant state of the development of in vitro production techniques for mass production of the entomogenous protozoa still limit the usefulness of this system. However, the production of contaminant-free spores and the ability to produce the protozoan in the absence of a host culture are significant advancements that will enhance the efficient use of in vitro production methods for entomogenous protozoa.

Fermentation Technology

The processes of surface or submerged fermentation have been extensively used in the mass production of many facultative pathogens of insects but the techniques have had little utility in the production of entomogenous protozoa. Most species of entomogenous protozoa are obligate parasites and few attempts have been made to grow them either on or in an artificial medium. However, considerable literature exists on the growth of some species of facultatively parasitic protozoa, primarily the entomogenous flagellates in the family Trypanosomatidae. Wallace[45] summarizes the techniques routinely used in the culture of insect trypanosomatids. These usually involve the use of a diphasic medium, i.e., blood agar to which a liquid component is added. Many of the entomogenous trypanosomatids have been cultured in such a manner and some, such as *Crithidia fasciculata*, have been routinely grown in a completely defined media for many years.[662,663] However, a review of these efforts is beyond the scope of this paper and is unwarranted due to their limited potential as microbial control agents.

Similarly, a tremendous volume of literature exists on the culture, nutrition, and physiology of the tetrahymenid ciliates, some of which are parasitic in insects. The voluminous literature on *T. pyriformis* is well covered in the book *Biology of Tetrahymena*[664] but, again, the facultative or accidental nature of the species as a parasite of insects does not justify an extensive review here. Kidder and Dewey[665] provided an earlier review of ciliate biochemistry in pure cultures. It is interesting to note, however, that only two species, *Tetrahymena dimorpha*[260b] and *T. sialidos*,[260i] of the tetrahymenids intimately associated with insects have been successfully cultured in vitro. Both of these species were easily cultured in a protease-peptone, yeast-extract medium commonly used for the culture of tetrahymenid ciliates. In addition, Batson[260b] found that *T. dimorpha* was able to grow well in Mitsuhashi and Maramorosch's insect tissue culture medium supplemented with fetal calf serum but growth could not be sustained in unsupplemented medium or in unsupplemented or supplemented Grace's insect tissue culture medium. While the limited success in culturing the other entomogenous species may be largely due to a lack of effort, Lynn et al.[260a] were only successful in prolonging the life of individual ciliates, *T. rotunda*, for a few hours in a variety of media. If such ciliates are to be pursued as control agents of mosquitoes or other aquatic insects, increased efforts will have to be devoted to their mass production. Although it would also be highly desirable to produce other species of the obligate-parasitic protozoa by fermentation, the intracellular development of these protozoa and their often complex life cycles indicates that this would be an arduous task.

BIOCONTROL OF ARTHROPOD PESTS

Relatively few attempts have been made to evaluate the field efficacy of various entomogenous protozoa.[1-5,39-41] Most attempts have involved a few microsporidia and neogregarines against lepidopterous, coleopterous, dipterous, and orthopteroid pests of forests, field crops, rangelands, or those pests found in aquatic habitats. Emphasis has usually been on assessing the efficacy of the protozoan as a short-term control agent despite the fact that protozoa are probably best suited as agents for long-term establishment or introduction programs.[1,3,5,39] Formulations of protozoa have been applied as sprays or baits; little effort has been made to determine their compatibility with various adjuvants or insecticides. Emphasis herein will be primarily placed on attempts to evaluate the field efficacy of the most promising candidates (Table 4) including some general comments on other entomogenous protozoa.

Flagellates

Few of the entomogenous flagellates are pathogenic for their hosts; thus there has been no attempt to use them as microbial control agents. Most entomogenous flagellates are either commensals or mutualists in insects; and even among the strictly entomogenous trypanosomatids, most species do not cause deleterious effects in their hosts.[10,14] Heavily infected hosts may be less active and diarrheic but even hemocoelic infections are not necessarily fatal. *Herpetomonas muscarum*, a frequent inhabitant of the gut of muscoid and calliphorid flies, is not harmful when confined to the alimentary tract. However, it was found in the hemocoels of moribund and dead larvae of *Musca domestica*,[666] and it was considered to be a primary etiological agent in eye gnat larvae which eventually succumbed to bacterial septicema.[461,462]

Mortality due to *Herpetomonas swainei* in the jack pine sawfly, *Neodiprion swainei*, did not exceed 20% in larvae exposed as 1st and 2nd instars. A reduction in the fat body was the only effect noted in 2nd- or 3rd-stage larvae; infected eonymphs and adults were asymptomatic.[64] Probably the most promising candidate for further study as a control agent is the recently described species *Blastocrithidia caliroae*.[46] Field observations indicated that this species caused high mortality of the pear slug, *Caliroa cerasi*. An epizootic of this flagellate was responsible for the collapse of a population of *C. cerasi* in the upper Rhine Valley in Germany. Larvae usually die during the 4th to 6th instars, following color changes associated with the drying of a mucous coating and a cessation of feeding. Although this flagellate and perhaps others are significant, naturally occurring mortality factors, their general low virulence, and the lack of resistant life-cycle stages may limit their potential usefulness as microbial control agents.

Amoebae

The amoeba, *Malameba locustae*, was probably the first entomogenous protozoan used as a microbial control agent. Taylor and King[24] mixed cyst-containing feces from infected grasshoppers with bran and molasses and distributed it along roads and fences over an area of approximately 10 m^2 at three different places near Iowa City, Iowa. Grasshoppers were collected 8 weeks later from the treated area and examined for infection. They concluded, based on an infection rate of 2.7% of 422, that the incidence of infection in grasshoppers could be increased. This work still represents the only field trial conducted with an entomogenous amoeba. Henry[667] considers *M. locustae* to be a particularly useful agent for long-term pest management of grasshoppers. This amoeba also is frequently a serious pathogen of laboratory colonies of grasshoppers (see *Malameba locustae*, Natural Occurrence) and may require control through prophylactic use of antibiotics.[84,550] Heavily infected grasshoppers may exhibit molting difficulties, prolonged nymphal instars, reproductive failures, and premature death.[14] While infections are usually more limited under field conditions, Venter[280] attributed the collapse of a potential outbreak of grasshoppers in South Africa to an epizootic of *M. locustae*. As with many protozoa, the chronic nature of *M. locustae* in grasshoppers probably inhibits further study of this species as a microbial control agent. Its wide host range, the presence of a resistance cyst in its life cycle, and the potential for dissemination through the use of baits, however, enhances it potential as a microbial control agent. If cysts could be produced in vitro, the amoeba might be an effective microbial control agent for the management of grasshoppers.

Other entomogenous amoebae are less well known and none exhibits any real potential as microbial control agents for pest species.[14] Most are encountered infrequently, poorly described, and may be more commensalistic than parasitic. Purrini[80] postulated that *Malameba scolyti* is lethal to bark beetles due to extensive infection of Malpighian tubules. However, in a later study the only external sign or symptom of infection in adults or pupae was a change in the coloration of infected pupae (despite extensive hypertrophy of infected organs and cellular changes in the Malpighian tubules).[582] As with entomogenous flagellates, most amoebae are of questionable merit as microbial control agents because they are essentially avirulent and they produce chronic infections.

Eugregarines

This is also a group with little potential as microbial control agents. Most species are relatively harmless parasites or commensals in the intestinal tract or hemocoel of their host. The lack of merogony in their life cycle is associated with nonvirulence. Even when present in large numbers, most eugregarine infections do not produce obvious signs and symptoms. A few species will produce adverse effects when their hosts are reared or held under suboptimal conditions.[459,460,460a] Lipa[422] indicated that many eugregarines damage the epithelium of the gut and gastric ceca both mechanically and physiologically. However, as damaged cells are usually regenerated rather quickly, little lasting effects were noted. Damage caused by coelomic eugregarines is probably more serious. Larvae of the scarab, *Hoplia* sp., usually die before pupation when infected with an undescribed species of *Monocystis* as do the grubs of *Melolontha melolontha* infected with a similar gregarine.[12]

The only eugregarine to have received attention as a potential microbial control agent is *Ascogregarina culicis*, a parasite of the yellow fever mosquito, *Aedes aegypti*. This eugregarine has been implicated as causing stunting or death in mosquitoes[298,306,313,316] and was shown by Barrett[306] and Barrett et al.[310] to severely damage the Malpighian tubules of pupae and adult mosquitoes. In an extensive evaluation of this parasite, McCray et al.[296] found that only the life span of infected adult *Ae. aegypti* was reduced. No significant effect of the parasite was demonstrated on larval development, size, mortality, pupation, pupal weight, adult emergence, survival, or adult fecundity. Their conclusion that *A. culicis* possesses a low and questionable potential as a microbial control agent for *Ae. aegypti* has been supported by field observations by Stapp and Casten,[308] Hayes and Haverfield,[309] and Gentile et al.[307] in naturally infected populations of mosquitoes in the U.S. More recently, however, Walsh and Olson[311] found that *A. culicis* caused high mortality (from 77 to 86%) in several other species of mosquitoes, although only 11% mortality was obtained for *Ae. aegypti*. These alternate hosts died during the larval stage and the parasite was usually unable to establish itself in the Malpighian tubules. Development in the midguts of these hosts elicited an encapsulation reaction to the gregarine. While agreeing that *A. culicis* has little or no effect on *Ae. aegypti*, they indicated that this gregarine is lethal to *Culiseta inornata*, *Psorophora columbiae*, and four other species of *Aedes*. A further indication of the potential of this species was shown by Sneller.[668] Heavily infected larvae produced pupae which weighed significantly less than noninfected pupae and, more importantly, the vector potential of infected *Ae. aegypti* was decreased for the filaria heartworm parasite, *Dirofilaria immitis*. Apparently fewer filariae were able to develop in infected hosts due to the extensive destruction of the Malpighian tubules and the concurrent lack of suitable sites within the tubules for development of the obligatory intracellular phase of the heartworm microfilariae.

Several other species of *Ascogregarina* are also known in various species of mosquitoes (Table 2) but there is little indication of any real possibility for their use as microbial control agents. For example, *A. clarki* probably also destroys some cells of the midgut and Malpighian tubules of *Ae. sierrensis* but a lasting effect was not suspected.[112]

Neogregarines

The neogregarines are more serious parasites of insects than the eugregarines. Weiser[12] indicates that the presence of a secondarily acquired merogonic phase in their life cycle compensates for the low number of spores produced during sporogony. With this mechanism acting to increase the number of parasites concurrent with their intracellular development in various organs of a host, most of the neogregarines are capable of causing serious debilitating injuries. Infection of the host is usually most noticeable at critical stages of growth such as molting, pupation, or adult emergence. Deformities or death result from destruction of cells, such as fat body cells, thus depriving the host of nutritional reserves critical to normal growth and cellular differentiation. Species which develop in the gut, Malpighian tubules, or hypodermal cells are usually less virulent but cause severe problems in laboratory colonies of insects. Species (*Mattesia dispora*, *M. povolnyi*, and *Farinocystis tribolii*) that develop in the fat body probably offer the best potential as microbial control agents. They are highly virulent pathogens and are significant in natural suppression of several economically important insects. More general information of neogregarines as pathogens of insects is presented by Weiser[12] and Brooks.[14] The remainder of this section, on efficacy, will be focused on use of *M. grandis* and *M. trogodermae* as microbial control agents.

Mattesia grandis

The destructive potential of this neogregarine was demonstrated by an extensive and devastating epizootic in a large laboratory colony of the boll weevil, *Anthonomus grandis*, in Mississippi.[129] The symptomatology of disease is associated with a progressive development and destruction of the fat body that impairs larval molting, pupation, ecdysis, and egg synthesis in the infected females.[130,669] Adults also may be deformed, the fecundity and longevity of females may be reduced, and infected adults are progressively more susceptible to insecticides.[513a] Larval mortality peaks about 2 weeks after exposure to the pathogen.[130]

Field tests, with *M. grandis* against adult weevils, were initiated in the early 1960s. A bait formulation containing spores and a feeding stimulant was used to induce ingestion, since adults do not normally feed on cotton plants.[29] Weevils were introduced into small field cages on bait-treated plants at rates varying from 1×10^7 to 2×10^8 spores per plant. Infection rates varied from 18.3 to 93.3% while only 1 of 162 weevils became infected on plants treated with an aqueous suspension of 1×10^8 spores per plant. Field cage tests carried out in 1965 also demonstrated the effectiveness of a bait formulation as a method of introducing *M. grandis* into a weevil population.[30] Fifty-five percent of the weevils were infected at the end of the test, and the treated weevils produced about half as many adults as the untreated population. Additional field-cage tests[32] and a large field test,[33] with a bait formulation containing improved feeding stimulants, also demonstrated the effectiveness of the bait principle for boll weevil control. In these tests the bait formulations contained either

FIGURE 16. Adult males of *Trogoderma glabrum* attracted to disk treated with a synthetic sex pheromone and spores of the neogregarine, *Mattesia trogodermae*. In attempting to copulate with the disk, the beetles become contaminated with spores which they subsequently transmit to females beetles during copulation. (Courtesy of W. E. Burkholder, U.S. Department of Agriculture, Agricultural Research Service, Madison, Wis.)

spores of *M. grandis* or spores of the microsporidium *N. gasti* at a rate of 1×10^6 spores per mℓ of formulated bait. In 1966, both pathogens were equally effective in reducing the weevil populations. However, rain limited the overall effectiveness of the bait and only about 30 to 40% of the adults were infected.[32] Another field of weevils treated late in 1966 received a single application of spores of *N. gasti*, followed shortly by two applications of spores of *M. grandis* and several applications of the bait without the pathogens.[32] Adults, which were entering diapause, carried the disease into the overwintering population of weevils. None of the overwintering weevils, however, was found infected upon emergence from hibernation sites the following spring. Since the 1967 tests were conducted exclusively with spores of *N. gasti*, its applicability to probable effectiveness of *M. grandis* is mostly conjecture. Certainly the potential of *M. grandis* (in bait formulations) as a microbial control agent has been demonstrated. The high cost of production,[554] however, may limit its use for suppression or control of boll weevils.[2]

Mattesia trogodermae

This species is a good example of a neogregarine that regulates natural populations of stored-product pests. The high larval mortalities, associated with infected cultures of various species of *Trogoderma* beetles, were noted by several authors[331-333] prior to its description by Canning.[106] Heavily infected larvae exhibit premature mortality, body distention, sluggish movement, and unusually moist fecal pellets voided in long strands. Schwalbe et al.[670] determined that the weight of infected larvae and glycogen and protein reserves rapidly diminished. Severity of infection by *M. trogodermae* varies with host species. Highly susceptible hosts include *Trogoderma glabrum*, *T. simplex*, *T. sternale*,[331,333,661] and *T. inclusum*.[661] Less severely affected hosts include *T. granarium*[331,333] and *T. teukton*.[331] Marzke and Dicke[331] found *T. inclusum* to be less severely affected than did Schwalbe et al.[661] while *T. grassmani* and *T. variabile* were refractory to infection,[661] even though natural populations were heavily infected.[330]

The potential of *M. trogodermae* as a microbial agent for controlling dermestids in stored products was demonstrated by Schwalbe et al.[661] and Shapas et al.[671] with *T. glabrum*. Adult-to-adult transmission of spores, which occurs naturally during mating, was enhanced by use of a spore-transfer site (Figure 16) treated with a synthetic sex pheromone and spores of *M. trogodermae*. Male beetles were contaminated with spores, after they attempted to mate with the transfer site, and then transmitted the spores to females during copulation. Additional transmission was through larval ingestion of spores from dead adults or from ingestion of surface-contaminated food. One application of pheromone-baited, spore-transfer sites, under simulated warehouse conditions, resulted in substantial suppression of beetle populations.[671] Under heavier population pressure the treated populations increased only 4-fold in the first generation (vs. a 24-fold increase in the controls) and fell below pretreatment levels by the second generation (vs. a 100-fold increase in the control population). The treatment, however, had no effect on the first generation of a low-density population. Although *M. trogodermae* has still not been tested under field conditions, its specificity for dermestids of the genus *Trogoderma* (Table 30), its capacity for being mass produced,[132] and its harmlessness to vertebrates,[329] indicate it is potentially useful as a microbial control agent. The concept of the use of pheromones as lures for inoculation devices and pathogen dissemination has been discussed in several reviews by Burkholder[672,673] and Burkholder and Shapas.[674]

Coccidia

The occurrence of coccidia in insects generally is considered as accidental.[12] In many respects, however, the coccidia are similar to the neogregarines in that their development in the fat body or alimentary tract can be detrimental. A few species cause mortality of stored grain pests or other insects. *Adelina tribolii* often causes epizootics in laboratory and natural populations of the confused flour beetle, *Tribolium confusum*,[1,12] and is infectious for other species of flour pests. A more recently described species, *A. sericesthis*, is active in the natural control of the scarabaeid grub *Sericesthis pruinosa* in Australia.[140] Infected larvae usually die at metamorphosis due to the depletion of fat body reserves by the developing coccidian. Since most species have not been extensively studied and little is known of their host specificity, no attempt has yet been made to field test their efficacy.

Microsporidia

The large number of microsporidia precludes a detailed general discussion of their efficacy as insect pathogens. They range from innocuous parasites to virulent pathogens. Most microsporidia produce chronic infections characterized by abnormal feeding, irregular growth or retarded larval development, color changes in localized areas of the body, incomplete metamorphosis, formation of deformed pupae or adults, or lowered adult fecundity and longevity. Death occurs as a result of destruction of cells or is associated with septicemia produced by the entry and multiplication of other pathogens (particularly bacteria) in the hemocoel via the disrupted gut epithelium. Symptoms vary depending upon degree and extent of infection or the stage of the host at infection.

The microsporidia include most of the protozoa regulating natural populations of insects and are considered the most promising candidates as microbial control agents. Species such as *Nosema pyrausta*, *N. fumiferanae*, and *N. tortricis* are often responsible for extensive epizootics and population collapses of their hosts and are important biotic factors limiting natural populations of insects. Eight species of microsporidia listed in Table 4 are potential microbial control agents. Discussion of efficacy of the eight species of microsporidia (Table 4) will illustrate the role of microsporidia as potential microbial control agents. In a recent review Canning[675] also discussed the efficacy of microsporidia, with emphasis on parameters affecting the selection of species, with the greatest potential as microbial control agents.

Nosema algerae

This microsporidium has received extensive consideration as a potential microbial control agent of mosquitoes. Destructive epizootics have been reported in laboratory colonies of anopheline mosquitoes,[192,193,337,338,343,553] and it is pathogenic to many other mosquitoes

(Table 31). Anopheline mosquitoes are generally more susceptible than species of *Aedes* or *Culex* but there is variability even within a species. For example, larval or pupal mortality by *N. algerae* was not obtained in some cultures of *Anopheles stephensi*[193,477] while extensive mortalities were obtained in other cultures of this same species.[192,553,676] Extensive mortality also was reported in cultures of other anophelines such as *An. gambiae*[193,337,338] and *An. quadrimaculatus*.[336,677] Surviving anopheline adults exhibit reduced fecundity and longevity[343,477,676-679] and a reduced capacity to transmit malaria.[338,678,680-683] In culicine and other mosquitoes, *N. algerae* can develop in a wide range of tissues.[319,336] Pathogenic effects range from little or none in *Aedes aegypti*,[192,193,319] *Culex territans*, and *Cu. tritaeniorhynchus*[319] to extensive destruction of tissues and organs in adults of moderately susceptible species (*Cu. nigripalpus* and *Cu. tarsalis*[319]). In some hosts, pathogenic effects also may vary with the strain of *N. algerae*. Savage[319] found that an El Salvador isolate of *N. algerae* produced higher infection rates and higher larval and pupal mortality in *Ae. aegypti* than did the Walter Reed or Panama isolates.

Pathogenicity for a wide range of hosts and its ease of mass production make *N. algerae* a promising microbial agent for the control of mosquitoes; however, efforts to evaluate this species under field conditions have been few. The lack of an effective formulation to keep viable spores in the feeding horizon of mosquito larvae has resulted in a virtual abandonment of this agent. In the only extensive test carried out under field conditions, Anthony et al.[562] applied spores of *N. algerae* to six test plots in Panama to control *An. albimanus*. The plots ranged in size from 37 to 1950 m² and were treated with one to eight applications of spores at intervals of 2 to 3 or 3 to 4 days. Application rates varied from 1.1×10^7 to 2.2×10^9 spores per m². Infection rates in each plot were dosage dependent and ranged from 16% (in one test receiving a single application of spores) to 86% (in a plot that received four applications at the highest rate of 2.2×10^9 spores per m²). In the latter test plot applications were made on day 0, 4, 7, and 11. Infected larvae (about 38%) were first found on day 5. Peak infection (about 86%) was obtained on days 12 and 15, and remained relatively high (about 20 to 60%) for the following 2 weeks. One infected larvae was found 70 days after treatment. Estimates of overall population control were not attained since the plots were not sufficiently isolated. However, high mortality occurred among the test larvae brought back to the laboratory. These tests conclusively demonstrated that high infection rates could be attained following application of spores to natural breeding areas of *An. albimanus*. Dense mats of *Hydrilla*, in the test plots, probably kept the spores, which do not float, near the surface so that they could be ingested by surface-feeding larvae. The authors concluded that a formulation designed to hold the spores at the water surface would likely increase infection rates and also extend the residual activity of spores. This could extend the time between applications and also reduce the spore rates.

The loss of spore activity, as a result of the spores rapidly settling to the bottom of treated pools, was probably the biggest factor limiting success of field tests against *An. stephensi* with *N. algerae* in Pakistan.[563] Field tests at Lahore consisted of nine freshly-dug m² pools (Figure 17) laid out in a 3×3 Latin square (two treatments and a control). Each pool was seeded with 2000 newly hatched larvae: treated pools received a single application of 10^7 or 10^9 spores per pool. Slight reductions were obtained in the number of early instar larvae collected (5 days after treatment) and in the number of pupae (10 days after treatment) at both spore rates. Presence of a similar species of *Nosema*, in some of the mosquito stock material used to seed the test pools, complicated the comparison of infection rates. In spite of this, increases of 31 and 46%, in the prevalences of infections over that of the control were obtained in pools treated with the low and high rates of spores of *N. algerae*, respectively. These increases favorably compared with a site at Karachi which showed a 40 to 50% infection rate, 4 to 5 days after treatment. At Karachi, spores were applied to several mosquito-breeding sites around the city at a rate of 10^9 spores per m². The investigators

FIGURE 17. Test pools in Lahore, Pakistan used for the field evaluation of spores of *Nosema algerae* for control of larvae of *Anopheles stephensi*. (Courtesy of J. V. Maddox and N. E. Alger.)

concluded that no significant long term reduction in larval populations of *An. stephensi* was obtained. In the laboratory they[563] found that spores settled out of the feeding zone of mosquitoes within 24 hr. Subsequent lab and field work, aimed at developing a suitable formulation for floating spores, has been unsuccessful. The results of a small field test in 1978 (in Pakistan), with a formulation that had given good laboratory results, were disappointing. Wave action and disturbances from frogs and birds caused the spores to sink before any significant extension in spore activity was achieved.[597] Future efforts to evaluate the efficacy of this microsporidium will likely be curtailed unless an effective formulation is designed that will keep the spores in the feeding horizon of mosquitoes for several days.

Nosema fumiferanae

This microsporidium is one of only a few species whose potential as a microbial control agent for insect pests of forests has been studied. It is a significant endemic pathogen of populations of the spruce budworm, *Choristoneura fumiferana*. Chronic and debilitating infections caused by *N. fumiferanae* are typical of many species of microsporidia. Although heavily infected budworms may die as late instar larvae, infections are usually nonlethal in nature. Many infected larvae, pupae, and adults lack overt signs and symptoms of disease.[684] Thomson,[684] in an extensive study, found that the microsporidium retards larval and pupal development and reduces pupal weight and adult longevity and fecundity. These effects were usually more pronounced in females. Wilson[685-685b] corroborated these results and also demonstrated that deleterious effects could be increased by feeding spores to naturally infected budworms.[686] Mortality is dosage dependent[684,687] but even heavily infected larvae can develop into adults. Infected females transmit the microsporidium transovarially to all their progeny. As indicated previously, *N. fumiferanae* commonly occurs in natural populations of *C. fumiferana* and is often epizootic. In at least two periods (1955 to 1959[351] and 1973 to 1978[353]) the prevalence of infection annually increased in natural populations of *C. fumiferana* from the Uxbridge Forest area of southern Ontario. Thomson[351] attributed the reductions in egg numbers in 1959 and the subsequent low larval populations in 1960 to microsporidian effects on larval survival and adult fecundity. Wilson[353] also noted slight reductions in larval populations following an increase in prevalences of infection from 1973 to 1978.

Efforts to evaluate the efficacy of *Nosema fumiferanae* under field conditions were limited

FIGURE 18. Field application of spores of *Nosema fumiferanae* on white spruce trees using a packsack-type mist sprayer. (Courtesy of G. G. Wilson, Canadian Forestry Service, Sault Ste. Marie, Ontario.)

to two preliminary tests in Ontario. On Manitoulin Island in 1975, Wilson and Kaupp[652] sprayed individual white spruce trees (Figure 18) naturally infested with 3rd and 4th instar spruce budworms with approximately 1.8×10^{11} spores per tree. Counts at 11 and 25 days post-treatment indicated a significant increase in the prevalence of infection among larvae from sprayed trees over that of untreated trees. The prevalence of infection was also significantly higher among adults and their progeny from treated trees. A similar test was conducted in 1976 near Thessalon, Ontario in which applications were made on both white spruce and balsam fir trees including spruce trees with larvae at the needle-mining stage of development.[653] All of the balsam fir trees were treated once at a rate of 5×10^{10} spores per tree. A few of the white spruce trees were treated once at this same rate but other spruce trees were treated either once or twice with 2.5×10^{10} spores per tree. All applications contained 25% (v/v) molasses and a sunlight protectant. Spraying trees with budworms in the needle-mining stage on spruce trees was unsuccessful. Significant increases, on both spruce and balsam trees, in the levels of infection, however, were obtained when larvae were treated during the late instars. The twofold increase in spore concentration or the second application of spores did not significantly increase the prevalence of infection. Slightly higher infection rates were obtained on fir than spruce trees attributable to the earlier opening of the fir buds which made the larvae more accessible to the spores. Although data to determine extent of control were not collected in these preliminary tests, the results did demonstrate that *N. fumiferanae* could be successfully introduced into a population of spruce budworms. Examinations of the same spruce trees, on Manitoulin Island in 1975, revealed that the levels of *N. fumiferanae* remained essentially unchanged for 2 successive years while infection rates in untreated trees rose to the treated-tree levels.[688] Similarly, the prevalence of infection the following year, among larvae from the spruce trees treated in the 1976 test, remained the same while significantly rising in larvae from untreated trees. There was a reduction in the levels of *N. fumiferanae* in larvae from balsam fir but the prevalence was still significantly higher than in larvae from untreated trees. Thus, the levels of infection, in each of the treated populations, were advanced about 2 to 3 years due to applications of spores. Long-term effects (adult longevity and fecundity; larval and pupal development) of the microsporidium could influence subsequent future population levels of this pest.[41]

FIGURE 19. Application of spores of *Nosema gasti* in a cotton field by spinning disks mounted on a self-propelled spray machine. (Courtesy of R. E. McLaughlin, U.S. Department of Agriculture, Agricultural Research Service, State College, Miss.)

Nosema gasti

The detection of this microsporidium by McLaughlin and his associates, during their intensive study of *Mattesia grandis*, led to its immediate evaluation as a control agent for the boll weevil, *Anhthonomus grandis*. *N. gasti* was initially detected in several laboratory colonies of the boll weevil;[203] however, data were not presented on prevalence of infection in the colonies or on gross effects on the boll weevil. In an earlier report, McLaughlin[354] indicated that *N. gasti* caused chronic infections and that the adults had no symptoms, although pupal and adult mortality was obtained at high spore dosages.[554] Other effects of *N. gasti* (on adult longevity, fecundity, and overwintering survival) were derived from field tests with *N. gasti* and *M. grandis*.

Control attempts with *N. gasti* against the boll weevil were first conducted in 1966 in field cages.[32] In this and subsequent tests[2,33] an improved bait formulation, originally developed for *M. grandis*, was used at 1×10^6 spores per mℓ. In 1966, applications were made at frequent intervals over a period of 17 weeks beginning in early July. Although effectiveness of the baits was limited by frequent rains, as much as 30 to 40% of the adult weevils fed on the bait resulting in the partial suppression of weevil populations. Both *N. gasti* and *M. grandis* were equally effective in reducing weevil populations. A field test, conducted in 1967,[32] used a spore-formulation of *N. gasti* bait applied by spinning disks on a high clearance, self-propelled sprayer (Figure 19). This test demonstrated the potential usefulness of *N. gasti* for boll weevil control. Bait was again applied throughout the cotton-growing season beginning in June and continuing until the end of October. The overwintered weevils in June and early July responded well to the bait; 50 to 70% ingested the bait and as many as 80% became infected. The F_1 generation and ensuing generations in midsummer did not respond as well; only 2 to 19% ingested the bait and only 11 to 25% became infected. Weevils entering diapause in the fall again responded well to the bait and 50 to 70% of the weevils became infected. Over 60% of the overwintering weevils collected from woods trash were also infected and winter mortality in the treated area was 96% contrasted with 84% in nearby untreated areas. In subsequent tests, conducted during 1969—70 and 1970—71, fall treatments were followed by spring treatments.[2] Sufficient suppression was obtained in one test to delay chemical control measures until late season. In the other test, fields receiving no chemical treatments yielded more cotton than fields that received full chemical treatments. Although McLaughlin[2] concluded that the use of *N. gasti* or *M. grandis* for boll weevil control was not economically feasible (subsequent efforts on boll weevil control have been concentrated on eradication), those tests demonstrated the effectiveness of baits for introducing a protozoan into a host population.

Nosema locustae

This species has been extensively studied as a microbial control agent and is the only protozoan registered by the U.S. Environmental Protection Agency.[276] *N. locustae* is commonly found in dead and moribund grasshoppers in laboratory colonies and occurs at enzootic levels in the U.S. in natural populations of grasshoppers.[366] Early reports by Canning[205-207] established that the pathogenic effects of *N. locustae* on grasshoppers vary with the host age and species. For example, early instar nymphs of *Locusta migratoria migratorioides* are highly susceptible to *N. locustae*; most die prior to the final molt. Mid- to late-stage nymphs usually die at or just after the final molt and the few surviving adults are unable to fly.[207] While newly emerged adults are susceptible, 3-day-old adults are refractory to infection. The prevalence and level of infection also vary among different species of grasshoppers, both in natural populations of grasshoppers[366] as well as in field populations treated with spores.[34] In a recent review[321] Henry and Oma discussed the effects of *N. locustae* on cannibalism, development, and fecundity of grasshoppers, data derived from earlier, unpublished studies by Henry.[689] Cannibalism is increased among group-reared, infected grasshoppers, usually during molting. In addition to the delayed development with nymphs, persisting for prolonged periods at the penultimate instar, heavily infected grasshoppers feed less than healthy insects, are lethargic, and often exhibit deformities, i.e., twisted wings and legs and a protrusion of the foregut through the cervical membrane. Egg production decreases among grasshoppers when infected as nymphs or adults. More recent studies on the effect of *N. locustae* on food consumption by grasshoppers were published by Oma and Hewitt[478b] and Johnson and Pavlikova.[478c] In another study on infected grasshoppers,[478] significant increases in mortality, over that of the controls, occurred only in grasshoppers subjected to the stress of group rearing. Additional observations on field-treated grasshoppers have confirmed the chronic and severely debilitating nature of infections caused by *Nosema locustae* and the variable susceptibility of major grasshopper pests of rangelands.[34-36]

In the initial field study (ca. 4 ha) with *N. locustae*,[34] spores were sprayed on wheat bran and applied by ground equipment. Five concentrations, ranging from 7.75×10^8 to 1.3×10^{10} spores per ha in 2.24 kg bran per ha, were used. Hydroxymethyl cellulose (0.2% w/v) was added as a sticker. Each treatment was replicated four times. While significant differences in infection rates were not obtained between spore treatments (Table 43), the average prevalence of infection increased from about 3.6% at 3 weeks postapplication to 34.1% at 6 weeks postapplication. All differences were statistically significant when compared to the control plots. Densities of the three predominant species of grasshoppers were significantly reduced 4 to 6 weeks after applications, particularly for *Melanoplus gladstoni* (which was in an earlier stage of development at the time of application).

A series of three studies was conducted to provide a foundation for the use of *N. locustae* against grasshoppers. In the first of these studies,[35] attempts were made to determine the importance of application times, spore concentrations, and the level of spore carrier (bran) on grasshopper control. The experiment included 24 treatment combinations: (1) times of application: four applications spaced over 22 days at approximately 1 week intervals; (2) concentrations of spores: 1.5×10^8, 4.7×10^8, and 1.4×10^{10}/ha, each mixed with 0.2% hydroxymethyl cellulose as a sticker; and (3) levels of spore carrier: 1.12 and 4.49 kg wheat bran per ha. A replicate of 24 treated and 12 check plots (each about 4 ha) was established

Table 43
PERCENTAGE PREVALENCE OF INFECTION AMONG GRASSHOPPERS AFTER SPORE APPLICATIONS OF *NOSEMA LOCUSTAE*[34]

Treatment (No. spores/ha)	Interval posttreatment (weeks)			
	3	4	5	6
0	0	2.4	3.2	3.9
7.75×10^8	4.3	12.1	31.7	33.7
1.6×10^9	1.7	15.3	33.0	31.7
3.2×10^9	2.7	17.8	24.5	32.8
6.4×10^9	4.0	21.5	30.1	36.2
1.3×10^{10}	5.1	23.9	43.0	36.0

Table 44
PERCENTAGE PREVALENCE OF INFECTION OF *NOSEMA LOCUSTAE* AMONG SURVIVING GRASSHOPPERS RELATED TO APPLICATION TIME AND SPORE-BRAN CONCENTRATIONS[35]

		Percent infected related to spore concentration and bran/ha					
		1.5×10^8 spores		4.65×10^8 spores		1.4×10^{10} spores	
Application	Weeks post-application	1.12 kg bran	4.49 kg bran	1.12 kg bran	4.49 kg bran	1.12 kg bran	4.49 kg bran
First	4	1.1	1.1	7.3	5.0	19.4	16.2
	6	2.7	5.8	16.2	11.2	29.8	35.6
	9	20.6	9.5	13.2	25.6	5.9	28.6
Second	4	1.1	5.7	11.0	10.4	46.6	35.1
	6	8.0	5.9	22.2	15.8	49.6	43.8
	8	7.0	1.0	20.0	23.0	20.8	36.7
Third	4	9.5	5.9	20.7	12.8	31.1	31.6
	6	7.1	7.7	35.6	19.2	50.0	29.2
	7	16.0	6.0	38.4	12.5	39.7	48.4
Fourth	4	5.3	3.4	19.7	18.9	19.6	28.9
	6	9.9	7.0	29.2	18.1	30.4	40.5

at each of two locations about 1.6 km apart. The bran was applied by a buffalo turbine on a pickup truck. Reductions in grasshopper densities and the prevalence of infection among survivors (Table 44) at the last sampling date were primarily related to spore concentration. The average reduction in density ranged from 21% for the low rate to 73% for the high rate of spores. At the median spore dose (4.7×10^8) the applications appeared most efficient when the spores were applied at 1.1 kg bran per ha. The greatest reductions in grasshoppers occurred for the first and second applications while the third and fourth applications resulted in the highest prevalence of infection. When all factors were analyzed, the data suggested that an application of 1.6×10^9 to 2.3×10^9 spores on 1.12 to 1.68 kg bran per ha, (applied at the times when the predominant summer species were 3rd-instar nymphs) would result in 50 to 60% reductions in density within 4 to 6 weeks and a 35 to 50% infection rate among survivors. Additional mortality and reduced fecundity also could be expected among the survivors. Thus, a standard formulation of 2.46×10^9 spores on 1.68 kg bran per ha was adopted for subsequent studies.

Table 45
**EFFECTIVENESS OF SPORES OF
NOSEMA LOCUSTAE AGAINST
GRASSHOPPERS WHEN APPLIED
ON WHEAT BRAN OR AS ULV
SPRAYS**[a][37]

Treatment		$\bar{x}$ no. grasshoppers/m²	
Method	No. spores/ha	4 weeks	7 weeks
Wheat bran	2.47×10^9	7.63a	8.14a
ULV spray	6.18×10^9	12.66b	12.79b
ULV spray	1.24×10^{10}	12.86b	12.95b
ULV spray	2.47×10^{10}	12.95b	11.27b
Check		13.75b	13.10b

[a] Values followed by the same letter do not differ significantly ($p = 0.05$).

A second test was designed to determine the effect of prolonged storage on spore viability when tested against several species of grasshoppers.[36] These tests were similar to previous ones as far as plot size, formulation procedures, application rate, and sampling procedures were concerned. Applications of fresh spores (stored in distilled water at about −10°C for up to 4 months) resulted in higher reductions in density of grasshoppers and higher prevalences of infection among survivors than spores stored at −10°C in water for 8 months to 3 years or harvested from cadavers stored for 1 year. Reductions in grasshoppers were less than had been previously obtained.[35] However, applications of fresh spores reduced grasshoppers in both complexes by 50% or more within 4 weeks.

The last study tested the relative effectiveness of a wheat bran formulation and an ultra-low volume (ULV) spray application of spores of *Nosema locustae*.[37] The standard application rate (2.47×10^9 spores on 1.68 kg wheat bran per ha) was compared with ULV applications (6.18×10^9, 1.24×10^{10}, and 2.47×10^{10} spores in 585 mℓ distilled water per ha) and an untreated check. As in previous tests, 0.2% hydroxymethyl cellulose was added to the spores prior to formulation as a bait or spray. The treatments (four replicates) were applied to plots (each 4 ha) with ground equipment when the predominant species were in the 3rd and 4th instars. The average densities of grasshoppers from plots treated with spores on wheat bran were significantly lower at 4 and 7 weeks postapplication than in plots treated with the ULV sprays or in untreated plots (Table 45). A higher prevalence of infection also was found in grasshoppers from plots treated with the bran than in the control plots or in sprayed plots. Despite the advantages of ULV sprays for grasshopper control, the high cost of producing the extra spores (10- to 15-fold increase) that is required to attain comparable results, make bran application cheaper and more efficient.

A recent attempt, to evaluate the use of spores of *N. locustae* applied by ground equipment against grasshoppers, was carried out in Canada.[690] A cyclone-type seeder was used to broadcast spore-treated bran in a cattle pasture at the rate of 2.5×10^9 spores or 5.0×10^9 spores per 1.68 kg wheat bran per ha. Applications were made when the three predominant species (*Melanoplus sanquinipes*, *M. packardii*, and *Camnula pellucida*) were primarily in the 3rd instar. About 50% of the grasshoppers were infected 4 to 5 weeks postapplication. The prevalence of infection among the surviving grasshoppers continued to rise, as their populations were reduced, reaching a peak of 95 to 100% near the end of the season (12 weeks after application). Population reductions of about 60% were attained (after 8 weeks) for *M. sanguinipes* and *M. packardii*; a reduction of only 28% (after 12 weeks) was detected in *C. pellucida*. Significantly lower egg production was also noted in the two species of *Melanoplus*.

Current attempts to exploit the potential of *Nosema locustae* for controlling grasshoppers have shifted from ground applications to aerial application of spores on bran or bran-spore combinations with insecticides. A large scale pilot test involving about 39,000 ha, was initiated in Montana in 1975 to determine the long-term debilitory effects of *N. locustae*.[38] Bait was applied aerially at a rate of 2.1×10^9 and 2.1×10^8 spores per ha, respectively. These rates were compared to a standard rangeland insecticide treatment (malathion) and an untreated check, and populations were monitored for 3 years. An extensive epizootic in the 2nd year of the test (by the fungus *Entomophaga grylli*) obscured the long term evaluation of the effects of *N. locustae*. Significant reductions in grasshoppers, however, were associated with the high rates in both the 1st and the 2nd year. Slight, but nonsignificant reductions in densities, were also recorded in the 2nd and 3rd year of the study in plots treated with the low rate.

A cooperative pilot pest management program was conducted in the western U.S. to evaluate the potential of using a spore bait and insecticide (carbaryl) combination for both short- and long-term suppression of grasshoppers.[516,517] This test was possible since it had been demonstrated that spores of *N. locustae* were compatible with malathion, carbaryl, dimethoate, and carbofuran.[514,515] Wheat bran was treated either with carbaryl (1.96 AI by weight) or with *N. locustae* (1.47×10^9 spores per kg of bran) and applied aerially at the rate of 1.68 kg of bait per ha. Large, nonreplicated plots were established at two locations. At one location the treatments included a 2:1 mixture of the carbaryl bait and *N. locustae* bait; a 1:2 mixture of carbaryl bait and *N. locustae* bait; and an untreated control. The other location included treatments of carbaryl bait only, *N. locustae* bait only, and an untreated check. Evaluations were conducted 48 hr postapplication to determine the acceptance of wheat bran and the effectiveness of carbaryl in reducing grasshoppers. Bait acceptance varied greatly among species and populations of most of the major grasshopper species were greatly reduced, particularly in the full dosage treatment with carbaryl.[516] These results also indicated that the grasshopper species that produce an abundance of *N. locustae* inoculum suffered less mortality from the carbaryl-treated bait than did those species with little value for inoculum production. Prevalence of infection at 4 weeks post-treatment was nil, and average infection levels at 6 weeks were only 12.7% for *N. locustae* alone, 8.1% for the 2:1 mixture, and 4.0% for the 1:2 mixture of spores to carbaryl.[517] The high reductions in longevity in treatments containing carbaryl were attributed to an enhancement of the predator-prey ratio by the selective elimination of grasshoppers. The reduced efficacy of *N. locustae* was probably associated with a loss in spore viability due to storing the formulated bait (for 11 days under unfavorable conditions) prior to application. Results of the pilot pest management program were sufficiently encouraging to foster additional tests with *N. locustae* alone and integrated with chemicals in Argentina and West Africa.

Fowler et al.[286a] reported on the results of field trials, in an unpublished summary, with *N. locustae* and carbaryl in Cape Verde and Mauritania, West Africa. (A more limited version of this study has also been recently published.[286b]) Tests at each location (four replications in a randomized complete block design) received the following treatments: (1) 5.25×10^9 spores per 2 kg of wheat bran per ha; (2) 3.5×10^9 spores on 1.3 kg of wheat bran plus 13.3 g of carbaryl on 0.67 kg wheat bran per ha; (3) 1.75×10^9 spores on 0.67 kg wheat bran plus 26.7 g or carbaryl on 1.33 kg of wheat bran per ha; (4) 40 g of carbaryl on 2 kg wheat bran per ha; and (5) untreated control. Bait formulations were applied using a cyclone seed spreader. The integrated treatments were applied separately in swaths to achieve the 2:1 and 1:2 ratios of treatments (2) and (3), respectively. Significant suppression of grasshoppers in Cape Verde was obtained in all carbaryl-treated plots for at least 14 days postapplication while the largest reduction (35%) in grasshoppers in the *Nosema*-treated plot was obtained 28 days postapplication. The percentage of grasshoppers infected 3 weeks postapplication varied from 1.4 in the carbaryl treatments to 12.1 in the *Nosema* treatment. At 6 weeks postapplication, the level of infection was similar in all treatments, varying from 12.0 to 15.5%. In the Mauritania plots at 4 weeks postapplication, infections in the *Nosema* and the 2:1 treatments were significantly higher than in the 1:2 and carbaryl treatments. Grasshopper movements between plots resulted in a dilution of infection levels at both locations and lower than expected reductions in grasshoppers (exposed to the *Nosema* treatment) in the Cape Verde plots. However, the study further demonstrated the value of an integrated control program for grasshoppers using bait formulations of *N. locustae* and chemical insecticides.

Although emphasis on using *N. locustae* continues to be placed on aerial applications of wheat-bran bair, Johnson and Henry[286d] recently used a truck-mounted, turbine-blower to apply *N. locustae*-treated bran to control grasshoppers in roadside vegetation. Low-level epizootics of *N. locustae* were initiated in plots at both the low (2 kg bait per ha) and high (4 kg bait per ha) rates of bait application while higher infection rates might have been obtained if the applications had been made earlier in the season, their work did prove that *N. locustae* could be introduced into and maintained in roadside grasshopper populations.

Efforts have also been made to use *N. locustae* to control the Mormon cricket, *Anabrus simplex*, on rangelands in Colorado.[286c] Bait (2.5×10^9 spores to 1 kg bran) was applied aerially and post-treatment samples were taken from both within and outside the treatment area. During the first season there was no evidence of a reduction in cricket density within the treated area but infected, moribund crickets were obtained 11 weeks postapplication. Infected crickets also were found the next season from within the treatment area, and the number and size of cricket bands appeared lower than the previous season. Grasshopper reductions could not be confirmed experimentally due to the use of insecticides by landowners outside the treatment area. However, the lower cricket populations suggested that the bait application had contributed to control within the treatment area.

Nosema pyrausta

The prominence of *N. pyrausta*, as a regulatory agent of natural populations of the European corn borer, is well documented.[323,376,376a,656,691-693] As pointed out previously, this species is a common, cosmopolitan parasite of *Ostrinia nubilalis* and is considered the most effective of all of the naturally occurring pathogens of this pest.[656] This species, typical of many lepidopterophilic microsporidia, is transmitted transovarially as well as perorally and generally causes sublethal infections characterized by the gradual debilitation of host larvae, pupae, or adults. In 1928 Paillot[209] speculated that infection by *N. pyrausta* reduced the vitality and favored the action of other destructive forces on the corn borer. However, it was not until 1954 that Zimmack et al.[26] showed that larval survival and growth rate as well as adult longevity and fecundity were adversely affected by infection. Kramer[322] confirmed the effects of *N. pyrausta* on survival of infected borers and also showed that mortality was related to the stresses of cold winter and hot summer.[371] Extensive laboratory studies also confirmed the deleterious effects of infection on adult longevity and fecundity[322,323,323a,435,692] and reduced oxygen consumption by infected borers.[694] Population reductions and reduced larval weight of infected borers have also been demonstrated in several studies.[375,376a,695,696] As with many microsporidia, the pathological effects of *N. pyrausta* on *O. nubilalis* are usually most severe when infected borers are exposed to additional stress factors such as overcrowding[323a,694] or resistant host plants.[695,696]

In their initial field tests with *N. pyrausta*, Zimmack et al.[26] demonstrated that the prevalence of infection in natural populations of borers was increased by spraying corn hills

Table 46
EFFECT OF FOLIAR APPLICATIONS OF *NOSEMA PYRAUSTA* SPORES ON EUROPEAN CORN BORER LARVAE[a] [656]

Treatment			Larval weight (mg)	Infected larvae (%)	No. cavities/ plant
Spores/ plant	mℓ water/ plant	No. larvae/ plant			
1974 — First-Generation Larvae (Whorl Stage Corn)					
22.5 × 10^7	25	1.6a	49.1a	62.1a	2.7a
Control	—	3.1b	63.3a	1.5b	5.1b
1975 — Second-Generation Larvae (Pollen-Shedding Stage Corn)					
24.3 × 10^7	27.5	5.3a	112.2a	99.2a	11.9a
24.3 × 10^7	15.3	5.5a	109.0a	95.2a	11.7a
Control	—	6.4a	115.4a	20.8b	13.8a

[a] Values followed by the same letter do not differ significantly ($p = 0.05$).

(with aqueous spore suspensions) but population reduction was not obtained. More extensive field tests were conducted by Lewis and Lynch[656] (Table 46). Foliar spray applications of vacuum-dried spores were made on either whorl-stage or pollen-shedding corn that had been artificially infested with corn borer egg masses so as to simulate first and second generation borer populations, respectively. In the simulated first-generation tests, the foliar applications resulted in lower numbers of larvae per plant, significant reductions in the number of cavities per plant, and significantly higher percentages of infected larvae. Significantly higher percentages of infected larvae also were obtained in the simulated second-generation tests as well as nonsignificant reductions in the number of larvae per plant, larval weight, and larval tunnels per plant. An average of 63.8 and 97.2% infected larvae were achieved, respectively, in the simulated first- and second-generation tests. Reductions of 48.4, 18.8, and 43.8% were obtained in three separate tests with first-generation larvae compared with reductions of 17.2 and 14.1% in two tests with second-generation larvae. Lewis and Lynch[656] suggested that the full potential of the suppression of corn borer populations would only be apparent in filial generations resulting from transovarial transmission of the microsporidium to survivors.

The possibility of utilizing *N. pyrausta* in combination with insecticides has also been investigated.[518,519] When spores were placed topically on artificial diet previously formulated with carbaryl, carbofuran, or the spore-crystal powder of *Bacillus thuringiensis (Bt)* var. *kurstaki*, higher percentage larval mortalities were obtained than with spores alone. Percent infection did not differ significantly among treatments but the intensity of infection was significantly reduced in the spore-insecticide combinations, especially in the *Nosema pyrausta* spore-*Bt* combination. In field tests where applications of spores were followed (24 hr later) by insecticidal treatments, the percentage of larvae infected and the intensity of infection did not differ significantly from larvae treated only with spores.[518] Significant reductions in larval numbers and stalk tunnels, as a result of additive effects of each material acting independently,[519] also were obtained in other field tests. Combinations of *Nosema pyrausta* with *B. thuringiensis* or with carbaryl were more effective than the combination with carbofuran.

The role of stress factors, in increasing the potential of *N. pyrausta* for population suppression of *O. nubilalis*, has also been demonstrated in preliminary field tests. These tests evaluated the response of infected and noninfected larvae to applications of spores of *Vairimorpha necatrix* and *N. pyrausta*, alone and in combination with each other.[697] Infected larvae were more adversely affected than noninfected larvae. Larvae only exposed to *N. pyrausta* were generally less affected than larvae exposed to *V. necatrix* alone or in combination with spores of *N. pyrausta*.

Despite these indications that *N. pyrausta* might have some potential as a short-term control agent, Canning[5] concluded that this microsporidium is probably not virulent enough on its own to serve as an effective short-term control agent for corn borers. Her assessment is in general agreement with the more recent study of Laing and Jaques[213a] who also evaluated the effectiveness of foliar spray applications of *N. pyrausta* spores on sweet and field corn in Ontario during the 6-year period from 1978 to 1983. Application rates ranged from 2×10^{12} to 1.3×10^{13} spores per ha and treatments were applied two to three times each season. Skim-milk powder was used as a UV protectant. Although the incidence of infection in *O. nubilalis* was higher in the treated plots than in the untreated controls, the treatments did not reduce damage or population levels of the borer.

Pleistophora schubergi

Infection by *P. schubergi*, a pathogen of midgut cells of a wide range of lepidopteran larvae, causes a variety of symptoms related to digestive disturbances. In highly susceptible species, early instar larvae die during the next larval stadium following spore ingestion. Earlier death may occur 1 to 3 days postexposure due to a bacterial septicemia that accompanies gut disruption.[381] Infected larvae, in later stages or in less susceptible hosts, are asymptomatic or typically exhibit symptoms characteristic of sublethal infections; however, most infected larvae eventually die in a late stage or as pupae or adults.[212,626,626a,698] Infected larvae show a loss of appetite, reduction in size, weight and vigor, prolonged or retarded larval development, and molting difficulties.[212,224,226,380,699] Simchuk[699] observed a stimulation of larval growth immediately after infection which was followed by retarded development and eventual death. In some species cocoon formation may be aberrant,[626] pupae may be malformed or produce deformed adults unable to mate and lay eggs,[212,626] or adult fecundity may be lessened.[685b] The intensity of pupal diapause also may be affected.[627]

The wide host range of *P. schubergi*, its common occurrence in natural populations of forest pest defoliators, and its efficacy under laboratory conditions demonstrate its potential as a microbial control agent. However, few field trials have been conducted. An attempt by Weiser[556] to use *P. schubergi* against *Euproctis chrysorrhoea* was unsuccessful possibly due to the removal of spores by rain and/or their inactivation by sunlight.

Higher levels of infection were obtained, in more recent field tests, following an application of aqueous sprays of spores against several lepidopterous pests of forest trees (Table 47). Wilson and Kaupp[653] attributed the higher infection rates (with larvae of *Choristoneura fumiferana* on balsam fir) to the open nature of spruce buds than those of white spruce at the time of spraying. Samples of larvae from the same trees revealed a very low carryover of infection (3.1%) the next year on white spruce and none on balsam fir or on either tree thereafter (2 to 3 years later).[325,688] Similarly, there was no carryover of infection after 1 year in *Malacosoma disstria* feeding on aspen trees.[41,655] While most field tests (Table 47) proved that *P. schubergi* will infect a high percentage of lepidopterous larvae and the hymenopteran, *Pristiphora erichsonii*,[700] reductions in populations have not been demonstrated. Thus, additional field testing will be required to determine whether *P. schubergi* actually will suppress or control forest defoliators.

Vairimorpha necatrix

The high virulence of this microsporidium, its activity against several lepidopterous pests (Table 40), and its ease of mass production[493] mark this species as one of relatively few

Table 47
SUMMARY OF FIELD TRIALS WITH *PLEISTOPHORA SCHUBERGI* USED AGAINST INSECT PESTS OF FOREST

Host species	Tree species	Dosage	Adjuvants	Days post-application	Infection (%)	Ref.
Anisota senatoria	Northern red oak	2×10^7 spores/ml (to point of runoff)	0.1% Nu-Film 17	14	72.0	557
		2×10^8 spores/ml (to point of runoff)	0.1% Nu-Film 17	14	95.8	
Symmerista canicosta	Northern red oak	2×10^7 spores/ml (to point of runoff)	0.1% Nu-Film 17	14	100.0	
		2×10^8 spores/ml (to point of runoff)	0.1% Nu-Film 17	14	100.0	
Choristoneura fumiferana	White spruce	5×10^{10} spores/tree	25% (v/v) molasses, 30 g/l IMC90-001	19 / 26	64.8 / 48.6	653
	Balsam fir	5×10^{10} spores/tree	25% (v/v) molasses, 30 g/l IMC90-001	19 / 26	96.3 / 95.2	
Malacosmoa disstria	Trembling aspen	1.8×10^{11} spores/tree (96% spores of *P. schubergi* and 4% *Nosema disstria*)	50% (v/v) molasses, 30 g/ℓ IMC90-001	17	85.0ª	655
		1.8×10^{11} spores/tree (50% spores of *P. schubergi* and 50% *N. disstria*)	50% (v/v) molasses, 30 g/ℓ IMC90-001	17	76.1a	
Pristiphora erichsonii	Larch	5×10^8 spores/tree	30 g/ℓ IMC90-001	14	40.0	700

ª Represents % infected by *P. schubergi* only.

entomogenous protozoa considered to possess good potential as a short-term, microbial control agent.[4,326] As initially demonstrated by Maddox[385] and confirmed by Chu and Jaques[701] and Fuxa,[635] infection by *V. necatrix* is characterized by two types of mortality: relatively quick death due to gut damage caused by high spore doses and death at pupation at low doses due to a chronic microsporidiosis from extensive development in fat body and other tissues. In a quick death scenario the extrusion of polar tubes (from large numbers of spores) damages the gut epithelium and larvae die (within 3 to 6 days) from this damage or from a bacterial septicemia. This first type of mortality makes *V. necatrix* attractive as a short-term microbial control agent. As expected of a highly virulent pathogen, transmission of *V. necatrix* is primarily *per os*. Infected larvae usually die before pupation. The number of spores necessary to infect one half the population (ID_{50}) equals the number required to cause death in one half of the population (LD_{50}).[326,385] Since few larvae survive to become adults, there is no opportunity for the microsporidium to be transmitted transovarially. Spores have been found in adults of a few species of Lepidoptera[385,635,702] but evidence of an active infection was lacking. However, there is evidence that the microsporidium may be transovarially transmitted. Tanabe,[386] in laboratory tests reported transovarial transmission by adults fed spores in honey solution. Infection rates by *V. necatrix* in field populations of crop pests are usually low but it has been epizootic among forest pests[385,392] and in *Pseudaletia unipuncta* from grasslands in Hawaii.[384,390] Attempts to use spores of *V. necatrix* as a microbial control agent have progressed from initial evaluations in aqueous sprays to the more recent successful use in bait formulations.

Maddox,[385] in his initial field trials with *V. necatrix*, injected spore suspensions in the silk channel of sweet corn ears in an attempt to control *H. zea*. Spore rates ranged from 1

$\times$ 10^6 to 4 $\times$ 10^8 spores per ear. Although 70% of the live larvae were infected and many larvae were killed by either a bacterial septicemia or microsporidiosis, significant crop protection was not obtained. Other tests conducted by Maddox et al. against noctuid pests of vegetables also were unsuccessful despite high levels of infection and mortality.[326] In most of these tests, unacceptable "cosmetic" crop damage to marketable vegetables occurred before death of the pests.

Other tests with spores applied as sprays were more encouraging.[629,633] At an application rate of 2.2 $\times$ 10^{12} spores per ha (without a UV protectant), Mistric and Smith[633] obtained significant control and leaf protection of *Heliothis virescens* feeding on tobacco. However, results were not as good as several other microbial agents or the commonly used insecticide, methomyl. Jaques[629] sprayed spores (4.5 $\times$ 10^{10} and 4.5 $\times$ 10^{11} spores per ha plus the spreader-sticker Plyac) on cabbage infested with the cabbage looper *T. ni* and the imported cabbageworm *P. rapae*. Since the imported cabbageworm was not susceptible to *V. necatrix*, the spores were combined with a granulosis virus. Populations of *T. ni* were reduced to 15% and feeding damage was 64% of that of the untreated plot. In addition, the proportion of undamaged cabbage and the weight of marketable heads were significantly higher in treated than in check plots. Successful control of *T. ni* may be due to its relative susceptibility to *Vairimorpha necatrix*. Fuxa[635] found that *T. ni* was the most susceptible of several species of noctuids as evidenced by gut damage and microsporidiosis.

More recent and extensive field trials with various formulations of spores of *V. necatrix* were conducted by Fuxa and Brooks[630] on tobacco, soybeans, and sorghum in attempts to control *Heliothis zea* and *H. virescens*. Adequate crop protection was not obtained with sprays containing up to 2.5 $\times$ 10^{13} spores per ha plus a UV protectant and spreader-sticker (SAN 285 ADWP 66). High infection rates (up to 63.2% of *H. virescens* on tobacco and 77 and 99% of *H. zea* on sorghum and soybeans, respectively) were obtained but significant reductions in the number of larvae or in feeding damage were obtained in only a few treatments on tobacco. Favorable results on tobacco may be related to the fact that *H. virescens* is up to 11 times more susceptible than *H. zea* to gut damage by *V. necatrix*.[635]

Since 2.5 $\times$ 10^{13} spores per ha represent spores produced in ca. 2500 larvae, they concluded that control of a fruit or blossom feeding pest such as *H. zea* with *V. necatrix* was impractical. However, foliage feeders such as the green cloverworm, *Plathypena scabra*, may be controlled since significant reductions in green cloverworms were obtained on soybeans with a spore rate as low as 2.5 $\times$ 10^{12} per ha. In contrast to poor protection with spray applications, hand applications of corn meal formulations of spores (Figure 20A) on tobacco at 2.5 $\times$ 10^{12} spores per ha resulted in excellent control of tobacco budworms. Budworm densities and damages, in several tests, were reduced to levels comparable to a commercial corn meal formulation of *Bacillus thuringiensis* and the insecticide methomyl. Moribund and dead 2nd and 3rd instar larvae were found frequently on plants 4 to 7 days after treatment (Figure 20B). Fuxa and Brooks[630] suggested that the corn meal formulation was an attractant or bait, protected the spores, or was simply a way of concentrating the spores directly on tobacco plants.

Brooks[703] carried out two field tests in 1980 on soybeans to further evaluate the potential of using corn meal formulations of *V. necatrix* against *H. zea*. Dry corn meal or corn grit formulations containing 2.47 $\times$ 10^{12} spores per ha were applied with a hand-crank, row-crop duster, or as a spray either with or without a UV protectant or a feeding stimulant (Coax). Spores were vacuum dried in cellulose prior to mixing as a dry powder with corn meal or water. Despite high infection rates (87.5 and 98.3%) in all the treatments, significant reduction in populations was not obtained at 8 to 11 days postapplication. Only the insecticide methomyl provided significant reductions (at 2 to 4 days postapplication). Even methomyl did not significantly reduce larval populations to below the control plots at the conclusion of the tests. Thus field results substantiate the conclusion of Fuxa and Brooks[630] that foliage

FIGURE 20. Field tests with a cornmeal formulation of spores of *Vairimorpha necatrix*. (A) Appearance of tobacco plant immediately after hand application of spore formulation to bud area of plant; (B) moribund, third-stage larvae of the tobacco budworm, *Heliothis virescens*, in bud area of plant about 5 days post-treatment. (Courtesy of J. R. Fuxa.)

applications of spores of *V. necatrix* are unlikely to provide significant control when *H. zea* predominantly feeds on fruit or blossoms. However, the high infection rates obtained demonstrate that control may be possible in the next generation in a suitable crop-pest system.

Richter and Fuxa[385b] conducted similar field tests that emphasized that a suitable crop-pest system must involve a pest species that is at least moderately susceptible to this microsporidium. The results of their evaluation of aqueous and corn grit formulations of spores of *V. necatrix* for control of the velvetbean caterpillar, *Anticarsia gemmatalis*, on soybean were discouraging. In three separate tests the highest percentage of infection obtained was only 10.7%,[385b] and another study indicated that *V. necatrix* was not virulent to the velvetbean caterpillar.[385a]

Further efforts to evaluate the efficacy of bait formulations of *V. necatrix* were carried out by Grundler[704,704a] against larvae of the black cutworm, *Agrotis ipsilon*, feeding on corn. Spores were combined with a grape pomace and tested under both greenhouse and field conditions. At the optimal pomace rate of 22.5 kg per ha containing 3.8 to 4.5×10^{13} spores, significant reductions in damage (plants cut) were obtained. While this represented a significant reduction over the number of damaged plants in the control plots, it still exceeded the economic threshold level (8% damage for above-ground cutting and 4% for below-ground cutting). Infection rates were independent of pomace and spore rates but damage was lowest at the highest spore rate. Despite the inadequacy of the pomace-spore bait in controlling cutworms in postemergence corn, Grundler suggested that a preplant or planting time application of the bait might allow corn to outgrow cutworms.

Considerable interest also has been shown for use of *V. necatrix* to control *Ostrinia nubilalis* on corn.[560,697,702] The corn borer is susceptible to (both types of mortality) *V. necatrix*[702] and spores remain viable for up to 12 days in whorl- and pollen-shedding-stage corn.[560] The results of laboratory and preliminary field tests with *V. necatrix* against both *N. pyrausta*-infected and noninfected larvae also suggest that this microsporidium might successfully be used to control field populations of *O. nubilalis*.[697] However, Laing and Jaques[213a] found that spray applications of 3.9×10^{12} to 7.7×10^{13} spores per ha of

V. necatrix on sweet and field corn did not reduce crop damage or population levels of the corn borer in field tests conducted in Ontario during 1981 to 1983.

As with most microsporidia, field tests with *V. necatrix* demonstrated that high infection rates in pest populations can be obtained with appropriate spore formulations. However, the use of this species as a short-term microbial control agent will likely depend on the development of suitable spore-bait formulations for use only against lepidopterous pests occupying unusual ecological niches. Plant damage can be minimized if a large number of spores can be concentrated at the feeding site, i.e., in the vegetative buds, crowns, or leaf axiles of plants. Plant protection against foliage feeders will probably be more difficult to obtain.

Vavraia culicis

Recent demonstrations of the mass production potential of *V. culicis* in surrogate larvae[328,404a,639] and its broad geographic and host range (Table 41) indicate its promise as a microbial control agent for mosquitoes. As with *Nosema algerae*, host susceptibility as well as pathological effects vary with the species.[328,394,404a] Initial observations indicated that *V. culicis* was essentially avirulent to *Anopheles gambiae* despite extensive destruction of infected tissues.[401] Reynolds[396] found that the larval life of infected mosquitoes was prolonged, and, as suggested by Garnham[404] and confirmed by Bano,[403] infection (in adult *An. stephensi*) resulted in a partial inhibition of the sporogonic cycle of *Plasmodium*. In more extensive studies, Reynolds[705] found that *V. culicis* reduced the fecundity and longevity of infected adults of *Culex pipiens quinquefasciatus*, and Kelly et al.[328] reported between 50 and 60% mortality in larvae of *Cx. salinarus* exposed to spores at a 5×10^4 spores per mℓ. They[328] also obtained severe infections in *Cx. tarsalis*, *An. albimanus*, *Cx. pipiens quinquefasciatus*, and *Aedes taeniorhynchus*. As one of few truly perorally infectious microsporidia, *V. culicis* will likely be extensively investigated as a microbial control agent for mosquitoes. However, there has been only one attempt to evaluate a field application of spores against mosquitoes.[706]

In 1967, Reynolds[706] introduced spores of *V. culicis* into artificial containers and abandoned cisterns containing natural populations of *Cx. pipiens quinquefasciatus* on the South Pacific island of Nauru. Spores, previously propagated in larvae and pupae of *An. stephensi*, were transported to Nauru just prior to application. Applications of 6000 and 7000 spores per mℓ resulted in infection rates up to 30%. Infected larvae were present for up to 66 days at some sites. In a follow-up survey, 18 to 24 months later, low levels of infection were found at two sites. However, no significant reductions in mosquito populations or evidence of dispersal to other mosquito breeding sites were obtained. Although long-term control was not achieved in this test, similar and even more extensive tests can now be conducted using higher rates of spores, since spores are now readily obtainable using larvae of *H. zea* or *Mamestra brassicae* for production of *V. culicis*.

Ciliates

Only a few species of hymenostome ciliates are known as pathogens of various aquatic insects. These occur in *Tetrahymena* and *Lambornella* and include such species as *T. pyriformis*, an accidental or potential pathogen of various dipterous larvae, and *L. clarki*, a species that forms invasion cysts on the cuticle of larval *Aedes sierrensis*. Most work with entomogenous ciliates has been taxonomic and much more basic work is needed before attempts to use them as microbial control agents will be meaningful.[260m] However, studies such as those of Grassmick and Rowley[707] and LeBrun[260f,260g] may indicate that even *T. pyriformis* can cause mortality in some species of mosquitoes (at least under laboratory conditions) and further research may be warranted to evaluate potential as microbial control agents. In addition the usually lethal nature of ciliatoses[260a] and the culturability of most of the ciliates are significant characteristics that enhance their consideration as potential biological control agents of biomedically important insects, e.g., mosquitoes and blackflies.[259]

BIOTECHNOLOGY

In recent years the possibility of improving entomopathogens or developing new microbial control agents through genetic engineering has received considerable attention.[708-710] However, most of this interest has, again, focused on the bacteria and viruses as entomopathogens; the protozoa have been ignored as prospects for genetic engineering. This lack of attention is perhaps understandable when considered in respect to factors governing successful genetic engineering of microbial insecticides;[709] most of the entomogenous protozoa cannot be cultured in vitro and practically nothing is known about the genetics of their mode of action or factors comprising their effectiveness. In addition, the large number of undescribed species of protozoa and our incomplete knowledge of their life cycles indicate that much basic research needs to be done with the entomogenous protozoa before they will be good prospects for attempts at genetic engineering.

Despite these limitations, however, the spores of many species, especially microsporidia, can be produced in large quantities using live hosts and can be induced to germinate under controlled conditions. Thus, it should be possible to characterize their genomes using molecular biochemical techniques. The ability to culture some species in insect cell lines should also make them amenable to manipulation by classic selection techniques and recombinant DNA technology. The use of such techniques may allow the development of protozoa with genes for the production of various toxins or enhanced virulence, for increasing their tolerance to physical and chemical conditions, or for broading their host ranges. For example, microsporidia such as *Nosema pyrausta*, *N. fumiferanae*, and *N. heliothidis* are common, endemic pathogens of the lepidopterans *Ostrinia nubilalis*, *Choristoneura fumiferana*, and *Heliothis zea*, respectively, whose regulatory affects on their hosts might be greatly enhanced if one could introduce strains genetically engineered for increased virulence. The use of strains with increased tolerance to UV or to dessication could also increase their persistence under field conditions or their shelf life during storage.

The applicability of another biotechnological advancement in microsporidiology, the production of monoclonal antibodies by specific hydridoma cell lines, has already been demonstrated. Kawanishi and his co-workers[711] have produced monoclonal antibodies to the exospore proteins of *Nosema locustae*. When tested against purified spores of a number of microsporidian isolates, several clones were found to produce antibodies that reacted only with the spore proteins of *N. locustae*. One of these antibodies exhibited high avidity for a *N. locustae* exospore antigen and was utilized in an enzyme-linked immunosorbent assay for spore quantification. The assay was capable of detecting less than 50 spores. Other antibodies exhibited different reactivity patterns with the spores of heterologous species. As the patterns appeared to fall in general groupings, they indicate that monoclonal antibodies may have potential for classifying closely related microsporidia based on their spore antigens. The production of such monoclonal antibodies should eventually allow the development of immunodiagnostic techniques for the rapid and precise identification of spores in commercial formulations or in various target or nontarget organisms.

CONCLUSIONS AND FUTURE DEVELOPMENTS

The protozoa represent a group of pathogens that are in their infancy as microbial control agents when compared to other groups of entomopathogens. However, the size and diversity of the protozoa and the role that many species play as highly significant, natural mortality factors demonstrate their potential as microbial control agents. In addition to candidate species emphasized in this review, many other species also may be candidates for use against insect pests. Perhaps as significant, many new as yet undescribed species, particularly the microsporidia, will also be available as future candidate control agents. Basic research on life cycle patterns and efforts to achieve peroral infections may provide new approaches to successfully using protozoa to control mosquitoes, blackflies, and other medically important insects. The prospects of some species as microbial control agents may also be improved through genetic engineering.

Since most of the protozoa that are candidate microbial control agents are obligate parasites, spores will have to be produced for use in live hosts. Limited quantities of some protozoa may be produced in vitro (in tissue or cell cultures) but in vitro mass production of protozoa is still a long way off. The use of substitute or surrogate hosts for the production of some species is currently the most effective and economical means of producing spores of species in hosts difficult to rear or from which only small yields of spores are obtainable.

The relatively low virulence of most of the pathogenic protozoa for their hosts indicates that few species will likely be used as chemical insecticides are currently used. Special formulation and application techniques will need to be developed that will increase their ingestion by foliage-feeding pests and those which occur in aquatic habitats. Bait formulations appear to offer the best immediate approach for achieving this goal. Various formulations can also be used in combination with insecticides to ensure better short-term control or to predispose pests to the pathogenic effects of protozoa. Novel application techniques, e.g., the application of spores in water through overhead irrigation systems,[712] also may increase the efficacy of some species under field conditions. The increasing concern for environmental quality and the development of integrated pest management programs will also provide more opportunities for use of microbials that cannot presently compete with less expensive and more broad-spectrum chemical insecticides. Pest management programs also may offer the best opportunity for use of pathogens for long-term control. Thus, protozoa which cannot provide immediate population reduction or plant protection may provide significant long-term relief by reducing the vigor, longevity, and fecundity of the infected hosts.

There do not appear to be any real hazards of protozoa for nontarget organisms, particularly humans or plants; however, the host range and potential hazard of only relatively few species have been examined to date. Thus, host specificity and safety of candidate species must be determined prior to extensive field evaluation and development as commercial microbial control agents. Techniques to rapidly and precisely identify species of protozoa also will be needed to detect the presence of spores in commercial formulations or their occurrence in nature in target or nontarget organisms.

Considerable progress has been made in recent years to demonstrate the potential usefulness of protozoa as microbial control agents. With continued research, protozoa can be a worthwhile addition to our available supply of biorational pesticides for use in controlling populations of insect pests and for increasing the quality of the environment. We have significant challenges, in the years ahead, to exploit and fully realize the potential for protozoa as microbial control agents of insect pests.

ACKNOWLEDGMENTS

My sincere gratitude is expressed to C. B. Moore, C. M. Ignoffo, and J. E. Henry for ther review of this chapter, to Diane Jones for typing the manuscript, and to S. Jaronski for assistance in figure preparation. Appreciation is also extended to the following colleagues who supplied photographs or unpublished information for this chapter: J. E. Henry, J. V. Maddox, N. E. Alger, R. E. McLaughlin, E. U. Canning, G. G. Wilson, T. C. Andreadis, W. E. Burkholder, J. R. Fuxa, E. D. Rowton, D. L. Hostetter, J. J. Germida, S. W. Avery, and the late E. I. Hazard.

REFERENCES

1. **McLaughlin, R. E.,** Use of protozoans for microbial control of insects, in *Microbial Control of Insects and Mites,* Burges, H. D. and Hussey, N. W., Eds., Academic Press, New York, 1971, 151,
2. **McLaughlin, R. E.,** Protozoa as microbial control agents, *Misc. Publ. Entomol. Soc. Am.,* 9, 95, 1973.
3. **Pramer, D. and Al-Rabiai, S.,** Regulation of insect populations by protozoa and nematodes, *Ann. N.Y. Acad. Sci.,* 217, 85, 1973.
4. **Brooks, W. M.,** Production and efficacy of protozoa, *Biotechnol. Bioeng.,* 22, 1415, 1980.
5. **Canning, E. U.,** An evaluation of protozoal characteristics in relation to biological control of pests, *Parasitology,* 84, 119, 1982.
6. **Paillot, A.,** *L'Infection chez les Insects,* G. Patissier, Trévoux, 1933.
7. **Steinhaus, E. A.,** *Insect Microbiology,* Comstock, Ithaca, N.Y., 1947.
8. **Steinhaus, E. A.,** *Principles of Insect Pathology,* McGraw-Hill, New York, 1949.
9. **Aoki, K.,** *Konchu Byorigaku,* Gihodo, Tokyo, 1957.
10. **Lipa, J. J.,** Infections caused by protozoa other than sporozoa, in *Insect Pathology, An Advanced Treatise,* Vol. 2, Steinhaus, E. A., Ed., Academic Press, New York, 1963, 335.
11. **Lipa, J. J.,** *Zarys Patologii Owadow,* PWRiL, Warszawa, Poland, 1967.
12. **Weiser, J.,** Sporozoan infections, in *Insect Pathology, An Advanced Treatise,* Vol. 2, Steinhaus, E. A., Ed., Academic Press, New York, 1963, 291.
13. **Weiser, J.,** *Nemoci Hmyzu,* Academia, Praha, Czechoslovakia, 1966.
14. **Brooks, W. M.,** Protozoan infections, in *Insect Diseases,* Vol. 1, Cantwell, G. E., Ed., Marcel Dekker, New York, 1974, 237.
15. **Steinhaus, E. A.,** Microbial control — the emergence of an idea, *Hilgardia,* 26, 107, 1956.
16. **Wallace, F. G.,** The trypanosomatid parasites of insects and arachnids, *Exp. Parasitol.,* 18, 124, 1966.
17. **Bassi, A.,** *Del mal del Segno Calcinaccio o Moscardino Malattia che Affligge i Bachi da Seta,* Parte Prima, Dalla Tipografia Orcesi, Lodi, 1835.
18. **Pasteur, L.,** *Études sur la Maladie des vers à Soie,* Gauthier-Villars, Paris, 1870.
19. **Naegeli, C.,** Die neue Krankheit des Seidenspinners, *Nosema bombycis,* Naegeli., *Bot. Ztg.,* 15, 760, 1857.
20. **Balbiani, E. G.,** Sur les microsporidies ou psorospermies des articulés, *C. R. Acad. Sci.,* 95, 1168, 1882.
21. **Stempell, W.,** Über *Nosema bombycis,* Naegeli, *Arch. Protistenkd.,* 16, 281, 1909.
22. **Pasteur, L.,** On the use of fungi against Phylloxera, *C. R. Acad. Sci.,* 79, 1233, 1874.
23. **Pérez, C.,** Microsporidies parasites des crabes d'Arcachon, *Soc. Sci. Arcachon Trav. Lab.,* 8, 15, 1905.
24. **Taylor, A. B. and King, R. L.,** Further studies on the parasitic amoebae found in grasshoppers, *Trans. Am. Micros. Soc.,* 56, 172, 1937.
25. **Hall, I. M.,** Studies of microorganisms pathogenic to the sod webworm, *Hilgardia,* 22, 535, 1954.
26. **Zimmack, H. L., Arbuthnot, K. D., and Brindley, T. A.,** Distribution of the European corn borer parasite *Perezia pyraustae* and its effect on the host, *J. Econ. Entomol.,* 47, 641, 1954.
27. **Weiser, J. and Veber, J.,** Moznosti biologickeho boje s prastevnickem americkym (*Hyphantria cunea* Drury). II. *Cesk. Parazitol.,* 2, 191, 1955.
28. **Weiser, J. and Veber, J.,** Die Mikrosporidie *Thelohania hyphantriae* Weiser des weissen Bärenspinners und anderer Mitglieder seiner Biocoenose, *Z. Angew. Entomol.,* 40, 55, 1957.
29. **McLaughlin, R. E.,** Infection of the boll weevil with *Mattesia grandis* induced by a feeding stimulant, *J. Econ. Entomol.,* 59, 909, 1966.
30. **McLaughlin, R. E.,** Development of the bait principle for boll-weevil control. II. Field-cage tests with a feeding stimulant and the protozoan *Mattesia grandis,* *J. Invertebr. Pathol.,* 9, 70, 1967.
31. **Daum, R. J., McLaughlin, R. E., and Hardee, D. D.,** Development of bait principle for boll weevil control: cottonseed oil, a source of attractants and feeding stimulants for the boll weevil, *J. Econ. Entomol.,* 60, 321, 1967.
32. **McLaughlin, R. E., Daum, R. J., and Bell, M. R.,** Development of bait principle for boll-weevil control. III. Field-cage tests with a feeding stimulant and the protozoans *Mattesia grandis* (Neogregarinida) and a microsporidian, *J. Invertebr. Pathol.,* 12, 168, 1968.
33. **McLaughlin, R. E., Cleveland, T. C., Daum, R. J., and Bell, M. R.,** Development of the bait principle for boll weevil control. IV. Field tests with a bait containing a feeding stimulant and the sporozoans *Glugea gasti* and *Mattesia grandis,* *J. Invertebr. Pathol.,* 13, 429, 1969.
34. **Henry, J. E.,** Experimental application of *Nosema locustae* for control of grasshoppers, *J. Invertebr. Pathol.,* 18, 389, 1971.
35. **Henry, J. E., Tiahrt, K, and Oma, E. A.,** Importance of timing, spore concentrations, and levels of spore carrier in applications of *Nosema locustae* (Microsporida: Nosematidae) for control of grasshoppers, *J. Invertebr. Pathol.,* 21, 263, 1973.
36. **Henry, J. E. and Oma, E. A.,** Effect of prolonged storage of spores on field applications of *Nosema locustae* (Microsporida: Nosematidae) against grasshoppers, *J. Invertebr. Pathol.,* 23, 371, 1974.

37. **Henry, J. E., Oma, E. A., and Onsager, J. A.**, Relative effectiveness of ULV spray applications of spores of *Nosema locustae* against grasshoppers, *J. Econ. Entomol.*, 71, 629, 1978.
38. **Henry, J. E. and Onsager, J. A.**, Large-scale test of control of grasshoppers on rangeland with *Nosema locustae*, *J. Econ. Entomol.*, 75, 31, 1982.
39. **Tanada, Y.**, Epizootiology and microbial control, in *Comparative Pathobiology*, Vol. 1, Bulla, L. A. and Cheng, T. C., Eds., Plenum Press, New York, 1976, 247.
40. **Henry, J. E.**, Natural and applied control of insects by protozoa, *Annu. Rev. Entomol.*, 26, 49, 1981.
41. **Wilson, G. G.**, Protozoans for insect control, in *Microbial and Viral Pesticides*, Kurstak, E. Ed., Marcel Dekker, New York, 1982, 587.
42. **Honigberg, B. M., Balamuth, W., Bovee, E. C., Corliss, J. O., Gojdics, M., Hall, R. P., Kudo, R. R., Levine, N. D., Loeblich, A. R., Weiser, J., and Wenrich, D. H.**, A revised classification of the phylum Protozoa, *J. Protozool.*, 11, 7, 1964.
43. **Levine, N. D., Corliss, J. O., Cox, F. E. G., Deroux, G., Grain, J., Honigberg, B. M., Leedale, G. F., Loeblich, A. R., Lom, J., Lynn, D., Merinfeld, E. G., Page, F. C., Poljansky, G., Sprague, V., Vávra, J., and Wallace, F. G.**, A newly revised classification of the protozoa, *J. Protozool.*, 27, 37, 1980.
44. **Poinar, G. O. and Thomas, G. M.**, *Diagnostic Manual for the Identification of Insect Pathogens*, Plenum Press, New York, 1978.
44a. **Poinar, G. O. and Thomas, G. M.**, *Laboratory Guide to Insect Pathogens and Parasites*, Plenum Press, New York, 1984.
45. **Wallace, F. G.**, Biology of the Kinetoplastida of arthropods, in *Biology of the Kinetoplastida*, Vol. 2, Lumsden, W. H. R. and Evans, D. A., Eds., Academic Press, New York, 1979, 213.
46. **Lipa, J. J., Carl, K. P., and Valentine, E. W.**, *Blastocrithidia caliroae* sp. n., a flagellate parasite of *Caliroa cerasi* (L.) (Hymenoptera: Tenthredinidae) and notes on its epizootics in host field populations, *Acta Protozool.*, 16, 121, 1977.
47. **Patton, W. S.**, The life cycle of a species of *Crithidia* parasitic in the intestinal tract of *Gerris fossarum*, *Arch. Protistenk.*, 12, 131, 1908.
48. **Laird, M.**, *Blastocrithidia* n.g. (Mastigophora; Protomonadina) for *Crithidia* (in part), with a subarctic record for *B. gerridis* (Patton), *Can. J. Zool.*, 37, 749, 1959.
49. **McCulloch, I.**, An outline of the morphology and life history of *Crithidia leptocoridis*, sp. nov., *Univ. Calif. Publ. Zool.*, 16, 1, 1915.
50. **Wallace, F. G. and Dyer, M. I.**, The culture of *Blastocrithidia leptocoridis* (McCulloch), *J. Parasitol.*, 46, 43, 1960.
51. **Lipa, J. J.**, *Blastocrithidia raabei* sp. n., a flagellate parasite of *Mesocerus marginatus* L. (Hemiptera: Coreidae), *Acta Protozool.*, 4, 19, 1966.
52. **Hanson, W. L. and McGhee, R. B.**, The biology and morphology of *Crithidia acanthocephali* n. sp., *Leptomonas leptoglossi* n. sp., and *Blastocrithidia euschisti* n. sp., *J. Protozool.*, 8, 200, 1961.
53. **Léger, L.**, Sur un flagelle parasite de l' *Anopheles maculipennis*, *C. R. Soc. Biol.*, 54, 354, 1902.
54. **Wallace, F. G.**, Flagellate parasites of mosquitoes with special reference to *Crithidia fasiculata* Léger, 1902, *J. Parasitol.*, 29, 196, 1943.
55. **Strickland, C.**, Description of a *Herpetomonas* parasitic in the alimentary tract of the common green-bottle fly, *Lucilia* sp., *Parasitology*, 4, 222, 1911.
56. **Wallace, F. G. and Clark, T. B.**, Flagellate parasites of the fly, *Phaenicia sericata* (Meigen), *J. Protozool.*, 6, 58, 1959.
57. **Noguchi, H. and Tilden, E. B.**, Comparative studies of herpetomonads and leishmanias. I. Cultivation of herpetomonads from insects and plants, *J. Exp. Med.*, 44, 307, 1926.
58. **Hanson, W. L. and McGhee, R. B.**, Experimental infection of the hemipteron *Oncopeltus fasciatus* with Trypanosomatidae isolated from other hosts, *J. Protozool.*, 10, 233, 1963.
59. **Patton, W. S. and Strickland, C.**, A critical review of the relation of blood sucking invertebrates to the life cycles of trypanosomes of vertebrates, with a note on the occurrence of a species of *Crithidia*, *C. ctenophthalmi*, in the alimentary tract of *Ctenophthalmus agyrtes*, Heller, *Parasitology*, 1, 322, 1909.
60. **Kramar, J.**, Parasiti larvy *Tipula maxima* Poda, *Vestn. Cesk. Spol. Zool.*, 14, 55, 1950.
61. **Vickerman, K.**, *Herpetomonas ludwigi* (Kramar), 1950, n. comb., the trypanosomatid parasite of cranefly larvae (Diptera, Tipulidae), *Parasitology*, 50, 351, 1960.
62. **Leidy, J.**, A synopsis of Entozoa and some of their ectocongeners observed by the author, *Proc. Acad. Natl. Sci. Philadelphia*, 8, 42, 1856.
63. **Rogers, W. E. and Wallace, F. G.**, Two new subspecies of *Herpetomonas muscarum* (Leidy, 1856) Kent, 1880, *J. Protozool.*, 18, 645, 1971.
64. **Smirnoff, W. A. and Lipa, J. J.**, *Herpetomonas swainei* sp. n., a new flagellate parasite of *Neodiprion swainei* (Hymenoptera: Tenthredinidae), *J. Invertebr. Pathol.*, 16, 187, 1970.
65. **Fantham, H. B.**, Some insect flagellates and the problem of the transmission of *Leishmania*, *Br. Med. J.*, 2, 1196, 1912.

66. **Wenyon, C. M.**, *Protozoology, A Manual for Medical Men, Veterinarians and Zoologists,* Vol. 1 and 2, William Wood, New York, 1926.
67. **McGhee, R. B. and Hanson, W. L.**, Growth and reproduction of *Leptomonas oncopelti* in the milkweed bug, *Oncopeltus fasciatus, J. Protozool.,* 9, 488, 1962.
68. **Zotta, G.**, Sur un flagellé du type *Herpetomonas* chez *Pyrrhocoris apterus,* (Note préliminaire), *Ann. Sci. Univ. Jassy,* 7, 211, 1912.
69. **Zotta, G.**, Sur la culture en milieu N. N. N. du *Leptomonas pyrrhocoris, C. R. Soc. Biol.,* 84, 822, 1921.
70. **Zotta, G.**, Sur la transmission expérimentale du *Leptomonas pyrrhocoris* Z. chez des insectes divers, *C. R. Soc. Biol.,* 85, 135, 1921.
71. **Gibbs, A. J.**, *Leptomonas serpens* n. sp. parasitic in the digestive tract and salivary glands of *Nezara viridula* (Pentatomidae) and in the sap of *Solanum lycopersicum* (tomato) and other plants, *Parasitology,* 47, 297, 1957,
72. **Wallace, F. G.**, *Leptomonas seymouri* sp. n. from the cotton stainer *Dysdercus suturellus, J. Protozool.,* 24, 483, 1977.
73. **Chatton, E. and Alilaire, E.**, Coexistence d'un *Leptomonas (Herpetomonas)* et d'un *Trypanosoma* chez un muscide non vulnérant, *Drosophila confusa* Staeger, *C. R. Soc. Biol.,* 64, 1004, 1908.
74. **Chatton, E.**, Position systématique et signification phylogénique des trypanosomes malpighiens des muscides. Le genre *Rhynchoidomonas* Patton, *C. R. Soc. Biol.,* 74, 551, 1913.
75. **Patton, W. S.**, *Rhynchomonas luciliae,* nov. gen., nov. spec. A new flagellate parasitic in the malpighian tubes of *Lucilia serenissima* Walk., *Bull. Soc. Pathol. Exot.,* 3, 300 and note of correction, 433, 1910.
76. **Patton, W. S.**, Studies on the flagellates of the genera *Herpetomonas, Crithidia* and *Rhynchoidomonas* No. 3. The morphology and life history of *Rhynchoidomonas siphunculinae* sp. nov., parasitic in the malpighian tubes of *Siphunculina funicola* de Meijere, *Indian J. Med. Res.,* 8, 603, 1921.
76a. **Purrini, K. and Halperin, J.**, Studies on some protozoan parasites of the bark beetle (Scolytidae, Coleoptera) in Israel, *Boll. Lab. Agric. Fillippo Silvestri,* 40, 61, 1983.
77. **King, R. L. and Taylor, A. B.**, *Malpighamoeba locustae,* n. sp. (Amoebidae), a protozoan parasitic in the malpighian tubes of grasshoppers, *Trans. Am. Microsc. Soc.,* 55, 6, 1936.
78. **Stoll, N. R., Dollfus, R. P., Forest, J., Niley, N. D., Sabrosky, C. W., Wright, C. V., and Melville, R. V.**, Editorial committee, *International Code of Zoological Nomenclature Adopted by the XV International Congress of Zoology,* Int. Trust Zool. Nomenature, London, 1964.
79. **Harry, O. G. and Finlayson, L. H.**, The life-cycle, ultrastructure and mode of feeding of the locust amoeba *Malpighamoeba locustae, Parasitology,* 72, 127, 1976.
80. **Purrini, K.**, *Malamoeba scolyti* sp. n. (Amoebidae, Rhizopoda, Protozoa) parasitizing the bark beetles, *Drycoetes autographus* Ratz., and *Hylurgops palliatus* Gyll. (Scolytidae, Coleoptera), *Arch. Protistenk.,* 123, 358, 1980.
81. **Porter, A.**, *Amoeba chironomi,* nov. sp., parasitic in the alimentary trace of the larva of *Chironomus, Parasitology,* 2, 32, 1909.
82. **Keilin, D.**, Une nouvelle entamibe, *Entamoeba mesnili* n. sp., parasitic intestinale d'une larve d'un diptère, *C. R. Soc. Biol.,* 80, 133, 1917.
83. **Bishop, A. and Tate, P.**, The morphology and systematic position of *Dobellina mesnili* nov. gen. (*Entamoeba mesnili* Keilin, 1917), *Parasitology,* 31, 501, 1939.
84. **Henry, J. E.**, *Malameba locustae* and its antibiotic control in grasshopper cultures, *J. Invertebr. Pathol.,* 11, 224, 1968.
85. **Evans, W. A. and Elias, R. G.**, The life cycle of *Malamoeba locustae* (King et Taylor) in *Locusta migratoria migratoides* (R. et F.), *Acta Protozool.,* 7, 229, 1970.
86. **Prell, H.**, The amoeba-disease of adult bees: a little-noticed springtime disease, *Bee World,* 8, 10, 1926.
87. **Prell, H.**, Beiträge zur Kenntniss einer Amöbenseuche der Honigbiene, *Z. Angew. Entomol.,* 12, 163, 1927.
88. **Minchin, H. A.**, On some parasites observed in the rat-flea *(Ceratophyllus fasciatus), Festschr. 60, Geburstag R. Hertwigs,* 1, 289, 1910.
89. **Levine, N. D.**, Checklist of the species of the aseptate gregarine families Aikinetocystidae, Diplocystidae, Allanocystidae, Schaudinnellidae, Ganymedidae, and Enterocystidae, *J. Invertebr. Pathol.,* 29, 175, 1977.
90. **Watson, M. E.**, Studies on gregarines, *Ill. Biol. Monogr.,* 2, 215, 1916.
91. **Kamm, M. W.**, Studies on gregarines II, *Ill. Biol. Monogr.,* 7, 1, 1922.
92. **Levine, N. D.**, Taxonomy of the Archigregarinorida and Selenidiidae (Protozoa, Apicomplexa), *J. Protozool.,* 18, 704, 1971.
93. **Levine, N. D.**, Revision and checklist of the species of the aseptate gregarine genus *Lecudina, Trans. Am. Microsc. Soc.,* 95, 695, 1976.
94. **Levine, N. D.**, Revision and checklist of the species (other than *Lecudina*) of the aseptate gregarine family Lecudinidae, *J. Protozool.,* 24, 41, 1977.
95. **Levine, N. D.**, Checklist of the species of the aseptate gregarine family Urosporidae, *Int. J. Parasitol.,* 7, 101, 1977.

96. **Levine, N. D.**, Revision and check-list of the species of the aseptate gregarine family Monocystidae, *Folia Parasitol.*, 24, 1, 1977.
97. **Levine, N. D.**, New genera and higher taxa of septate gregarines (Protozoa, Apicomplexa), *J. Protozool.*, 26, 532, 1979.
98. **Grassé, P. P.**, Classe des grégarinomorphes (Gregarinomorpha n. nov., Gregarinae Haeckel, 1866; Gregarinidea Lankester, 1885; grégarines des anteurs), in *Traité de Zoologie*, Vol. 1, Grassé, P. P., Ed., Masson et Cie, Paris, 1953, 550.
99. **Manwell, R. D.**, Gregarines and haemogregarines, in *Parasitic Protozoa*, Vol. 3, Kreier, J. P., Ed., Academic Press, New York, 1977, 1.
100. **Weiser, J.**, A new classification of the Schizogregarina, *J. Protozool.*, 2, 6, 1955.
101. **Weiser, J.**, A new classification of the Schizogregarina: a correction, *J. Protozool.*, 2, 88, 1955.
102. **Ashford, R. W.**, *Lymphotropha tribolii* gen. nov., sp. nov., Neogregarinida, Schizocystidae, from the haemocoele of *Tribolium castaneum* (Herbst), *J. Protozool.*, 12, 609, 1965.
103. **Ormières, R.**, Une grégarine paradoxale, *Gigaductus anchi* Tuz. Orm., 1966: ultrastructure des stages schizogoniques et position systématique des Gigaductidae Filipponi 1948, *Protistologica*, 7, 261, 1971.
104. **Weiser, J.**, Schizogregariny z hmyzer skodiciho zasobam mouky II. *Mattesia dispora* Naville 1930 a *Coelogregarina ephestiae* Ghelelovitch 1947, *Vestn. Cesk. Spol. Zool.*, 17, 73, 1954.
105. **Weiser, J.**, Zur systematischer Stellung der Schizogregariner der Mehlmotte, *Ephestia kühniella* Z., *Arch. Protistenk.*, 100, 127, 1954.
106. **Canning, E. U.**, Observations on the life history of *Mattesia trogodermae* sp. n., a schizogregarine parasite of the fat body of the Khapra beetle, *Trogoderma granarium* Everts, *J. Insect Pathol.*, 6, 305, 1964.
107. **Ghélélovitch, S.**, *Coelogregarina ephestiae*, schizogrégarine parasite d' *Ephestia kühniella* Z. (Lépidoptére), *Arch. Zool. Exp. Gen.*, 85, 155, 1948.
108. **Weiser, J. and Briggs, J. D.**, Identification of pathogens, in *Microbial Control of Insects and Mites*, Burges, H. D. and Hussey, N. W., Eds. Academic Press, New York, 1971, 13.
109. **Lien, S. M. and Levine, N. D.**, Three new species of *Ascocystis* (Apicomplexa, Lecudinidae) from mosquitoes, *J. Protozool.*, 27, 147, 1980.
110. **Vávra, J.**, *Lankesteria barretti* n. sp. (Eugregarinida, Diplocystidae), a parasite of the mosquito *Aedes triseriatus* (Say) and a review of the genus *Lankesteria* Mingazzini, *J. Protozool.*, 16, 546, 1969.
110a. **Purrini, K.**, *Ascoystis brachyceri* n. sp. (Lecudinidae, Eugregarinida, Sporozoa) parasitizing *Megaselia subnitida* Lundbeck (Phoridae, Brachycera, Diptera), *Arch. Protistenk.*, 123, 192, 1980.
111. **Adler, W. and Mayrink, W.**, A gregarine, *Monocystis chagasi* n. sp., of *Phlebotomus longipalpis*. Remarks on the accessory glands of *P. longipalpis*, *Rev. Inst. Med. Trop. Sao Paulo*, 3, 230, 1961.
112. **Sanders, R. D. and Poinar, G. O.**, Fine structure and life cycle of *Lankesteria clarki* sp. n. (Sporozoa: Eugregarinida) parasitic in the mosquito *Aedes sierrensis* (Ludlow), *J. Protozool.*, 20, 594, 1973.
113. **Ross, R.**, Some observations on the crescent-sphere-flagella metamorphosis of the malaria-parasite within the mosquito, *Trans. South Indian Branch Br. Med. Assoc.*, 6, 334, 1895.
114. **Garnham, P. C. C.**, The zoogeography of *Polychromophilus* and description of a new species of a gregarine (*Lankesteria galliardi*), *Ann. Parasitol.*, 48, 231, 1973.
115. **Blanchard, L. F.**, Grégarine coelomique chez un coleoptère, *C. R. Acad. Sci.*, 135, 1123, 1902.
116. **Shortt, H. E. and Swaminath, C. S.**, *Monocystis mackiei* n. sp. parasitic in *Phlebotomus argentipes* Ann. and Brun., *Indian J. Med. Res.*, 15, 539, 1927.
117. **Guenther, K.**, Üeber eine Gregarine in *Ficalbia dofleini* Guenther, *Zool. Anz.*, 44, 264, 1914.
118. **Keilin, D.**, Une nouvelle schizogrégarine *Caulleryella aphiochaetae*, n.g., n. sp., parasite intestinal d'une larve d'un diptère cyclorhaphe (*Aphiochaeta rufipes* Meig.), *C. R. Soc. Biol.*, 76, 768, 1914.
119. **Bresslau, E. and Buschkiel, M.**, Parasiten der Stechmückenlarven (IV. Mitteilung der Beitrage zur Kenntnis der Libensweise unserer Stechmücken), *Biol. Zentrabl.*, 39, 325, 1919.
120. **Dasgupta, B.**, A new schizogregarine, *Mattesia orchopiae* n. sp. in a flea of squirrels in England, *Parasitology*, 48, 375, 1958.
121. **Weiser, J.**, Schizogregariny z hmyzu skodiciho zasobam mouky, I, *Vestn. Cesk. Spol. Zool.*, 16, 199, 1953.
122. **Dederichs, P. J. and Scholtysick, E.**, New gregarines of the genus *Gigaductus* from Carabidae, *Acta Protozool.*, 16, 5, 1977.
123. **Tuzet, O. and Ormières, R.**, *Gigaductus anchi* n. sp., grégarine parasite d'*Anchus ruficornis* Goeze (Coleoptera Caraboidea) et le problème des Gigaductidae, *Protistologica*, 2, 43, 1966.
124. **Massot, M. and Ormières, R.**, *Gigaductus steropi* n. sp., néogrégarine parasite de *Steropus madidus* Fabr. (Coléopt. Caraboidea) ultrastructure des stages intracellulaires, *Protistologica*, 15, 495, 1979.
125. **Grell, K. G.**, Untersuchungen an Schizogregarinen. I. *Lipocystis polyspora* n. g., n. sp., eine neue Schizogregarine aus dem Fettkörper von *Panorpa communis* L., *Arch. Protistenk.*, 91, 526, 1938.
126. **Keilin, D.**, The structure and life history of *Lipotropha* n.g., a new type of schizogregarine parasitic in the fat body of a dipterous larva *(Systenus)*, *Proc. Cambridge Phil. Soc. Biol. Sci.*, 1, 18, 1923.

127. **Reichenow, E.**, *Machadoella triatomae* n.g., n. sp., eine Schizogregarine aus *Triatoma dimidiata*, *Arch. Protistenk.*, 84, 431, 1935.
128. **Naville, A.**, Recherches cytologiques sur les Schizogregarines, I. Le cycle évolutif de *Mattesia dispora* n. g., n. sp., *Z. Zellforsch. Mikrosk. Anat.*, 11, 375, 1930.
129. **McLaughlin, R. E.**, *Mattesia grandis* n. sp., a sporozoan pathogen of the boll weevil, *Anthonomus grandis* Boheman, *J. Protozool.*, 12, 405, 1965.
130. **McLaughlin, R. E.**, Some relationships between the boll weevil, *Anthonomus grandis* Boheman, and *Mattesia grandis* McLaughlin (Protozoa: Neogregarinida), *J. Invertebr. Pathol.*, 7, 464, 1965.
131. **Vávra, J. and McLaughlin, R. E.**, The fine structure of some developmental stages of *Mattesia grandis* McLaughlin (Sporozoa, Neogregarinida), a parasite of the boll weevil *Anthonomus grandis* Boheman, *J. Protozool.*, 17, 483, 1970.
132. **Nara, J. M.**, Life History and Spore Production of *Mattesia trogodermae* Canning (Neogregarinida: Ophryocystidae) a Pathogen of *Trogoderma glabrum* (Herbst) (Coleoptera: Dermestidae), Ph.D. thesis, University of Wisconsin, Madison, 1975.
133. **Weiser, J.**, Prispevek k znalosti cizopasniku kurovce *Ips typographus*. II, *Vestn. Cesk. Spol. Zool.*, 19, 374, 1955.
134. **Weiser, J.**, Three new pathogens of the Douglas fir beetle, *Dendroctonus pseudotsugae*: *Nosema dentroctoni* n. sp., *Ophryocystis dendroctoni* n. sp., and *Chytridiopsis typographi* n. comb., *J. Invertebr. Pathol.*, 16, 436, 1970.
135. **McLaughlin, R. E. and Myers, J.**, *Ophryocystis elektroscirrha* sp. n., a neogregarine pathogen of the monarch butterfly *Danaus plexippus* (L.) and the Florida queen butterfly *D. glippus berenice* Cramer, *J. Protozool.*, 17, 300, 1970.
136. **Léger, L.**, Sur un nouveau sporozoaire des larves de Diptères, *Schizocystis*, *C. R. Acad. Sci.*, 131, 722, 1900.
137. **Keilin, D.**, On a new schizogregarine, *Schizocystis legeri* n. sp., an intestinal parasite of dipterous larvae (*Systenus*), *Parasitology*, 15, 103, 1923.
138. **Weiser, J.**, Vyvoj schizogregariny *Syncystis mirabilis* (A. Schneider) ve splestuli blative, *Cesk. Parasitol.*, 2, 181, 1955.
139. **Yarwood, E. E.**, The life cycle of *Adelina cryptocerci*, a coccidian parasite of the roach *Cryptocercus punctulatus*, *Parasitology*, 29, 370, 1937.
139a. **Larsson, R.**, Ultrastructural study and description of *Klossia aphodii* n. sp. (Apicomplexa, Adeleidae), a coccidian parasite of *Aphodius fimetarius* and *A. contaminatus* (Coleoptera: Scarabaeidae) in Sweden, *Zool. Anz.*, 214, 241, 1985.
140. **Weiser, J. and Beard, R.**, *Adelina sericesthis* n. sp., a new coccidian parasite of scarabaeid larvae, *J. Insect Pathol.*, 1, 99, 1959.
141. **Carini, A.**, Sobre uma *Barrouxia* parasito do tubo intestinal de Hemipteros do genero *Belostoma*, *Arch. Biol. Sao Paolo*, 26, 251, 1942.
142. **Reyer, W.**, Infektionsversuche mit *Barrouxia* Schneider an *Lithobius forficatus*, insbesondere zur Frage der Sexualität der Coccidien-Sporozoiten, *Z. Parasitenk.*, 9, 478, 1937.
143. **Machado, A.**, Citolojia e ciclo evolutivo da *Chagasella alydi*, moco coccidio parazito d'um Hemiptero do genero *Alydus*, *Mem. Inst. Oswaldo Cruz*, 5, 32, 1913.
143a. **Chagas, C.**, Cytologische Studien über *Adelea hartmanni*, ein neues Coccidium aus dem Darm von *Dysdercus ruficollis* L., *Mem. Inst. Oswaldo Cruz*, 2, 168, 1910.
144. **Ludwig, F. W.**, Studies on the protozoan fauna of the larva of the crane fly, *Tipula abdominalis*, II. The life history of *Ithania wenrichi* n. g., n. sp., a coccidian, found in the caeca and mid-gut and a diagnosis of Ithaiinae n. subfam., *Trans. Am. Microsc. Soc.*, 45, 22, 1947.
145. **Vincent, M.**, On *Legerella hydropori* n. sp., a coccidian parasite of the malpighian tubules of *Hydroporus palustris* L. (Col.), *Parasitology*, 19, 394, 1927.
146. **Nöller, W.**, Die Übertragungsweise der Rattentrypanosomen, *Arch. Protistenk.*, 34, 295, 1914.
147. **Beesley, J. E.**, The life cycle of *Rasajeyna nannyla* n. gen., n. sp., a coccidian pathogen of *Tipula paludosa* Meigen, *Parasitology*, 74, 273, 1977.
148. **Sprague, V.**, Classification and phylogeny of the Microsporidia, in *Comparative Pathobiology*, Vol. 2, Bulla, L. A. and Cheng, T. C., Eds., Plenum Press, New York, 1977, 1.
149. **Weiser, J.**, Contribution to the classification of microsporidia, *Vestn. Cesk. Spol. Zool.*, 41, 308, 1977.
149a. **Weiser, J.**, Phylum Microspora Sprague, 1969, in *Illustrated Guide to the Protozoa*, Lee, J. J., Hutner, S. H., and Bovee, E. C., Eds., Soc. Protozoologists, Allen Press, Lawrence, Kan., 1985, 375.
149b. **Weiser, J.**, Taxonomy of the order Chytridiopsida (Microsporidia), *J. Protozool.*, 30, 75A, 1983.
149c. **Issi, I. V.**, Microsporidia as a phylum of parasitic protozoa, *Acad. Sci. U.S.S.R. Soc. Protozool. Ser. Protozool.*, 10, 6, 1986.
149d. **Larsson, R.**, Ultrastructural investigation of two microsporidia with rod-shaped spores, with descriptions of *Cylindrospora fasciculata* sp. nov. and *Resiomeria odonatae* fen. et sp. nov. (Microspora, Thelohaniidae), *Protistologica*, 22, 379, 1986.

149e. **Larsson, R.**, Ultrastructure, function, and classification of microsporidia, *Prog. Protistol.*, 1, 325, 1986.
150. **Sprague, V.**, Microspora, in *Synopsis and Classification of Living Organisms*, Vol. 1, Parker, S. P., Ed., McGraw-Hill, New York, 1982, 589.
151. **Hazard, E., Ellis, E. A., and Joslyn, D. J.**, Identification of microsporidia, in *Microbial Control of Pests and Plant Diseases 1970—1980*, Burges, H. D., Ed., Academic Press, New York, 1981, 163.
152. **Sprague, V.**, Annotated list of species of microsporidia, in *Comparative Pathobiology*, Vol. 2, Bulla, L. A. and Cheng, T. C., Eds., Plenum Press, New York, 1977, 31.
153. **Sprague, V.**, Characterization and composition of the genus *Nosema*, *Misc. Publ. Entomol. Soc. Am.*, 11, 5, 1978.
154. **Kellen, W. R. and Lipa, J. J.**, *Thelohania californica* n. sp., a microsporidian parasite of *Culex tarsalis* Coquillett, *J. Insect Pathol.*, 2, 1, 1960.
155. **Hazard, E. I. and Oldacre, S. W.**, Revision of microsporida (Protozoa) close to *Thelohania*, with descriptions of one new family, eight new genera, and thirteen new species, *U.S. Dept. Agric. Tech. Bull.*, 1530, 1, 1975.
155a. **Larsson, R.**, Ultrastructural study and description of *Chapmanium dispersus* n. sp. (Microspora, Thelohaniidae), a microsporidian parasite of *Endochironomus* larvae (Diptera: Chironomidae), *Protistologica*, 20, 547, 1984.
156. **Kellen, W. R. and Wills, W.**, New *Thelohania* from Californian mosquitoes (Nosmatidae: Microsporidia), *J. Insect Pathol.*, 4, 41, 1962.
157. **Kudo, R. R.**, A biologic and taxonomic study of the Microsporidia, *Ill. Biol. Monogr.*, 9, 1, 1924.
158. **Kudo, R. R.**, Studies on microsporidia parasitic in mosquitoes. II. On the effect of the parasites upon the host body, *J. Parasitol.*, 8, 70, 1922.
159. **Weiser, J. and Purrini, K.**, Seven new microsporidian parasites of springtails (Collembola) in the Federal Republic of Germany, *Z. Parasitenk.*, 62, 75, 1980.
159a. **Weiser, J.**, Zur Kenntnis der Mikrosporidien aus Chironomiden-Larven III, *Zool. Anz.*, 170, 226, 1963.
159b. **Larsson, R.**, On the cytology, development and systematic position of *Thelohania asterias* Weiser, 1963, with creation of the new genus *Bohuslavia* (Microspora, Thelohaniidae), *Protistologica*, 21, 235, 1985.
159c. **Maurand, J. and Manier, J. P.**, Une Microsporidie nouvelle pour les larves de Simulies, *Protistologica*, 3, 445, 1967.
160. **Jouvenaz, D. P. and Hazard, E. I.**, New family, genus, and species of microsporida (Protozoa: Microsporida) from the tropical fire ant, *Solenopsis geminata* (Fabricius) (Insecta: Formicidae), *J. Protozool.*, 25, 24, 1978.
161. **Jouvenaz, D. P., Lofgren, C. S., and Allen, G. E.**, Transmission and infectivity of spores of *Burenella dimorpha* (Microsporida: Burenellidae), *J. Invertebr. Pathol.*, 37, 265, 1981.
162. **Larsson, R.**, Insect pathological investigations on Swedish Thysanura II. A new microsporidian parasite of *Petrobius brevistylis* (Microcoryphia, Machilidae); description of the species and creation of two new genera and a new family, *Protistologica*, 16, 85, 1980.
162a. **Issi, I. V., Radishcheva, D. F., and Dolzhenko, V. I.**, Microsporidia of flies of the genus *Delia* (Diptera, Muscidae) harmful to agriculture, *Bull. Vses. Nauchno-Issled. Inst. Zashch. Rast.*, 55, 3, 1983.
163. **Beaudoin, R. and Wills, W.**, A description of *Caudospora pennsylvanica* sp. n. (Caudosporidae, Microsporidia), a parasite of the larvae of the black fly, *Prosimulium magnum* Dyar and Shannon, *J. Invertebr. Pathol.*, 7, 152, 1965.
164. **Jamback, H. A.**, *Caudospora* and *Weiseria*, two genera of microsporidia parasitic in blackflies, *J. Invertebr. Pathol.*, 16, 3, 1970.
165. **Strickland, E. H.**, Some parasites of *Simulium* larvae and their effects on the development of the host, *Biol. Bull.*, 21, 302, 1911.
166. **Maurand, J.**, Les Microsporidies des larves de Simulies: systématique, données cytochimiques, pathologiques et écologiques, *Ann. Parasitol. Hum. Comp.*, 50, 371, 1975.
167. **Weiser, J.**, Studie o microsporidiich z larev hmyzu nasich vod, *Vestn. Cesk. Spol. Zool.*, 10, 245, 1946.
167a. **Larsson, R.**, A revisionary study of the taxon *Tuzetia* Maurand, Fize, Fenwick and Michel, 1971, and related forms (Microspora, Tuzetiidae), *Protistologica*, 19, 323, 1983.
167b. **Loubès, C. and Maurand, J.**, Étude ultrastructurale de *Pleistophora debaisieuxi* Jirovec, 1943 (Microsporida): son transfert dans la genre *Tuzetia* Maurand, Fize, Michel et Fenwick, 1971, et remarques sur la structure et la genèse du filament polaire, *Protistologica*, 12, 577, 1976.
168. **Weiser, J.**, *Caudospora simulii*, n. g., n. sp., Microsporidie parasite des larves de *Simulium*, *Ann. Parasitol. Hum. Comp.*, 22, 11, 1947.
169. **Schneider, A.**, Sur le développment du *Stylorhynchus longicollis*, *Arch. Zool. Exp. Gen.*, 2, 1, 1884.
170. **Manier, J. and Ormières, R.**, Ultrastructure de quelques stades de *Chytridiopsis socius* Schn. parasite de *Blaps lethifera* Marsh. (Coleopt. Tenebr.), *Protistologica*, 4, 181, 1968.
171. **Weiser, J.**, Prispevek k znalosti cizopasniku kurovce *Ips typographus* L., I., *Vestn. Cesk. Spol. Zool.*, 18, 217, 1954.

172. **Weiser, J.**, Zwei Mikrosporidien aus Köcherfliegen-Larven, *Zool. Anz.*, 175, 229, 1965.
173. **Baudoin, J.**, Nouvelles espèces de microsporidies chez des larves de Trichoptères, *Protistologica*, 5, 441, 1969.
174. **Kudo, R. R.**, On the structure of some microsporidan spores, *J. Parasitol.*, 6, 178, 1920.
175. **Hazard, E. I. and Savage, K. E.**, *Stempellia lunata* sp. n. (Microsporida, Nosematidae) in larvae of the mosquito *Culex pilosus* collected in Florida, *J. Invertebr. Pathol.*, 15, 49, 1970.
175a. **Canning, E. U., Barker, R. J., Nicholas, J. P., and Page, A. M.**, The ultrastructure of three microsporidia from winter moth, *Operophtera brumata* (L.), and the establishment of a new genus *Cystosporogenes* n. g. for *Pleistophora operophterae* (Canning, 1960). *Syst. Parasitol.*, 7, 213, 1985.
176. **Voronin, V. N.**, Two new species of microsporidians (Protozoa, Microsporidia) from the mosquitoes of the family Chironomidae, *Parasitologiya (Leningrad)*, 9, 373, 1975.
177. **Kalavati, C. and Narasimhamurti, C. C.**, A microsporidian parasite, *Duboscqia coptotermi* sp. n., from the gut of *Coptotermes heimi* (Wasm.), *J. Parasitol.*, 62, 323, 1976.
178. **Pérez, C.**, Sur *Duboscqia legeri*, microsporidie nouvelle parasite du *Termes lucifugus*, et sur la classification des microsporidies, *C. R. Soc. Biol.*, 65, 631, 1908.
179. **Kudo, R. R.**, On the microsporidian *Duboscqia legeri* Pérez 1908, parasitic in *Reticulitermes flavipes*, *J. Morphol.*, 71, 307, 1942.
179a. **Larsson, R.**, Ultracytology of a tetrasporoblastic microsporidium of the caddisfly *Holocentropus picicornis* (Trichoptera, Polycentropodidae), with description of *Episeptum inversum* gen. et sp. nov. (Microspora, Gurleyidae), *Arch. Protistenkd.*, 131, 257, 1986.
180. **Golberg, A. M.**, Mikrosporidiozy komarov *Culex pipiens* L., *Med. Parasitol. Bolezn.*, 2, 204, 1971.
181. **Fantham, H. B., Porter, A., and Richardson, L. R.**, Some microsporidia found in certain fishes and insects in eastern Canada, *Parasitology*, 33, 186, 1941.
182. **Loubès, C. and Maurand, J.**, Étude ultrastructurale de *Gurleya chironomi* n. sp. Microsporidie parasite des larves d'*Orthocladius* (Diptera-Chironomidae), *Protistologica*, 11, 233, 1975.
183. **Hesse, H.**, Sur une nouvelle Microsporidie tetrasporée du genre *Gurleya*, *C. R. Soc. Biol.*, 55, 495, 1903.
184. **Hazard, E. I. and Fukuda, T.**, *Stempellia milleri* sp. n. (Microsporida: Nosematidae) in the mosquito *Culex pipiens quinquefasciatus* Say, *J. Protozool.*, 21, 497, 1974.
185. **Larsson, R.**, Cytology and taxonomy of *Helmichia aggregata* gen. et sp. nov. (Microspora, Thelohaniidae), a parasite of *Endochironomus* larvae (Diptera, Chironomidae), *Protistologica*, 18, 355, 1982.
186a. **Ormières, R. and Sprague, V.**, A new family, new genus, and new species allied to the microsporida, *J. Invertebr. Pathol.*, 21, 224, 1973.
186b. **Larsson, R.**, Description of *Hyalinocysta expilatoria* n. sp., a microsporidian parasite of the blackfly *Odagmia ornata*, *J. Invertebr. Pathol.*, 42, 348, 1983.
186c. **Issi, I. V. and Pankova, T. F.**, A new microsporidian, *Issia globulifera* n. sp. (Nosematidae), from *Anopheles maculipennis*, *Parazitologiya (Leningrad)*, 17, 189, 1983.
186d. **Batson, B. S.**, A light and electron microscopic study of *Hirsutusporos austrosimulii* gen. n., sp. n., (Microspora: Nosematidae), a parasite of *Austrosimulium* sp. (Diptera: Simuliidae) in New Zealand, *Protistologica*, 19, 263, 1983.
187. **Léger, L. and Hesse, E.**, *Mrazekia*, genre nouveau de microsporidies à spores tubuleuses, *C. R. Soc. Biol.*, 79, 345, 1916.
188. **Codreanu, R.**, On the occurrence of spore or sporont appendages in the microsporidia and their taxonomic significance, in *Proc. 1st Int. Cong. Parasitol.*, Corradetti, A., Ed., Pergamon Press, New York, 1966, 602.
189. **Knell, J.**, *Microsporidium goeldichironomi* n. sp. and *Microsporidium chironomi* n. sp. (Microsporida: Apansporoblastina): two new microsporidia from Florida chironomids, *J. Invertebr. Pathol.*, 37, 129, 1981.
190. **Weiser, J.**, Prispevek k znalosti parasitu prastevnika americkeho (*Hyphantria cunea* Drury), *Vestn. Cesk. Spol. Zool.*, 17, 228, 1953.
191. **Issi, I. V. and Voronin, V. N.**, The contemporary state of the problem on bispore genera of microsporidians, *Parazitologiya (Leningrad)*, 13, 150, 1979.
191a. **Hostounský, Z. and Zizka, Z.**, Defense reactions in an unusual host against microsporidan infections, *J. Protozool.*, 30, 32A, 1983.
192. **Vávra, J. and Undeen, A. H.**, *Nosema algerae* n. sp. (Cnidospora, Microsporida) a pathogen in a laboratory colony of *Anopheles stephensi* Liston (Diptera, Culicidae), *J. Protozool.*, 17, 240, 1970.
193. **Canning, E. U. and Hulls, R. H.**, A microsporidan infection of *Anopheles gambiae* Giles, from Tanzania, interpretation of its mode of transmission and notes on *Nosema* infections in mosquitoes, *J. Protozool.*, 17, 531, 1970.
194. **Canning, E. U. and Sinden, R. E.**, Ultrastructural observations on the development of *Nosema algerae* Vávra and Undeen (Microsporida, Nosematidae) in the mosquito *Anopheles stephensi* Liston, *Protistologica*, 9, 405, 1973.
195. **Zander, E.**, Tierische Parasiten als Krankheitserreger bei der Biene, *Leipz. Bienenztg.*, 24, 147, 1909.

196. **Fantham, H. B. and Porter, A.**, The morphology and life history of *Nosema apis* and the significance of its various stages in the so-called 'Isle of Wight' disease in bees (Microsporidiosis), *Ann. Trop. Med. Parasitol.*, 6, 163, 1912.
197. **Cali, A.**, Morphogenesis in the genus *Nosema*, in *Proc. IVth Int. Colloq. Insect Pathol.*, College Park, Md., 1971, 431.
198. **Youssef, N. N. and Hammond, D. M.**, The fine structure of the developmental stages of the microsporidian *Nosema apis* Zander, *Tissue Cell*, 3, 283, 1971.
199. **Ishihara, R.**, The life cycle of *Nosema bombycis* as revealed in tissue culture cells of *Bombyx mori*, *J. Invertebr. Pathol.*, 14, 316, 1969.
200. **Ishihara, R.**, Fine structure of *Nosema bombycis* (Microsporidia, Nosematidae) developing in the silkworm (*Bombyx mori*)-I., *Bull. Coll. Agric. Vet. Med. Nikon Univ.*, 27, 84, 1970.
201. **Thomson, H. M.**, *Perezia fumiferanae* n. sp., a new species of microsporidia from the spruce budworm *Choristoneura fumiferana* (Clem.), *J. Parasitol.*, 41, 416, 1955.
202. **Wilson, G. G.**, Studies on *Nosema fumiferanae* a Microsporidian Parasite of *Choristoneura fumiferana* (Clem.) (Lepidoptera: Tortricidae), Ph.D. thesis, Cornell University, Ithaca, N.Y., 1972.
203. **McLaughlin, R. E.**, *Glugea gasti* sp. n., a microsporidan pathogen of the boll weevil *Anthonomus grandis*, *J. Protozool.*, 16, 84, 1969.
204. **Streett, D. A., Sprague, V., and Harman, D. M.**, Brief study of microsporidian pathogens in the white pine weevil *Pissodes strobi*, *Chesapeake Sci.*, 16, 32, 1975.
205. **Canning, E. U.**, A new microsporidian, *Nosema locustae* n. sp. from the fat body of the African migratory locust, *Locusta migratoria migratorioides* R. & F., *Parasitology*, 43, 287, 1953.
206. **Canning, E. U.**, The life cycle of *Nosema locustae* Canning in *Locusta migratoria migratorioides* (Reiche and Fairmaire) and its infectivity to other hosts, *J. Insect Pathol.*, 4, 237, 1962.
207. **Canning, E. U.**, The pathogenicity of *Nosema locustae* Canning, *J. Insect Pathol.*, 4, 248, 1962.
208. **Paillot, A.**, Sur deux protozoaires nouveaux parasites des chenilles de *Pyrausta nubilalis* Hb., *C. R. Acad. Sci.*, 185, 673, 1927.
209. **Paillot, A.**, On the natural equilibrium of *Pyrausta nubilalis* Hb., *Int. Corn Borer Invest. Sci. Rep.*, 1, 77, 1928.
210. **Hall, I. M.**, Observations on *Perezia pyraustae* Paillot, a microsporidian parasite of the European corn borer, *J. Parasitol.*, 38, 48, 1952.
211. **Kramer, J. P.**, Studies on the morphology and life history of *Perezia pyraustae* Paillot (Microsporidia: Nosematidae), *Trans. Am. Microsc. Soc.*, 78, 336, 1959.
212. **Weiser, J.**, Die Mikrosporidien als Parasiten der Insekten, *Monogr. Angew. Entomol.*, 17, 1, 1961.
213. **Wenn, C. T.**, On a microsporidian parasite of the European corn borer in China, *Acta Zool. Sinica*, 17, 60, 1965.
213a. **Laing, D. R. and Jaques, R. P.**, Microsporidia of the European corn borer (Lepidoptera: Pyralidae) in southwestern Ontario: natural occurrence and effectiveness as microbial insecticides, *Proc. Entomol. Soc. Ontario*, 115, 13, 1984.
214. **Kramer, J. P.**, *Octosporea carloschagasi* n. sp., a microsporidian associate of *Trypanosoma cruzi* in *Panstrongylus megistus*, *Z. Parasitenk.*, 39, 221, 1972.
215. **Flu, P. C.**, Studien über die im Darm der Stubenfliege, *Musca domestica*, vorkommenden protozoären Gebilde, *Zentralbl. Bakteriol. Parasitenk. Infektionskr. Hyg. Abt. I Orig.*, 57, 522, 1911.
216a. **Kramer, J. P.**, The microsporidian *Octosporea muscaedomesticae* Flu, a parasite of calypterate muscoid flies in Illinois, *J. Insect Pathol.*, 6, 331, 1964.
216b. **Canning, E. U.**, Two new microsporidian parasites of the winter moth, *Operophtera brumata* (L.), *J. Parasitol.*, 46, 755, 1960.
216c. **Canning, E. U., Wigley, P. J., and Barker, R. J.**, The taxonomy of three species of microsporidia (Protozoa: Microspora) from an oakwood population of winter moths *Operophtera brumata* (L.) (Lepidoptera: Geometridae), *Syst. Parasitol.*, 5, 147, 1983.
217. **Hazard, E. I. and Anthony, D. W.**, A redescription of the genus *Parathelohania* Codreanu 1966 (Microsporida: Protozoa) with a reexamination of previously described species of *Thelohania* Henneguy 1892 and descriptions of two new species of *Parathelohania* from anopheline mosquitoes, *U.S. Dept. Agric. Tech. Bull.*, 1505, 1, 1974.
218. **Hesse, E.**, *Thelohania legeri* n. sp., microsporidie nouvelle, parasite des larves d' *Anopheles maculipennis* Meig., *C. R. Soc. Biol.*, 57, 570, 1904.
219. **Hesse, E.**, Sur le développement de *Thelohania legeri* Hesse, *C. R. Soc. Biol.*, 57, 571, 1904.
220. **Kudo, R. R.**, Studies on microsporidia parasitic in mosquitoes. IV. Observations upon the microsporidia found in the mosquitoes of Georgia, U.S.A., *Zentralbl. Bakteriol. Parasitenk. Infektionskr. Hyg. Abt. I. Orig.*, 96, 428, 1925.
221. **Steinhaus, E. A. and Hughes, K. M.**, Two newly described species of microsporidia from the potato tuberworm, *Gnorimoschema operculella* (Zeller) (Lepidoptera, Gelechiidae), *J. Parasitol.*, 35, 67, 1949.

222. **Sprague, V. and Ramsey, J.**, A preliminary note on *Plistophora kudoi* n. sp., a microsporidian parasite of the cockroach, *Anat. Rec.* 81, 132, 1941.
223. **Sprague, V. and Ramsey, J.**, Further observations on *Plistophora kudoi*, a microsporidian of the cockroach, *J. Parasitol.*, 28, 399, 1942.
224. **Zwölfer, W.**, Die Pebrine des Schwammspinners (*Porthetria dispar* L.) und Goldafters (*Nygmia phaeorrhoea* Don. = *Euproctis chrysorrhoea* L.) eine neue wirtschaftlich bedeutungsvolle Infektionskrankheit, *Dtsch. Ges. Angew. Entomol.*, 6, 98, 1927.
224a. **Milner, R. J. and Briese, D. T.**, Identification of the microsporidian *Pleistophora schubergi* infecting *Anaitis efformata* (Lepidoptera: Geometridae), *J. Invertebr. Pathol.*, 48, 100, 1986.
225. **Zwölfer, W.**, Die Pebrine des Schwammspinners und Goldafters, eine neue wirtschaftlich bedeutungsvolle Infektionskrankheit, *Z. Angew. Entomol.*, 12, 498, 1927.
226. **Lipa, J. J.**, Studia inwazjologiczne i epizootiologiczne nad kilkoma gatunkami peirwotniaków z rzedu Microsporidia pasózytaujoacymi w owadach, *Pr. Nauk Inst. Ochr. Rosl.*, 5, 103, 1963.
227. **Wilson, G. G.**, Occurrence of *Thelohania* sp. and *Pleistophora* sp. (Microsporida: Nosematidae) in *Choristoneura fumiferana* (Lepidoptera: Tortricidae), *Can. J. Zool.*, 53, 1799, 1975.
227a. **Maurand, J. and Loubès, C.**, Les microsporidies des larves de simulies: données ultrastructurales, *Z. Parasitenkd.*, 56, 131, 1978.
228. **Léger, L. and Hesse, E.**, Cnidosporidies des larves d'Éphemérès, *C. R. Acad. Sci.*, 150, 411, 1910.
229. **Desportes, I.**, Ultrastructure de *Stempellia mutabilis* Léger et Hesse, microsporidie parasite de l'émphémère *Ephemera vulgata* L., *Protistologica*, 12, 121, 1976.
230. **Codreanu, R. and Vávra, J.**, The structure and ultrastructure of the microsporidan *Telomyxa glugeiformis* Léger and Hesse, 1910, parasite of *Ephemera danica* (Müll.) nymphs, *J. Protozool.*, 17, 374, 1970.
230a. **Hazard, E. I. and Federici, B. A.**, Ultrastructure and description of a new species of *Telomyxa* (Microspora: Telomyxidae) from the semiaquatic beetle, *Ora texana* Champ. (Coleoptera: Helodidae), *J. Protozool.*, 32, 189, 1985.
231. **Strickland, E. H.**, Further observations on the parasites of *Simulium* larvae, *J. Morphol.*, 24, 43, 1913.
232. **Debaisieux, P. and Gastaldi, L.**, Les microsporidies parasites des larves de *Simulium* II, *Cellule*, 30, 187, 1919.
233. **Smirnoff, W. A.**, *Thelohania pristiphorae* sp. n., a microsporidian parasite of the larch sawfly, *Pristiphora erichsonii* (Hymenoptera: Tenthredinidae), *J. Invertebr. Pathol.*, 8, 360, 1966.
234. **Debaisieux, P.**, Deux microsporidies nouvelle de larves de *Chironomus*, *C. R. Soc. Biol.*, 107, 913, 1931.
235. **Jírovec, O.**, Studien über Microsporidien, *Vestn. Cesk. Spol. Zool.*, 4, 1, 1936.
236. **Larsson, R.**, Ultrastructural study of *Toxoglugea variabilis* n. sp. (Microspora: Thelohaniidae), a microsporidian parasite of the biting midge *Bezzia* sp. (Diptera: Ceratopogonidae), *Protistologica*, 16, 17, 1980.
237. **Léger, L. and Hesse, E.**, Microsporidies bactériformes et essai de systématique du groupe, *C. R. Acad. Sci.*, 174, 327, 1922.
238. **Léger, L. and Hesse, E.**, Microsporidies nouvelles parasites des animaux d'eau douce, *Trav. Lab. Hydrobiol. Pisc. Univ. Grenoble*, 14, 49, 1924.
239. **Léger, L.**, Une microsporidie nouvelle à sporontes épineux, *C. R. Acad. Sci.*, 182, 727, 1926.
240. **Léger, L.**, Sur "*Trichoduboscqia epeori*" Léger, microsporidie parasite des larvae d'Éphémèrides, *Trav. Lab. Hydrobiol. Pisc. Univ. Grenoble*, 18, 9, 1926.
241. **Batson, B. S.**, A light and electron microscopical study of *Trichoduboscqia epeori* Léger (Microspora:Duboscqiidae), *J. Protozool.*, 29, 202, 1982.
242. **Codreanu, R. and Codreanu-Balcesu, D.**, Caractérestiques ultrastructurales d'une Microsporidie nouvelle du genre *Tuzetia* parasite chez *Ecdyonurus* (Ephemeroptera), *J. Protozool.*, 29, 300, 1982.
243. **Knell, J. D. and Allen, G. E.**, Morphology and ultrastructure of *Unikaryon minutum* sp. n. (Microsporida: Protozoa), a parasite of the southern pine beetle, *Dendroctonus frontalis*, *Acta Protozool.*, 17, 271, 1978.
243a. **Toguebaye, B. S. and Marchand, B.**, Développement d'une microsporidie du genre *Unikaryon* Canning, Lai et Lie, 1974 chez un Coléoptère Chrysomelidae, *Euryope rubra* (Latreille, 1807): étude ultrastructurale, *Protistologica*, 19, 371, 1983.
244. **Kramer, J. P.**, *Nosema necatrix* sp. n. and *Thelohania diazoma* sp. n., microsporidians from the armyworm *Pseudaletia unipuncta* (Haworth), *J. Invertebr. Pathol.*, 7, 117, 1965.
245. **Pilley, B. M.**, A new genus, *Vairimorpha* (Protozoa: Microsporida), for *Nosema necatrix* Kramer 1965: pathogenicity and life cycle in *Spodoptera exempta* (Lepidoptera: Noctuidae), *J. Invertebr. Pathol.*, 28, 177, 1976.
246. **Kellen, W. R. and Lindegren, J. E.**, Biology of *Nosema plodiae* sp. n., a microsporidian pathogen of the Indian-meal moth, *Plodia interpunctella* (Hübner), (Lepidoptera: Phycitidae), *J. Invertebr. Pathol.*, 11, 104, 1968.
247. **Maddox, J. V. and Sprenkel, R. K.**, Some enigmatic microsporidia of the genus *Nosema*, *Misc. Publ. Entomol. Soc. Am.*, 11, 65, 1978.

286c. **Henry, J. E. and Onsager, J. A.**, Experimental control of the Mormon cricket, *Anabrus simplex*, by *Nosema locustae* (Microspora: Microsporida), a protozoan parasite of grasshoppers (Ort.: Acrididae), *Entomophaga*, 27, 197, 1982.
286d. **Johnson, D. L. and Henry, J. E.**, Low rates of insecticides and *Nosema locustae* (Microsporidia: Nosematidae) on baits applied to roadside for grasshopper (Orthoptera: Acrididae) control., *J. Econ. Entomol.*, 80, 685, 1987.
287. **Ernst, H. P. and Baker, G. L.**, *Malameba locustae* (King and Taylor) (Protozoa: Amoebidae) in field populations of Orthoptera in Australia, *J. Aust. Entomol. Soc.*, 21, 295, 1982.
288. **Ross, R.**, Report on a preliminary investigation into malaria in the Sigur Ghat, Ootacamund, *Indian Med. Gaz.*, 33, 133, 1898.
289. **Ross, R.**, Notes on the parasites of mosquitoes found in India between 1895 and 1899, *J. Hyg.*, 6, 101, 1906.
290. **Manson, P.**, On the life-history of the malaria germ outside the human body, The Goulstonian Lectures, No. 2. *Br. Med. J.*, 1, 712, 1896.
291. **Wenyon, C. M.**, Oriental sore in Baghdad, together with observations on a gregarine in *Stegomyia fasciata*, the haemogregarine of dogs and the flagellates of house flies, *Parasitology*, 4, 273, 1911.
292. **Ward, R. A., Levine, N. D., and Craig, G. B.**, *Ascogregarina* nom. nov. for *Ascocystis* Grassé, 1953 (Apicomplexa, Eugregarinorida), *J. Parasitol.*, 68, 331, 1982.
293. **Ganapati, P. N. and Tate, P.**, On the gregarine *Lankesteria culicis* (Ross), 1898, from the mosquito *Aëdes* (Finlaya) *geniculatus* (Olivier), *Parasitology*, 39, 291, 1949.
294. **Kramar, J.**, Hromadinka *Lankesteria culicis* Ross, parasit komara *Aëdes* (Finlaya) *geniculatus* Oliv., *Vestn. Cesk. Spol. Zool.*, 16, 43, 1952.
295. **Walsh, R. D. and Callaway, C. S.**, The fine structure of the gregarine *Lankesteria culicis* parasitic in the yellow fever mosquito *Aedes aegypti*, *J. Protozool.*, 16, 536, 1969.
296. **McCray, E. M., Fay, R. W., and Schoof, H. F.**, The bionomics of *Lankesteria culicis* and *Aedes aegypti*, *J. Invertebr. Pathol.*, 16, 42, 1970.
297. **Sheffield, H. G., Garnham, P. C. G., and Shiroishi, T.**, The fine structure of the sporozoite of *Lankesteria culicis*, *J. Protozool.*, 18, 98, 1971.
298. **Hati, A. K. and Ghosh, S. M.**, On the gregarine (*Lankesteria culicis* Ross) infection in *Aedes aegypti* mosquito in Calcutta, *Bull. Calcutta Sch. Trop. Med.*, 11, 7, 1963.
299. **Marchoux, E., Salimbeni, A., and Simond, P. L.**, La fiévre jaune. Rapport de la mission francaise, *Ann. Inst. Pasteur*, 17, 712, 1903.
300. **Stevenson, A. C. and Wenyon, C. M.**, Note on the occurrence of *Lankesteria culicis* in West Africa, *J. Trop. Med. Hyg.*, 18, 196, 1915.
301. **Bacot, A. W.**, Report of the entomological investigation undertaken for the Yellow Fever (West Africa) Commission for the year 1914—1915. IV, *Parasites*, 3, 1, 1916.
302. **Macfie, J. W. S.**, The identification of insects collected at Accra during the year 1916, and other entomological notes, Abstract in *Rev. Appl. Entomol.*, 6, 16, 1917.
303. **Sellards, A. W. and Siler, J. F.**, The occurrence of rickettsia in mosquitoes *(Aedes aegypti)* infected with the virus of dengue fever, *Am. J. Trop. Med.*, 8, 299, 1928.
304. **Laird, M.**, Parasites of Singapore mosquitoes, with particular reference to the significance of larval epibionts as an index of habitat population, *Ecology*, 40, 206, 1959.
305. **Dzerzhinsky, V. A., Nam, E. A., and Dubitsky, A. M.**, The finding of *Lankesteria culicis* and *Tetrahymena stegomyiae* in larvae of *Aedes aegypti*, *Parazitologiya*, 10, 381, 1976.
305a. **Kilochitskii, P. Y. and Bryginskii, S. A.**, Finding of *Lankesteria culicis* Ross (Sporozoa, Diplocystidae) in blood-sucking mosquitoes of the Ukraine, *Vestn. Zool.*, 1979, 87, 1979.
306. **Barrett, W. L.**, Damage caused by *Lankesteria culicis* (Ross) to *Aedes aegypti* (L.), *Mosq. News*, 28, 441, 1968.
307. **Gentile, A. G., Fay, R. W., and McCray, E. M.**, The distribution, ethology and control potential of the *Lankesteria culicis* (Ross)-*Aedes aegypti* (L.) complex in southern United States, *Mosq. News*, 31, 12, 1971.
308. **Stapp, R. R. and Casten, J.**, Field studies on *Lankesteria culicis* and *Aedes aegypti* in Florida, *Mosq. News*, 31, 18, 1971.
309. **Hayes, G. R. and Haverfield, L. E.**, Distribution and density of *Aedes aegypti* (L.) and *Lankesteria culicis* (Ross) in Louisiana and adjoining areas, *Mosq. News*, 31, 28, 1971.
310. **Barrett, W. L., Miller, F. M., and Kliewer, J. W.**, Distribution in Texas of *Lankesteria culicis* (Ross), a parasite of *Aedes aegypti*, *Mosq. News*, 31, 23, 1971.
311. **Walsh, R. D. and Olson, J. K.**, Observations on the susceptibility of certain culicine mosquito species to infection by *Lankesteria culicis* (Ross), *Mosq. News*, 36, 154, 1976.
312. **Feng, L. C.**, *Lankesteria culicis* a parasite of *Aedes koreicus*, *Ann. Trop. Med. Parasitol.*, 24, 361, 1930.
313. **Feng, L. C.**, Some parasites of mosquitoes and flies found in China, *Lingnan Sci. J.*, 12(Suppl.), 23, 1933.

313a. **Yu, H. S., Cho, H. W., and Pillai, J. S.**, Microbial pathogens of mosquitoes infected with protozoan parasites *Tetrahymena* and *Lankesteria, Korean J. Entomol.*, 9, 86, 1979.
314. **Ray, H.**, On the gregarine, *Lankesteria culicis* (Ross), in the mosquito, *Aedes (Stegomyia) albopictus* Skuse., *Parasitology*, 25, 392, 1933.
315. **Mathis, M. and Baffet, O.**, Purification d'un élevage d'*Aëdes argenteus* parasité par une Grégarine, *Lankesteria culicis* (Ross), *Bull. Soc. Pathol. Exot.*, 5, 435, 1934.
316. **Villacarlos, L. T. and Gabriel, B. P.**, Some microbial pathogens of four species of mosquitoes, *Kalikasan Phillip, J. Biol.*, 3, 1, 1974.
317. **Pillai, J. S., Neill, H. J. C., and Sone, P. F.**, *Lankesteria culicis* a gregarine parasite of *Aedes polynesiensis* in Western Samoa, *Mosq. News*, 36, 150, 1976.
318. **Laird, M.**, Gregarine and microsporidan protozoa in *Aedes polynesiensis*, Tokelu Islands. Recent accidental importations?, *Can. J. Zool.*, 60, 1922, 1982.
319. **Savage, K. E.**, *Nosema algerae* Vávra and Undeen, 1970 (Protozoa: Microsporida): its Bionomics and Development for Use as a Biological Control Agent of Mosquitoes, M.S. thesis, University of Florida, Gainesville, 1975.
320. **Wilson, G. G.**, *Nosema fumiferanae*, a natural pathogen of a forest pest: potential for pest management, in *Microbial Control of Insects, Mites, and Plant Diseases 1970—1980*, Burges, H. D., Ed., Academic Press, New York, 1981, 595.
321. **Henry, J. E. and Oma, E. A.**, Pest control by *Nosema locustae*, a pathogen of grasshoppers and crickets, in *Microbial Control of Pests and Plant Diseases 1970—1980*, Burges, H. D., Ed., Academic Press, New York, 1981, 573.
322. **Kramer, J. P.**, Some relationships between *Perezia pyraustae* Paillot (Sporozoa, Nosematidae) and *Pyrausta nubilalis* (Hübner) (Lepidoptera, Pyralidae), *J. Insect Pathol.*, 1, 25, 1959.
323. **Windels, M. B., Chiang, H. C., and Furgala, B.**, Effects of *Nosema pyrausta* on pupa and adult stages of the European corn borer *Ostrinia nubilalis, J. Invertebr. Pathol.*, 27, 239, 1976.
323a. **Siegel, J. P., Maddox, J. V., and Ruesink, W. G.**, Lethal and sublethal effects of *Nosema pyrausta* on the European corn borer (*Ostrinia nubilalis*) in central Illinois, *J. Invertebr. Pathol.*, 48, 167, 1986.
324. **Wilson, G. G.**, Detrimental effects of feeding *Pleistophora schubergi* (Microsporida) to spruce budworm (*Choristoneura fumiferana*) naturally infected with *Nosema fumiferanae, Can. J. Zool.*, 56, 578, 1978.
325. **Wilson, G. G.**, The potential of *Pleistophora schubergi* in microbial control of forest insects, *Environ. Can. For. Serv. Inf. Rep.*, FPM-X-49, Ottawa, Ontario, Canada, 1981.
326. **Maddox, J. V., Brooks, W. M., and Fuxa, J. R.**, *Vairimorpha necatrix* a pathogen of agricultural pests: potential for pest control, in *Microbial Control of Pests and Plant Diseases 1970—1980*, Burges, H. D., Ed., Academic Press, New York, 1981, 587.
327. **Canning, E. U.**, On the occurrence of *Plistophora culicis* Weiser in *Anopheles gambiae, Riv. Malariol.*, 36, 39, 1957.
328. **Kelly, J. F., Anthony, D. W., and Dillard, C. R.**, A laboratory evaluation of the microsporidian *Vavraia culicis* as an agent for mosquito control, *J. Invertebr. Pathol.*, 37, 117, 1981.
329. **Pounds, J. G.**, Safety and Potential Hazards of the Entomopathogen *Mattesia trogodermae* to Non-Target Species, Ph.D thesis, University of Wisconsin, Madison, 1977.
330. **Hall, I. M., Stewart, F. D., Arakawa, K. Y., and Strong, R. G.**, Protozoan parasites of species of *Trogoderma* in California, *J. Invertebr. Pathol.*, 18, 252, 1971.
331. **Marzke, F. O. and Dicke, R. J.**, Disease-producing protozoa in species of *Trogoderma, J. Econ. Entomol.*, 51, 916, 1958.
332. **Burkholder, W. E. and Dicke, R. J.**, Detection by ultraviolet light of stored-product insects infected with *Mattesia dispora, J. Econ. Entomol.*, 57, 818, 1964.
333. **Beal, R. S. and Spitler, G. H.**, Report on crossbreeding experiments in *Trogoderma, Proc. Entomol. Soc. Wash.*, 61, 1, 1959.
334. **Schwalbe, C. P., Boush, G. M., and Burkholder, W. E.**, Factors influencing the pathogenicity and development of *Mattesia trogodermae* infecting *Trogoderma glabrum* larvae, *J. Invertebr. Pathol.*, 21, 176, 1973.
335. **Laudani, H.**, Biology and habits of dermestids, *Pest Control*, 29, 58, 1961.
336. **Hazard, E. I. and Lofgren, C. S.**, Tissue specificity and systematics of a *Nosema* in some species of *Aedes, Anopheles,* and *Culex, J. Invertebr. Pathol.*, 18, 16, 1971.
337. **Hazard, E. I.**, Microsporidian diseases in mosquito colonies: *Nosema* in two *Anopheles* colonies, in *Proc. IVth Int. Colloq. Insect Pathol.*, College Park, Md., 1971, 267.
338. **Fox, R. M. and Weiser, J.**, A microsporidian parasite of *Anopheles gambiae* in Liberia, *J. Parasitol.*, 45, 21, 1959.
339. **Undeen, A. H.**, Growth of *Nosema algerae* in pig kidney cell cultures, *J. Protozool.*, 22, 107, 1975.
340. **Avery, S. W. and Anthony, D. W.**, Ultrastructural study of early development of *Nosema algerae* in *Anopheles albimanus, J. Invertebr. Pathol.*, 42, 87, 1983.

341. **Streett, D. A., Ralph, D., and Hink, W. F.**, Replication of *Nosema algerae* in three insect cell lines, *J. Protozool.*, 27, 113, 1980.
342. **Hazard, E. I.**, Investigation of pathogens of anopheline mosquitoes in the vicinity of Kaduna, Nigeria, mimeographed document, World Health Organization, VBC/72.384, Rome, 1972.
343. **Bai, M. G., Das, P. K., Gajanana, A., and Rajagopalan, P. K.**, Host-parasite relationship of *Nosema algerae*, a parasite of mosquitoes, *Indian J. Med. Res.*, 70, 620, 1979.
344. **Anon.**, Data sheet on the biological control agent *Nosema algerae* (Vávra and Undeen 1970), mimeographed document, *World Health Organization*, VBC/80.760, Rome, 1980.
345. **Jafri, R. H., Asif, M., and Aslamkhan, M.**, *Nosema* infection in *Anopheles stephensi* larvae from Lahore, Pakistan, *Pakistan J. Zool.*, 8, 232, 1976.
346. **Maddox, J. V.**, unpublished data, 1977.
347. **Percy, J.**, The intranuclear occurrence and fine structural details of schizonts of *Perezia fumiferanae* (Microsporida: Nosematidae) in cells of *Choristoneura fumiferana* (Clem.) (Lepidoptera: Tortricidae), *Can. J. Zool.*, 51, 553, 1973.
348. **Nordin, G. L. and Maddox, J. V.**, Microsporidia of the fall webworm, *Hyphantria cunea*. I. Identification, distribution, and comparison of *Nosema* sp. with similar *Nosema* spp. from other Lepidoptera, *J. Invertebr. Pathol.*, 24, 1, 1974.
349. **Graham, K.**, Insect pathology, *Can. Dept. Agric. Biomonthly Prog. Rep.*, 4, 2, 1948.
350. **Thomson, H. M.**, Some aspects of the epidemiology of a microsporidian parasite of the spruce budworm, *Choristoneura fumiferana* (Clem.), *Can. J. Zool.*, 36, 309, 1958.
350a. **Thomson, H. M.**, The effect of a microsporidian parasite of the spruce budworm, *Choristoneura fumiferana* (Clem.), on two internal hymenopterous parasites, *Can. Entomol.*, 90, 694, 1958.
350b. **Brooks, W. M.**, Transovarian transmission of *Nosema heliothidis* in the corn earworm, *Heliothidis zea*, *J. Invertebr. Pathol.*, 11, 510, 1968.
350c. **Kellen, W. R. and Lindegren, J. E.**, Transovarian transmission of *Nosema plodiae* in the Indian-meal moth, *Plodia interpunctella*, *J. Invertebr. Pathol.*, 21, 248, 1973.
351. **Thomson, H. M.**, The possible control of a budworm infestation by a microsporidian disease, *Can. Dept. Agric. Bimonthly Prog. Rep.*, 16, 1, 1960.
352. **Wilson, G. G.**, Incidence of microsporida in a field population of spruce budworm, *Environ. Can. For. Serv. Bimonthly Res. Notes*, 29, 35, 1973.
353. **Wilson, G. G.**, Observations on the incidence rates of *Nosema fumiferanae* (Microsporida) in a spruce budworm, *Choristoneura fumiferana*, (Lepidoptera: Tortricidae) population, *Proc. Entomol. Soc. Ontario*, 108, 144, 1977.
354. **McLaughlin, R. E.**, Laboratory techniques for rearing disease-free insect colonies: elimination of *Mattesia grandis* McLaughlin, and *Nosema* sp. from colonies of boll weevils, *J. Econ. Entomol.*, 59, 401, 1966.
355. **Purrini, K.**, Zwei Schizogregarinen-Arten (Protozoa, Sporozoa) bei vorratsschädlichen Insekten in jugoslawischen Mühlen, *Anz. Schaedlingskd. Pflanz. Umweltschutz*, 49, 83, 1976.
355a. **Ignoffo, C. M. and Garcia, C.**, Infection of the cabbage looper, bollworm, tobacco budworm, and pink bollworm with spores of *Mattesia grandis* McLaughlin collected from boll weevils, *J. Invertebr. Pathol.*, 7, 260, 1965.
355b. **Purrini, K.**, *Adelina tribolii* Bhatia und *A. mesnili* Pérez (Sporozoa, Coccidia) als Krankheitserreger bei vorratsschädlichen Insekten im Gebiet von Kosova, Jogoslawien. *Anz. Schaedlingskde. Pflanz. Umweltschutz*, 49, 51, 1976.
355c. **Purrini, K.**, Über zwei Protozoen-Arten, *Adelina tribolli* Bhatia comb. nov. (Coccidia) und *Nosema ptinidorum* n. sp. (Microsporidia) als Krankheitserreger der vorratsschädlichen Käfer *Ptinus pusillus* Strm. und *P. brunneus* Dft. (Col., Ptinidae), *Z. Angew. Entomol.*, 95, 477, 1983.
356. **Sprague, V. and Vernick, S. H.**, The ultrastructure of *Encephalitozoon cuniculi* (Microsporida, Nosematidae) and its taxonomic significance, *J. Protozool.*, 18, 560, 1971.
357. **Steinhaus, E. A.**, Report on diagnoses of diseased insects 1944—1950, *Hilgardia*, 20, 629, 1951.
358. **Goodwin, T. W.**, The biochemistry of locusts. II. Carotenoid distribution in solitary and gregarious phases of the African migratory locust *(Locusta migratoria migratorioides* R. & F.) and the desert locust *(Schistocerca gregaria* Forsk.), *Biochem. J.*, 45, 472, 1949.
359. **Goodwin, T. W.**, Insectorubin metabolism in the desert locust *(Schistocerca gregaria* Forsk.) and the African migratory locust *(Locusta migratoria migratorioides* R. & F.), *Biochem. J.*, 47, 554, 1950.
360. **Goodwin, T. W.**, The biochemistry of locust pigmentation, *Biol. Rev.*, 27, 439, 1952.
361. **Goodwin, T. W. and Srisukh, S.**, Insectorubin: the redox pigment present in the integument and eyes of the desert locust *(Schistocerca gregaria* Forsk.), the African migratory locust *(Locusta migratoria migratorioides* R. & F.) and other insects, *Biochem. J.*, 47, 549, 1950.
362. **Huger, A.**, Electron microscope study on the cytology of a microsporidian spore by means of ultrathin sectioning, *J. Insect Pathol.*, 2, 84, 1960.
363. **Henry, J. E.**, Extension of host range of *Nosema locustae* in Orthoptera, *Ann. Entomol. Soc. Am.*, 62, 452, 1969.

364. **Smith, R. W.**, A field population of *Melanoplus sanguinipes* (Fab.) (Orthoptera: Acrididae) and its parasites, *Can. J. Zool.*, 43, 179, 1965.
365. **Ewen, A. B.**, Extension of the geographic range of *Nosema locustae* (Microsporidia) in grasshoppers (Orthoptera: Acrididae), *Can. Entomol.*, 115, 1049, 1983.
366. **Henry, J. E.**, Epizootiology of infections by *Nosema locustae* Canning (Microsporida: Nosematidae) in grasshoppers, *Acrida*, 1, 111, 1972.
367. **Issi, I. V. and Lipa, J. J.**, Report on identification of protozoa pathogenic for insects in the Soviet Union (1961—1966), with descriptions of some new species, *Acta Protozool.*, 4, 281, 1968.
368. **Lewis, L. C. and Lynch, R. E.**, Lyophilization, vacuum drying, and subsequent storage of *Nosema pyrausta* spores, *J. Invertebr. Pathol.*, 24, 149, 1974.
369. **Kramer, J. P.**, Variations among the spores of the microsporidian *Perezia pyraustae* Paillot, *Am. Midl. Nat.*, 64, 485, 1960.
370. **Andreadis, T. G.**, *Nosema pyrausta* infection in *Macrocentrus grandii*, a braconid parasite of the European corn borer, *Ostrinia nubilalis*, *J. Invertebr. Pathol.*, 35, 229, 1980.
371. **Kramer, J. P.**, Observations on the seasonal incidence of microsporidiosis in European corn borer populations in Illinois, *Entomophaga*, 4, 37, 1959.
372. **Raun, E. S., York, G. T., and Brooks, D. L.**, Determination of *Perezia pyraustae* infection rates in larvae of the European corn borer, *J. Insect Pathol.*, 2, 254, 1960.
373. **York, G. T.**, Microsporidia in parasites of the European corn borer, *J. Insect Pathol.*, 3, 101, 1961.
373a. **Huger, A. M.**, Susceptibility of the egg parasitoid *Trichogramma evanescens* to the microsporidium *Nosema pyrausta* and its impact on fecundity, *J. Invertebr. Pathol.*, 44, 228, 1984.
374. **Peairs, F. B. and Lilly, J. H.**, *Nosema pyraustae* in populations of the European corn borer, *Ostrinia nubilalis* in Massachusetts, *Environ. Entomol.*, 3, 878, 1974.
375. **Hill, R. E. and Gary, W. J.**, Effects of the microsporidium, *Nosema pyrausta*, on field populations of European corn borers in Nebraska, *Environ. Entomol.*, 8, 91, 1979.
376. **Andreadis, T. G.**, Impact of *Nosema pyrausta* on field populations of *Macrocentrus grandii*, an introduced parasite of the European corn borer, *J. Invertebr. Pathol.*, 39, 298, 1982.
376a. **Andreadis, T. G.**, Epizootiology of *Nosema pyrausta* in field populations of the European corn borer (Lepidoptera: Pyralidae), *Environ. Entomol.*, 13, 882, 1984.
377. **Issi, I. V.**, On possible ways of speciation among Microsporidia, in *Prog. Protozool. Proc. 3rd Int. Congr. Protozool.*, Strelkov, A. A., Sukhanova, K. M., and Raikov, I. B., Eds., Nauka, Leningrad, 1969, 377.
378. **Issi, I. V.**, On polymorphism of the species *Plistophora schubergi* Zwolfer, 1927, possible ways of speciation in the order of Microsporidia, *Parazitologiya (Leningrad)*, 5, 297, 1971.
379. **Purrini, K.**, Light and electron microscopic studies on the microsporidian *Pleistophora schubergi neustriae* n. subsp. (Microsporida: Phylum Microspora) parasitizing the larvae of *Malacosoma neustriae* L. (Lymantriidae, Lepidoptera), *Arch. Protistenk.* 125, 345, 1982.
379a. **Canning, E. U. and Nicholas, J. P.**, Genus *Pleistophora* (Phylum Microspora): redescription of the type species, *Pleistophora typicalis* Gurley, 1893, and ultrastructural characterization of the genus, *J. Fish. Dis.*, 3, 317, 1980.
379b. **Maddox, J. V.**, personal communication, 1986.
380. **Günther, S.**, Zur Infektion des Goldafters (*Euproctis chrysorrhoea* L.) mit *Plistophora schubergi* Zwölfer (Microsporidia), *Z. Angew. Zool.*, 43, 397, 1956.
381. **Kaya, H. K.**, Pathogenicity of *Pleistophora schubergi* to larvae of the orange-striped oakworm and other lepidopterous insects, *J. Invertebr. Pathol.*, 22, 356, 1973.
382. **Wilson, G. G. and Burke, J. M.**, Microsporidian parasites of *Archips cerasivoranus* (Fitch) in the district of Algoma, Ontario, *Proc. Entomol. Soc. Ontario*, 109, 84, 1978.
383. **Tanada, Y. and Chang, G. Y.**, An epizootic resulting form a microsporidian and two virus infections in the armyworm, *Pseudaletia unipuncta* (Haworth), *J. Invertebr. Pathol.*, 4, 128, 1962.
384. **Tanada, Y.**, Effect of a microsporidian spore suspension on the incidence of cytoplasmic polyhedrosis in the alfalfa caterpillar, *Colias eurytheme* Boisduval, *J. Invertebr. Pathol.*, 4, 495, 1962.
385. **Maddox, J. V.**, Studies on a Microsporidiosis of the Armyworm, *Pseudaletia unipuncta* (Haworth), Ph.D. thesis, University of Illinois, Urbana, 1966.
385a. **Richter, A. R. and Fuxa, J. R.**, Pathogen-pathogen and pathogen-insecticide interactions in velvetbean caterpiller (Lepidoptera: Noctuidae), *J. Econ. Entomol.*, 77, 1559, 1984.
385b. **Richter, A. R. and Fuxa, J. R.**, Timing, formulation, and persistence of a nuclear polyhedrosis virus and a microsporidium for control of the velvetbean caterpillar (Lepidoptera: Noctuidae) in soybeans, *J. Econ. Entoml.*, 77, 1299, 1984.
385c. **Khan, A. R. and Selman, B. J.**, Effect of insecticide, microsporidian, and insecticide-microsporidian doses on the growth of *Tribolium castaneum* larvae, *J. Invertebr. Pathol.*, 44, 230, 1984.
385d. **Cossentine, J. E. and Lewis, L. C.**, Impact of *Vairimorpha necatrix* and *Vairimorpha* sp. (Microspora: Microsporida) on *Bonnetia comta* (Diptera: Tachinidae) within *Agrotis ipsilon* (Lepidoptera: Noctuidae) hosts, *J. Invertebr. Pathol.*, 47, 303, 1986.

386. **Tanabe, A. M.**, The Pathology of Two Microsporidia in the Armyworm *Pseudaletia unipuncta* (Haworth) (Lepidoptera, Noctuidae), Ph.D. thesis, University of California, Berkeley, 1971.
387. **Fowler, J. L. and Reeves, E. L.**, Spore dimorphism in a microsporidan isolate, *J. Protozool.*, 21, 538, 1974.
388. **Fowler, J. L. and Reeves, E.**, Microsporidian spore structure as revealed by scanning electron microscopy, *J. Invertebr. Pathol.*, 26, 1, 1975.
389. **Undeen, A.**, Spore-hatching processes in some *Nosema* species with particular reference to *N. algerae* Vávra and Undeen, *Misc. Publ. Entomol. Soc. Am.*, 11, 29, 1978.
389a. **Dall, D. J.**, A theory for the mechanism of polar filament extrusion in the microspora, *J. Theor. Biol.*, 105, 647, 1983.
390. **Tanada, Y.**, Incidence of microsporidiosis in field population of the armyworm, *Pseudaletia unipuncta* (Haworth), *Proc. Hawaiian Entomol. Soc.*, 18, 435, 1964.
391. **Tanada, Y.**, Field observations of the biotic factors regulating the population of the armyworm, *Pseudaletia unipuncta* (Haworth), *Proc. Hawaiian Entomol. Soc.*, 19, 302, 1966.
392. **Nordin, G. L., Rennels, R. G., and Maddox, J. V.**, Parasites and pathogens of the fall webworm in Illinois, *Environ. Entomol.*, 1, 351, 1972.
393. **Weiser, J. and Coluzzi, M.**, *Plistophora culisetae* n. sp., a new microsporidian (Protozoa, Cnidosporidia) in the mosquito *Culiseta longiareolata* (Maquart 1838), *Riv. Malariol.*, 43, 51, 1964.
394. **Weiser, J. and Coluzzi, M.**, The microsporidian *Plistophora culicis* Weiser, 1946 in different mosquito hosts, *Folia Parasitol.*, 19, 197, 1972.
395. **Weiser, J. and Coluzzi, M.**, *Plistophora culisetae* in the mosquito *Culiseta longiareolata* Maqu.-Further remarks, in *Proc. 1st Int. Cong. Parasitol.*, Corradetti, A., Ed., Pergamon Press, New York, 1966, 596.
396. **Reynolds, D. G.**, Infection of *Culex fatigans* with a microsporidian, *Nature (London)*, 210, 967, 1966.
397. **Yoeli, M. and Boné, G.**, Studies on *Anopheles dureni* Edwards, *Riv. Malariol.*, 46, 1, 1967.
398. **Chapman, H. C., Clark, T. B., Petersen, J. J., and Woodard, D. B.**, A two-year survey of pathogens and parasites of Culicidae, Chaoboridae, and Ceratopogonidae in Louisiana, *Proc. N.J. Mosq. Exterm. Assoc.*, 56, 203, 1969.
399. **Chapman, H. C.**, Biological control of mosquito larvae, *Annu. Rev. Entomol.*, 19, 33, 1974.
400. **Garnham, P. C. C.**, Microsporidia in laboratory colonies of *Anopheles*, *Bull. WHO*, 15, 845, 1956.
401. **Canning, E. U.**, *Plistophora culicis* Weiser (Protozoa, Microsporidia): its development in *Anopheles gambiae*, *Trans. R. Soc. Trop. Med. Hyg.*, 51, 8, 1957.
402. **Lainson, R. and Garnham, P. C. C.**, Stages of *Plistophora culicis* encountered during dissections of *Anopheles stephensi*, *Trans. R. Soc. Trop. Med. Hyg.*, 51, 6, 1957.
403. **Bano, L.**, Partial inhibitory effect of *Plistophora culicis* on the sporogonic cycle of *Plasmodium cynomolgi* in *Anopheles stephensi*, *Nature (London)*, 181, 430, 1958.
404. **Garnham, P. C. C.**, Some natural protozoal parasites of mosquitoes with special reference to Crithidia, Trans. 1st Int. Conf. Insect Pathol. and Biol. Control Praha, Czechoslovakia, 1958, 287.
404a. **Wang, B. T.**, The Pathobiology of the Mosquito Parasite: *Vavraia culicis* (Weiser), Ph.D. thesis, University of California, Los Angeles, 1982.
404b. **Weiser, J.**, Acute *Vavraia culicis* infection in a substitute host, *J. Protozool.*, 27, 73A, 1980.
405. **Weiser, J.**, Immunity of insects to protozoa, in *Immunity to Parasitic Animals*, Jackson, G. J., Herman, R., and Singer, I., Eds., Appleton-Century-Crofts, New York, 1969, 129.
406. **Stephens, J.**, Immunity in insects, in *Insect Pathology, An Advanced Treatise*, Vol. 1, Steinhaus, E. A., Ed., Academic Press, New York, 1963, 273.
407. **Briggs, J. D.**, Humoral immunity in lepidopterous larvae, *J. Exp. Zool.*, 138, 155, 1958.
408. **Briggs, J. D.**, Immunological responses, in *The Physiology of Insecta*, Vol. 3, Rockstein, M., Ed., Academic Press, New York, 1964, 259.
409. **Chadwick, J. S.**, Hemolymph changes with injection or induced immunity in insects and ticks, in *Invertebrate Immunity*, Maramorosch, K. and Shope, R. E., Eds., Academic Press, New York, 1975, 241.
410. **Jones, J. C.**, *The Circulatory System of Insects*, Charles C Thomas, Springfield, Ill., 1977.
411. **Hostounský, Z.**, *Nosema mesnili* (Paill.), a microsporidian of the cabbageworm *Pieris brassicae* (L.) in the parasites *Apanteles glomeratus* (L.), *Hyposoter ebeninus* (Grav.) and *Pimpla instigator* (F.), *Acta Entomol. Bohemoslav.*, 67, 1, 1970.
412. **Weiser, J.**, Mikrosporidien des Schwammspinners und der Goldafter, *Z. Angew. Entomol.*, 40, 509, 1957.
413. **Salt, G.**, The defense reactions of insects to metazoan parasites, *Parasitology*, 53, 527, 1963.
414. **Salt, G.**, The resistance of insect parasitoids to the defense reactions of their hosts, *Biol. Rev.*, 43, 200, 1968.
415. **Shapiro, M.**, Immunity of insect hosts to insect parasites, in *Immunity to Parasitic Animals*, Jackson, G. J., Herman, R., and Singer, I., Eds., Appleton-Century-Crofts, New York, 1969, 211.
416. **Brooks, W. M.**, Protozoan infections of insects with emphasis on inflammation, in *Proc. IVth Int. Colloq. Insect Pathol.*, College Park, Md., 1971, 11.

417. **Weiser, J.**, Protozoäre Infektionen in Kampfe gegen Insekten, *Z. Pflanzenpathol. Pflanzenschutz*, 63, 625, 1956.
417a. **Brooks, W. M., Hazard, E. I., and Becnel, J.**, Two new species of *Nosema* (Microsporida: Nosematidae) from the Mexican bean bettle *Epilachna varivestis* (Coleoptera: Coccinellidae), *J. Protozool.*, 32, 525, 1985.
418. **Jafri, R. H.**, Interaction of protozoan infections, toxins and radiation in insects, in *Prog. Protozool. Proc. 1st Int. Congr. Protozool.*, Ludvik, J., Lom, J., and Vávra, J., Eds., Academic Press, New York, 1963, 510.
419. **Spencer, J. P. and Olson, J. K.**, Evaluation of the combined effects of methoprene and the protozoan parasite *Ascogregarina culicis* (Eugregarinida, Diplocystidae), on *Aedes* mosquitoes, *Mosq. News*, 42, 384, 1982.
420. **Cuénot, L.**, Recherches sur l'évolution et la conjugaison des grégariens, *Archs Biol.*, 17, 581, 1900.
421. **Corbel, J. C.**, Role du kyste et des spores dans la fréquence et l'intensité des infestations grégariniennes chez les Orthoptères, *Protistologica*, 4, 19, 1968.
422. **Lipa, J. J.**, Studies on gregarines (Gregarinomorpha) of arthropods in Poland, *Acta Protozool.*, 5, 97, 1967.
423. **Henry, J. E.**, *Nosema acridophagus* sp.n., a microsporidian isolated from grasshoppers, *J. Invertebr. Pathol.*, 9, 331, 1967.
424. **Henry, J. E.**, Early morphogenesis of tumours induced by *Nosema acridophagus* in *Melanoplus sanguinipes*, *J. Insect Pathol.*, 15, 391, 1969.
425. **Brooks, W. M. and Cranford, J. D.**, Host-parasite relationships of *Nosema heliothidis* Lutz and Splendore, *Misc. Publ. Entomol. Soc. Am.*, 11, 51, 1978.
426. **Jordan, J. A. and Noblet, R.**, Pathological studies of *Heliothis* infected with *Nosema heliothidis* and *Vairimorpha necatrix*, *J. Ga. Entomol. Soc.*, 17, 183, 1982.
427. **Kellen, W. R. and Lindegren, J. E.**, *Nosema invadens* sp. n. (Microsporida: Nosematidae), a pathogen causing inflammatory response in Lepidoptera, *J. Invertebr. Pathol.*, 21, 293, 1973.
428. **Hesse, E.**, Microsporidies nouvelles des insectes, *Compt. Rend. Assoc. Franc.*, 33, 917, 1905.
429. **Poisson, R.**, Sur une infection à microsporidie chez la nèpe cendrées (Hémiptere-Hétéroptère) la réaction des tissus l'host vis-à-vis du parasite, *Arch. Zool. Exp. Gen.*, 67, 129, 1928.
430. **Brooks, W. M.**, *Nosema sphingidis* sp. n., a microsporidan parasite of the tobacco hornworm, *Manduca sexta*, *J. Invertebr. Pathol.*, 16, 390, 1970.
431. **Brooks, W. M.**, The inflammatory response of the tobacco hornworm, *Manduca sexta*, to infection by the microsporidian, *Nosema sphingidis*, *J. Invertebr. Pathol.*, 17, 87, 1971.
432. **Kellen, W. R., Hoffmann, D. F., and Collier, S. S.**, Studies on the biology and ultrastructure of *Nosema transitellae* sp. n. (Microsporidia: Nosematidae) in the navel orangeworm, *Paramyelois transitella* (Lepidoptera: Pyralidae), *J. Invertebr. Pathol.*, 29, 289, 1977.
433. **Pillai, J. S.**, *Pleistophora milesi*, a new species of microsporida from *Maorigoeldia argyropus* Walker (Diptera: Culicidae) in New Zealand, *J. Invertebr. Pathol.*, 24, 234, 1974.
434. **Kellen, W. R. and Lindegren, J. W.**, Inflammatory response of the Indian meal moth, *Plodia interpunctella*, to infection by *Nosema heterosporum*, *J. Invertebr. Pathol.*, 19, 418, 1972.
435. **Zimmack, H. L. and Brindley, T. A.**, The effect of the protozoan parasite *Perezia pyraustae* Paillot on the European corn borer, *J. Econ. Entomol.*, 50, 637, 1957.
435a. **Andreadis, T. G.**, Dissemination of *Nosema pyrausta* in feral populations of the European corn borer, *Ostrinia nubilalis*, *J. Invertebr. Pathol.*, 48, 335, 1986.
436. **Andreadis, T. G. and Hall, D. W.**, Development, ultrastructure, and mode of transmission of *Amblyospora* sp. (Microspora) in the mosquito, *J. Protozool.*, 26, 444, 1979.
436a. **Lord, J. C. and Hall, D. W.**, Sporulation of *Amblyospora* (Microspora) in female *Culex salinarius*: induction by 20-hydroxyecdysone, *Parasitology*, 87, 377, 1983.
436b. **Hall, D. W. and Washino, R. K.**, Sporulation of *Amblyospora californica* (Microspora: Amblyosporidae) in autogenous female *Culex tarsalis*, *J. Invertebr. Pathol.*, 47, 214, 1986.
436c. **Hall, D. W.**, The distribution of *Amblyospora* (Microspora) sp.-infected oenocytes in adult female *Culex salinarius*: significance for mechanism of transovarian transmission, *J. Am. Mosq. Control Assoc.*, 1, 514, 1985.
437. **Andreadis, T. G. and Hall, D. W.**, Significance of transovarial infections of *Amblyospora* sp. (Microspora: Thelohaniidae) in relation to parasite maintenance in the mosquito *Culex salinarius*, *J. Invertebr. Pathol.*, 34, 152, 1979.
438. **Hazard, E. I., Andreadis, T. C., Joslyn, D. J., and Ellis, E. A.**, Meiosis and its implications in the life cycles of *Amblysopora* and *Parathelohania* (Microspora), *J. Parasitol.*, 65, 117, 1979.
438a. **Hazard, E. I. and Brookbank, J. W.**, Karyogamy and meiosis in an *Amblyospora* sp. (Microspora) in the mosquito *Culex salinarius*, *J. Invertebr. Pathol.*, 44, 3, 1984.
438b. **Hazard, E. I., Fukuda, T., and Becnel, J. J.**, Gametogenesis and plasmogamy in certain species of Microspora, *J. Invertebr. Pathol.*, 46, 63, 1985.

438c. **Undeen, A. H. and Avery, S. W.**, Germination of experimentally nontransmissible microsporidia, *J. Invertebr. Pathol.*, 43, 299, 1984.

438d. **Sweeney, A. W., Hazard, E. I., and Graham, M. F.**, Intermediate host for an *Amblyospora* sp. (Microspora) infecting the mosquito, *Culex annulirostris*, *J. Invertebr. Pathol.*, 46, 98, 1985.

438e. **Andreadis, T. G.**, Experimental transmission of a microsporidian pathogen from mosquitoes to an alternate copepod host, *Proc. Natl. Acad. Sci. U.S.A.*, 82, 5574, 1985.

439. **Brooks, W. M.**, Protozoa: host-parasite-pathogen interrelationships, *Misc. Publ. Entomol. Soc. Am.*, 9, 105, 1973.

440. **Issi, I. V. and Maslennikova, V. A.**, Role of *Apanteles glomeratus* (L.) (Hymenoptera, Braconidae) in the transmission of *Nosema polyvora* (Blunk) (Protozoa, Microsporidia), *Entomol. Obozr.*, 45, 494, 1966.

441. **Laigo, F. M. and Tamashiro, M.**, Interactions between a microsporidian pathogen of the lawn-armyworm and the hymenopterous parasite *Apanteles marginiventris*, *J. Invertebr. Pathol.*, 9, 546, 1967.

442. **Larsson, R.**, Transmission of *Nosema mesnili* (Paillot) (Microsporida, Nosematidae), a microsporidian parasite of *Pieris brassicae* L. (Lepidoptera, Pieridae) and its parasite *Apanteles glomeratus* L. (Hymenoptera, Braconidae), *Zool. Anz.*, 203, 151, 1979.

442a. **Hamm, J. J., Nordlund, D. A., and Mullinix, B. G., Jr.**, Interaction of the microsporidium *Vairimorpha* sp. with *Microplitis croceipes* (Cresson) and *Cotesia marginiventris* (Cresson) (Hymenoptera: Braconidae), two parasitoids of *Heliothis zea* (Boddie) (Lepidoptera: Noctuidae), *Environ. Entoml.*, 12, 1547, 1983.

442b. **Own, O. S. and Brooks, W. M.**, Interactions of the parasite *Pediobius foveolatus* (Hymenoptera: Eulophidae) with two *Nosema* spp. (Microsporida: Nosematidae) of the Mexican bean beetle (Coleoptera: Coccinellidae), *Environ. Entomol.*, 15, 32, 1986.

442c. **Kellen, W. R. and Lindegren, J. E.**, Modes of transmission of *Nosema plodiae* Kellen and Lindegren, a pathogen of *Plodia interpunctella* (Hübner), *J. Stored Prod. Res.*, 7, 31, 1971.

442d. **Wilson, G. G.**, Transmission of *Nosema fumiferanae* (Microspora) to its host *Choristoneura fumiferana* (Clem.), *Z. Parasitenkd.*, 68, 47, 1982.

442e. **Siegel, J. P., Maddox, J. V., and Ruesink, W. G.**, Impact of *Nosema pyrausta* on a braconid, *Macrocentrus grandii*, in central Illinois, *J. Invertebr. Pathol.*, 47, 271, 1986.

443. **Tanada, Y.**, Field observations on a microsporidian parasite of *Pieris rapae* L. and *Apanteles glomeratus* (L.), *Proc. Hawaiian Entomol. Soc.*, 15, 609, 1955.

444. **McLaughlin, R. E. and Adams, C. H.**, Infection of *Bracon mellitor* (Hymenoptera: Braconidae) by *Mattesia grandis* (Protozoa: Neogregarinida), *Ann. Entomol. Soc. Am.*, 59, 800, 1966.

445. **Brooks, W. M. and Cranford, J. D.**, Microsporidoses of the hymenopterous parasites, *Campoletis sonorensis* and *Cardiochiles nigriceps*, larval parasites of *Heliothis* species, *J. Invertebr. Pathol.*, 20, 77, 1972.

446. **McNeil, J. N. and Brooks, W. M.**, Interactions of the hyperparasitoids *Catolaccus aeneoviridis* (Hym.: Pteromalidae) and *Spilochalcis side* (Hym.: Chalcididae) with the microsporidans *Nosema heliothidis* and *N. campoletidis*, *Entomophaga*, 19, 195, 1974.

447. **Muspratt, J.**, Observation on the larvae of tree-hole breeding Culicini (Diptera: Culicidae) and two of their parasites, *J. Entomol. Soc. S. Afr.*, 8, 13, 1945.

448. **Muspratt, J.**, Notes on a ciliate protozoon, probably *Glaucoma pyriformis*, parasitic in culicine mosquito larvae, *Parasitology*, 38, 107, 1974.

449. **Lom, J. and Vávra, J.**, The mode of sporoplasm extrusion in microsporidian spores, *Acta Protozool.*, 1, 81, 1963.

450. **Vávra, J.**, Structure of the microsporidia, in *Comparative Pathobiology*, Vol. 1, Bulla, L. A. and Cheng, T. C., Eds., Plenum Press, New York, 1976, 1.

451. **Ohshima, K.**, Stimulative or inhibitive substances to evaginate the filament of *Nosema bombycis* Nägeli. I. The case of artificial buffer solution, *Jpn. J. Zool.*, 14, 209, 1964.

452. **Ohshima, K.**, Effect of potassium ion on filament evagination of spores of *Nosema bombycis* as studied by the neutralization method, *Annot. Zool. Jpn.*, 37, 102, 1964.

453. **Ohshima, K.**, Substances stimulating or inhibitng the evagination of the filament of *Nosema bombycis* Nageli. III. Action of HCO_3^- in the digestive juice of starved silkworm larvae and the method effective in evaginating the filament *in vitro*, *Annot. Zool. Jpn.*, 38, 198, 1965.

454. **Ishihara, R.**, Stimuli causing extrusion of polar filaments of *Glugea fumiferanae* spores, *Can. J. Microbiol.*, 13, 1321, 1967.

455. **Weidner, E. and Trager, W.**, Structure and function of microsporidian spores, *Biol. Bull.*, 139, 433, 1970.

455a. **Undeen, A.**, The germination of *Vavraia culicis* spores, *J. Protozool.*, 30, 274, 1983.

455b. **Malone, L. A.**, Factors controlling in vitro hatching of *Vairimorpha plodiae* (Microspora) spores and their infectivity to *Plodia interpunctella*, *Heliothis virescens*, and *Pieris brassicae*, *J. Invertebr. Pathol.*, 44, 192, 1984.

456. **Weidner, E.**, The microsporidian spore invasion tube. The ultrastructure, isolation, and characterization of the protein comprising the tube, *J. Cell Biol.*, 71, 23, 1976.

457. **Weidner, E.**, The microsporidian spore invasion tube. III. Tube extrusion and assembly, *J. Cell. Biol.*, 93, 976, 1982.
458. **Weidner, E. and Byrd, W.**, The microsporidian spore invasion tube. II. Role of calcium in the activation of invasion tube discharge, *J. Cell Biol.*, 93, 970, 1982.
458a. **Weidner, E., Byrd, W., Scarborough, A., Pleshinger, J., and Sibley, D.**, Microspordian spore discharge and the transfer of polaroplast organelle membrane into plasma membrane, *J. Protozool.*, 31, 195, 1984.
459. **Harry, O. G.**, The effect of a eugregarine *Gregarina polymorpha* (Hammerschmidt) on the mealworm larva of *Tenebrio molitor* (L.), *J. Protozool.*, 14, 539, 1967.
460. **Dunkel, F. V. and Boush, G. M.**, Effect of starvation on the black carpet beetle, *Attagenus megatoma*, infected with the eugregarine *Pyxinia frenzeli*, *J. Invertebr. Pathol.*, 14, 49, 1969.
460a. **Zuk, M.**, The effects of gregarine parasites on longevity, weight loss, fecundity and developmental time in the field crickets *Gryllus veletis* and *G. pennsylvanicus*, *Ecol. Entomol.*, 12, 349, 1987.
461. **Bailey, C. H. and Brooks, W. M.**, Histological observations on larvae of the eye gnat, *Hippelates pusio* (Diptera: Chloropidae), infected with the flagellate *Herpetomonas muscarum*, *J. Invertebr. Pathol.*, 19, 342, 1972.
462. **Bailey, C. H. and Brooks, W. M.**, Effects of *Herpetomonas muscarum* on development and longevity of the eye gnat, *Hippelates pusio* (Diptera: Chloropidae), *J. Invertebr. Pathol.*, 20, 31, 1972.
463. **Weissenburg, R.**, Microsporidian interactions with host cells, in *Comparative Pathobiology*, Vol. 1, Bulla, L. A. and Cheng, T. C., Eds., Plenum Press, New York, 1976, 203.
464. **Weiser, J.**, The *Pleistophora debaisieuxi* xenoma, *Z. Parasitenk.*, 48, 263, 1976.
465. **Keyl, H. G.**, Erhöhung der chromosomalen Replikationsrate durch Mikrosporidieninfektion in Speicheldrüsenzellen von *Chironomus*, *Naturwissenschaften*, 47, 212, 1960.
466. **Diaz, M. and Pavan, C.**, Changes in chromosomes induced by microorganism infection, *Proc. Natl. Acad. Sci. U.S.A.*, 54, 1321, 1965.
467. **Pavan, C. and Basile, R.**, Chromosome changes induced by infections in tissues of *Rhynchosciara angelae*, *Science*, 151, 1556, 1966.
468. **Roberts, P. A., Kimball, R. F., and Pavan, C.**, Response of *Rhynchosciara* chromosomes to microsporidian infection, *Exp. Cell Res.*, 47, 408, 1967.
469. **Pavan, C. and DaCunha, A. B.**, Chromosomal activities in *Rhynchosciara* and other Sciaridae, *Annu. Rev. Genet.*, 3, 425, 1969.
470. **Pavan, C., Perondini, A. L. P., and Picard, T.**, Changes in chromosomes and in development of cells of *Sciara ocellaris* induced by microsporidian infections, *Chromosoma*, 28, 328, 1969.
471. **Pavan, C., Biesele, J., Riess, R. W., and Wertz, A. V.**, Changes in the ultrastructure of *Rhynchosciara* cells infected by microsporidia, *Stud. Genet.*, 6, 241, 1971.
472. **Jurand, A., Simoes, L. C. G., and Pavan, C.**, Changes in the ultrastructure of salivary gland cytoplasm in *Sciara ocellaris* (Comstock, 1882) due to microsporidian infection, *J. Insect Physiol.*, 13, 795, 1967.
473. **Wang, D. I. and Moeller, F. E.**, Ultrastructural changes in the hypopharyngeal glands of worker honey bees infected by *Nosema apis*, *J. Invertebr. Pathol.*, 17, 308, 1971.
474. **Liu, T. P. and Davies, D. M.**, Ultrastructure of the cytoplasm in fat-body cells of the blackfly, *Simulium vittatum*, with microsporidian infection; a freeze-etching study, *J. Invertebr. Pathol.*, 19, 208, 1972.
475. **Martins, R. R. and Perondini, A. L. P.**, Effects of microsporidia on the striated parietal muscle of *Rhynchosciara angelae* (Diptera: Sciaridae), *J. Invertebr. Pathol.*, 30, 422, 1977.
476. **Henry, J. E., Oma, E. A., Onsager, J. A., and Oldacre, S. W.**, Infection of the corn earworm, *Heliothis zea*, with *Nosema acridophagus* and *Nosema cuneatum* from grasshoppers: relative virulence and production of spores, *J. Invertebr. Pathol.*, 34, 125, 1979.
477. **Undeen, A. H. and Alger, H. E.**, The effect of the microsporidan, *Nosema algerae*, on *Anopheles stephensi*, *J. Invertebr. Pathol.*, 25, 19, 1975.
478. **Henry, J. E. and Oma, E. A.**, Effects of infections by *Nosema locustae* Canning, *Nosema acridophagus* Henry, and *Nosema cuneatum* Henry (Microsporida: Nosematidae) in *Melanoplus bivittatus* (Say) (Orthoptera: Acrididae), *Acrida*, 3, 223, 1974.
478a. **Erlandson, M. A., Mukerji, M. K., Ewen, A. B., and Gillott, C.**, Comparative pathogenicity of *Nosema acridophagus* Henry and *Nosema cuneatum* Henry (Microsporida: Nosematidae) for *Melanoplus sanguinipes* (Fab.) (Orthoptera: Acrididae), *Can. Entomol.*, 117, 1167, 1985.
478b. **Oma, E. A. and Hewitt, G. B.**, Effect of *Nosema locustae* (Microsporida: Nosematidae) on food consumption in the differential grasshopper (Orthoptera: Acrididae), *J. Econ. Entomol.*, 77, 500, 1984.
478c. **Johnson, D. L. and Pavlikova, E.**, Reduction of consumption by grasshoppers (Orthoptera: Acrididae) infected with *Nosema locustae* Canning (Microsporida: Nosematidae), *J. Invertebr. Pathol.*, 48, 232, 1986.
479. **Kramer, J. P.**, Longevity of microsporidian spores with special reference to *Octosporea muscaedomesticae* Flu, *Acta Protozool.*, 8, 217, 1970.
480. **Kramer, J. P.**, The extra-corporeal ecology of microsporidia, in *Comparative Pathobiology*, Vol. 1, Bulla, L. A. and Cheng, T. C., Eds., Plenum Press, New York, 1976, 127.
481. **Maddox, J. V.**, The persistence of the microsporida in the environment, *Misc. Publ. Entomol. Soc. Am.*, 9, 99, 1973.

482. **Maddox, J. V.**, Stability of entomophilic protozoa, *Misc. Publ. Entomol. Soc. Am.*, 10, 3, 1977.
483. **Kaya, H. K.**, Survival of spores of *Vairimorpha* (= *Nosema*) *necatrix* (Microsporida: Nosematidae) exposed to sunlight, ultraviolet radiation, and high temperature, *J. Invertebr. Pathol.*, 30, 192, 1977.
484. **Nara, J. M., Burkholder, W. E., and Boush, G. M.**, The influence of storage temperature on spore viability of *Mattesia trogodermae* (Protozoa: Neogregarinida), *J. Invertebr. Pathol.*, 38, 404, 1981.
485. **Allen, H. W.**, Nosema disease of *Gnorimoschema operculella* (Zeller) and *Macrocentrus ancylivorus* Rohwer, *Ann. Entomol. Soc. Am.*, 47, 407, 1954.
486. **Kulikov, N. S. and Akramovsky, M. N.**, Sroki ziznesposobnosti spor nosemy u pcel, *Pcelovodstvo*, 38, 46, 1961.
487. **Kudo, R. R.**, Microsporidia, in *Problems and Methods of Research in Protozoology*, Hegner, R. and Andrews, J., Eds., Macmillan, New York, 1930, 325.
488. **Weiser, J.**, Protozoan diseases in insect control, in Proc. 10th Int. Congr. Entomol., Montreal, Canada, 1956, 681.
489. **White, G. F.**, Nosema-disease, *U.S. Dept. Agric. Bull.*, 780, 1919.
490. **Burges, H. D., Canning, E. U., and Hurst, J. A.**, Morphology, development, and pathogenicity of *Nosema oryzaephili* n. sp. in *Oryzaephilus surinamensis* and its host range among granivorous insects, *J. Invertebr. Pathol.*, 17, 419, 1971.
491. **Milner, R. J.**, The survival of *Nosema whitei* spores stored at 4°C, *J. Invertebr. Pathol.*, 20, 356, 1972.
492. **Teetor-Barsch, G. E. and Kramer, J. P.**, The preservation of infective spores of *Octosporea muscaedomesticae* in *Phormia regina*, of *Nosema algerae* in *Anopheles stephensi*, and of *Nosema whitei* in *Tribolium castaneum* by lyophiliziation, *J. Invertebr. Pathol.*, 33, 300, 1979.
493. **Fuxa, J. R. and Brooks, W. M.**, Mass production and storage of *Vairimorpha necatrix* (Protozoa: Microsporida), *J. Invertebr. Pathol.*, 33, 86, 1979.
494. **Revell, I. L.**, Longevity of refrigerated nosema spores — *Nosema apis*, a parasite of honey bees, *J. Econ. Entomol.*, 53, 1132, 1960.
495. **Ohshima, K.**, Method of gathering and purifying active spores of *Nosema bombycis* and preserving them in good condition, *Annot. Zool. Jpn.*, 37, 94, 1964.
496. **Kamaski, H.**, Some pathogens and pests associated with tephritid flies in the laboratory, *J. Econ. Entomol.*, 63, 1353, 1970.
497. **Pilley, B. M.**, The storage of infective spores of *Vairimorpha necatrix* (Protozoa: Microsporida) in antibiotic solution at 4°C, *J. Invertebr. Pathol.*, 31, 341, 1978.
498. **Chu, W. H. and Jaques, R. P.**, Factors affecting infectivity of *Vairimorpha necatrix* (Microsporida: Nosematidae) in *Trichoplusia ni* (Lepidoptera: Noctuidae), *Can. Entomol.*, 113, 93, 1981.
499. **Vávra, J. and Maddox, J. V.**, Methods in microsporidiology, in *Comparative Pathobiology*, Vol. 1, Bulla, L. A. and Cheng, T. C., Eds., Plenum Press, New York, 1976, 281.
500. **Maddox, J. V. and Pilley, B. M.**, unpublished data, 1982.
501. **Moffett, J. O. and Wilson, W. T.**, The viability and infectivity of frozen *Nosema* spores, *Am. Bee J.*, 111, 55, 1971.
502. **Pilley, B. M.**, The preservation of infective spores of *Nosema necatrix* (Protozoa: Microsporida) in *Spodoptera exempta* (Lepidoptera: Noctuidae) by lyophilizaiton, *J. Invertebr. Pathol.*, 27, 349, 1976.
503. **Bailey, L.**, The preservation of infective microsporidan spores, *J. Invertebr. Pathol.*, 20, 252, 1972.
504. **Ignoffo, C. M. and Hostetter, D. L.**, Eds., Environmental stability of microbial insecticides, *Misc. Publ. Entomol. Soc. Am.*, 10, 1977.
505. **Ignoffo, C. M., Hostetter, D. L., Sikorowski, P. P., Sutter, G., and Brooks, W. M.**, Inactivation of representative species of entomopathogenic viruses, a bacterium, fungus, and protozoan by an ultraviolet light source, *Environ. Entomol.*, 6, 411, 1977.
506. **Kelly, J. F. and Anthony, D. W.**, Susceptibility of spores of the microsporidian *Nosema algerae* to sunlight and germicidal ultraviolet radiation, *J. Invertebr. Pathol.*, 34, 164, 1979.
507. **Wilson, G. G.**, The effects of temperature and ultraviolet radiation on the infection of *Choristoneura fumiferana* and *Malacosoma pluviale* by a microsporidian parasite, *Nosema (Perezia) fumiferanae* (Thom.), *Can. J. Zool.*, 52, 59, 1974.
508. **Sikorowski, P. P. and Lashomb, J. H.**, Effect of sunlight on the infectivity of *Nosema heliothidis* spores isolated from *Heliothis zea*, *J. Invertebr. Pathol.*, 30, 95, 1977.
509. **Teetor, G. E. and Kramer, J. P.**, Effect of ultraviolet radiation on the microsporidian *Octosporea muscaedomesticae* with reference to protectants provided by the host *Phormia regina*, *J. Invertebr. Pathol.*, 30, 348, 1977.
510. **Nilova, G. N. and Strelnikova, L. V.**, The effect of ultraviolet irradiation on the viability of spores of *Plistophora schubergi* Zwölfer and *Nosema arotidis* Lipa et Issi, *Parazitologiya*, 8, 463, 1974.
511. **Baribeau, M. F. and Burkhardt, C. C.**, Effect of heat and ultraviolet light on *Nosema apis* spores in relation to honey bee infection, *J. Kans. Entoml. Soc.*, 43, 455, 1970.
512. **Rosicky, B.**, Nosematosis of *Otiorrhynchus ligustici*. II. The influence of the parasitation by *Nosema otiorrhynchi* W. on the susceptibility of the beetles to insecticides, *Vestn. Cesk. Spol. Zool.*, 15, 219, 1951.

512a. **Hinks, C. F. and Ewen, A. B.**, Pathological effects of the parasite *Malameba locustae* in males of the migratory grasshopper *Melanoplus sanquinipes* and its interaction with the insecticide cypermethrin, *Entomol. Exp. Appl.*, 42, 39, 1986.

513a. **Bell, M. R. and McLaughlin, R. E.**, Influence on the protozoan *Mattesia grandis* McLaughlin on the toxicity to the boll weevil of four insecticides, *J. Econ. Entomol.*, 63, 266, 1970.

513b. **Listov, M. V. and Nesterov, V. A.**, Resistance of flourbeetle larvae (Coleoptera, Tenebrionidae) to methyl bromide in relation to their infection with microsporidia of *Nosema whitei* Weiser, 1953 and coccidia of *Adelina tribolii* Bhatia, 1937, *Entomol. Rev.*, 59, 13, 1980.

514. **Mussgnug, G. L. and Henry, J. E.**, Compatability of malathion and *Nosema locustae* Canning in *Melanoplus sanguinipes* (F.), *Acrida*, 8, 77, 1979.

515. **Mussgnug, G. L.**, Integration of *Nosema locustae* with Chemical Insecticides and Entomopoxvirus for Control of Grasshoppers, Ph.D. thesis, Montana State University, Bozeman, 1980.

515a. **Morris, O. N.**, Susceptibility of the migratory grasshopper, *Melanoplus sanguinipes* (Orthoptera: Acrididae), to mixtures of *Nosema locustae* (Microspora: Nosematidae) and chemical insecticides, *Can. Entomol.*, 117, 131, 1985.

516. **Onsager, J. A., Henry, J. E., Foster, R. N., and Staten, R. T.**, Acceptance of wheat bran bait by species of rangeland grasshoppers, *J. Econ. Entoml.*, 73, 548, 1980.

517. **Onsager, J. A., Rees, N. E., Henry, J. E., and Foster, R. N.**, Integration of bait formulations of *Nosema locustae* and carbaryl for control of rangeland grasshoppers, *J. Econ. Entomol.*, 74, 183, 1981.

517a. **Henry, J. E.**, Effect of grasshopper species, cage density, light intensity, and method of inoculation on mass production of *Nosema locustae* (Microsporida: Nosematidae), *J. Econ. Entoml.*, 78, 1245, 1985.

518. **Lublinkhof, J. and Lewis, L. C.**, Virulence of *Nosema pyrausta* to the European corn borer when used in combination with insecticides, *Environ. Entomol.*, 9, 67, 1980.

519. **Lublinkhof, J., Lewis, L. C., and Berry, E. C.**, Effectiveness of integrating insecticides with *Nosema pyrausta* for suppressing populations of the European corn borer, *J. Econ. Entomol.*, 72, 880, 1979.

520. **Goetze, G. and Zeutzschel, B.**, Nosema disease of honeybees, and its control with drugs: review of research work since 1954, *Bee World*, 40, 217, 1959.

521. **Gochnauer, T. A., Furgala, B., and Shimanuki, H.**, Diseases and enemies of the honey bee, in *The Hive and The Honey Bee*, Dadant, C. and Sons, Eds., Dadant & Sons, Hamilton, Illinois, 1975, 615.

522. **Furgala, B. and Mussen, E. C.**, Protozoa, in *Honey Bee Pests, Predators, and Diseases*, Morse, R. A., Ed., Comstock, Ithaca, N.Y., 1978, 62.

523. **Katznelson, H. and Jamieson, C. A.**, Control of nosema disease of honey bees with fumagillin, *Science*, 115, 70, 1952.

524. **Bailey, L.**, Effect of fumagillin upon *Nosema apis* (Zander), *Nature (London)*, 171, 212, 1953.

525. **Przelecka, A. and Hartwig, A.**, Cytochemical and autoradiographic investigation on nucleic acids in the intestine of *Apis mellifera* infected with *Nosema apis* Zander and treated with fumagillin DHC, *Proc. 21st Int. Agric. Congr.*, Washington, D.C., 1967, 488.

526. **Hartwig, A. and Przelecka, A.**, Nucleic acids in intestine of *Apis mellifera* infected with *Nosema apis* and treated with fumagillin DHC: cytochemical and autoradiographic studies, *J. Invertebr. Pathol.*, 18, 331, 1971.

527. **Liu, T. P.**, Effects of Fumidil B on the spore of *Nosema apis* and on lipids of the host cell as revealed by freeze-etching, *J. Invertebr. Pathol.*, 22, 364, 1973.

528. **Jaronski, S. T.**, Cytochemical evidence for RNA synthesis inhibition by fumagillin, *J. Antibiot.*, 25, 327, 1972.

529. **Moffett, J. O., Lackett, J. J., and Hitchcock, J. D.**, Compounds tested for control of nosema in honey bees, *J. Econ. Entomol.*, 62, 886, 1969.

530. **Mussen, E. C. and Furgala, B.**, Benomyl ineffective against *Nosema apis*, *Am. Bee J.*, 115, 478, 1975.

531. **Lewis, L. C. and Lynch, R. E.**, Treatment of *Ostrinia nubilalis* larvae with Fumidil B to control infections caused by *Perezia pyraustae*, *J. Invertebr. Pathol.*, 15, 43, 1970.

532. **Lynch, R. E. and Lewis, L. C.**, Reoccurrence of the microsporidian *Perezia pyraustae* in the European corn borer, *Ostrinia nubilalis*, reared on diet containing Fumidil B, *J. Invertebr. Pathol.*, 17, 243, 1971.

533. **Flint, H. M., Eaton, J., and Klassen, W.**, The use of Fumidil B to reduce microsporidian disease in colonies of the boll weevil, *Ann. Entomol. Soc. Am.*, 65, 942, 1972.

534. **Wilson, G. G.**, The use of Fumidil B to suppress the microsporidian *Nosema fumiferanae* in stock cultures of the spruce budworm, *Choristoneura fumiferana* (Lepidoptera: Tortricidae), *Can. Entomol.*, 106, 995, 1974.

535. **Shinholster, D. L.**, The Effects of X-Irradiation and Chemotherapy on the Host-Parasite Relationship Between *Tribolium castaneum* (Herbst) and Two Protozoan Parasites, *Nosema whitei* Weiser and *Adelina tribolii* Hesse, Ph.D. thesis, Cornell University, Ithaca, N.Y., 1974.

536. **Jaronski, S. T.**, Suppression of the microsporidan *Octosporea muscaedomesticae* in *Phormia regina* by fumagillin, in *Proc. 1st Int. Colloq. Invertebr. Pathol.*, Kingston, Ontario, 1976, 363.

537. **Armstrong, E.,** Fumidil B and benomyl: chemical control of *Nosema kingi* in *Drosophila willistoni, J. Invertebr. Pathol.,* 27, 363, 1976.
538. **Sohi, S. S. and Wilson, G. G.,** Effect of antimicrosporidian and antibacterial drugs on *Nosema disstriae* (Microsporida) infection in *Malacosoma disstria* (Lepidoptera: Lasiocampidae) cell cultures, *Can. J. Zool.,* 57, 1222, 1979.
539. **Bayne, C. J., Owczarzak, A., and Noonan, W. E.,** *In vitro* cultivation of cells and a microsporidian parasite of *Biomphalaria glabrata* (Pulmonata: Basommatophora), *Ann. N.Y. Acad. Sci.,* 266, 513, 1975.
540. **Kurtti, T. J. and Brooks, M. A.,** The rate of development of a microsporidan in moth cell culture, *J. Invertebr. Pathol.,* 29, 126, 1977.
541. **McCowen, M. C., and Callender, M. E., and Lawlis, J. F.,** Fumagillin (H$_3$), a new antibiotic with amoebicidal properties, *Nature (London),* 113, 202, 1951.
542. **Bailey, L.,** Control of amoeba disease by the fumigation of combs and by fumagillin, *Bee World,* 36, 162, 1955.
543. **Dunkel, F. V. and Boush, G. M.,** Biology of the gregarine *Pyxinia frenzeli* in the black carpet beetle, *Attagenus megatoma, J. Invertebr. Pathol.,* 11, 281, 1968.
544. **Hsiao, T. H. and Hsiao, C.,** Benomyl: a novel drug for controlling a microsporidan disease of the alfalfa weevil, *J. Invertebr. Pathol.,* 22, 303, 1973.
545. **Harvey, G. T. and Gaudet, P. M.,** The effects of benomyl on the incidence of microsporidia and the developmental performance of eastern spruce budworm (Lepidoptera: Tortricidae), *Can. Entomol.,* 109, 987, 1977.
546. **Brooks, W. M., Cranford, J. D., and Pearce, L. W.,** Benomyl: effectiveness against the microsporidian *Nosema heliothidis* in the corn earworm, *Heliothis zea, J. Invertebr. Pathol.,* 31, 239, 1978.
547. **Overstreet, R. M. and Whatley, E. C.,** Prevention of microsporidiosis in the blue crab with notes on natural infections, in *Proc. 6th Annu. Meet. World Maricult. Soc.,* Avoult, J. V. and Miller, R., Eds., Louisiana State University Press, Baton Rouge, 1976, 335.
548. **Overstreet, R. M.,** Buquinolate as a preventive drug to control microsporidosis in the blue crab, *J. Invertebr. Pathol.,* 26, 213, 1975.
548a. **Briese, D. T. and Milner, R. J.,** Effect of the microsporidian *Pleistophora schubergi* on *Anaitis efformata* (Lepidoptera: Geometridae) and its elimination from a laboratory colony, *J. Invertebr. Pathol.,* 48, 107, 1986.
549. **Bailey, L.,** *Infectious Diseases of the Honey-Bee,* Land Books, London, 1963.
550. **Henry, J. E. and Oma, E. A.,** Sulphonamide antibiotic control of *Malameba locustae* (King & Taylor) and its effect on grasshoppers, *Acrida,* 4, 217, 1975.
551. **Giordani, G.,** Amoeba disease of the honey bee, *Apis mellifera* Linnaeus, and an attempt at its chemical control, *J. Insect Pathol.,* 1, 245, 1959.
552. **Higby, G. C., Canning, E. U., Pilley, B. M., and Bush, P. J.,** Propagation of *Nosema eurytremae* (Microsporida: Nosematidae) from trematdoe larvae, in abnormal hosts and in tissue culture, *Parasitology,* 78, 155, 1979.
553. **Alger, N. E. and Undeen, A. H.,** The control of a microsporidian, *Nosema* sp., in an anopheline colony by an egg-rinsing technique, *J. Invertebr. Pathol.,* 15, 321, 1970.
554. **McLaughlin, R. E. and Bell, M. R.,** Mass production *in vivo* of two protozoan pathogens, *Mattesia grandis* and *Glugea gasti,* of the boll weevil, *Anthonomus grandis, J. Invertebr. Pathol.,* 16, 84, 1970.
555. **Nordin, G. L.,** Dietary effects of methoprene on *Vairimorpha necatrix* spore yield in *Heliothis virescens, J. Invertebr. Pathol.,* 37, 110, 1981.
556. **Weiser, J.,** Reported as personal communication by McLaughlin in Reference 1.
557. **Kaya, H. K.,** Persistence of spores of *Pleistophora schubergi* (Cnidospora: Microsporida) in the field, and their application in microbial control, *J. Invertebr. Pathol.,* 26, 329, 1975.
558. **Gardner, W. A., Sutton, R. M., and Noblet, R.,** Persistence of *Beauveria bassiana, Nomuraea rileyi,* and *Nosema necatrix* on soybean foliage, *Environ. Entomol.,* 6, 616, 1977.
559. **Fuxa, J. R. and Brooks, W. M.,** Persistence of spores of *Vairimorpha necatrix* on tobacco, cotton, and soybean foliage, *J. Econ. Entomol.,* 71, 169, 1978.
560. **Lewis, L. C.,** Persistence of *Nosema pyrausta* and *Vairimorpha necatrix* measured by microsporidiosis in the European corn borer, *J. Econ. Entomol.,* 75, 670, 1982.
561. **Germida, J. J.,** Persistence of *Nosema locustae* spores in soil as determined by fluorescence microscopy, *Appl. Environ. Microbiol.,* 47, 313, 1984.
561a. **Germida, J. J., Ewen, A. B., and Onofriechuk, E. E.,** *Nosema locustae* Canning (Microsporida) spore populations in treated field soils and resident grasshopper populations, *Can. Entomol.,* 119, 335, 1987.
562. **Anthony, D. W., Savage, K. E., Hazard, E. I., Avery, S. W., Boston, M. D., and Oldacre, S. W.,** Field tests with *Nosema algerae* Vávra and Undeen (Microsporida, Nosematidae) against *Anopheles albimanus* Wiedemann in Panama, *Misc. Publ. Entomol. Soc. Am.,* 11, 17, 1978.
563. **Alger, N. B. and Maddox, J. V.,** unpublished data, 1977.

563a. **Kucera, M. and Weiser, J.**, Different course of proteolytic inhibitory activity and proteolytic activity in *Galleria mellonella* larvae infected by *Nosema algerae* and *Vairimorpha heterosporum, J. Invertebr. Pathol.*, 45, 41, 1985.
564. **Ignoffo, C. M.**, Specificity of insect viruses, *Bull. Entomol. Soc. Am.*, 14, 265, 1968.
565. **Ignoffo, C. M.**, Evaluation of in vivo specificity of insect viruses, in *Baculoviruses for Insect Pest Control: Safety Considerations,* Summers, M., Engler, R., Falcon, L. A., and Vail, P. V., Eds., Am. Soc. Microbiol., Washington, D.C., 1975, 52.
566. **Weiser, J.**, Future requirements for laboratory research, Mimeographed document, World Health Organization, VBC/WP/73.8, Rome, 1973.
567. **Jaronski, S. T.**, Microsporida in cell culture, in *Advances in Cell Culture,* Vol. 3, Maramorosch, K., Ed., Academic Press, New York, 1984, 183.
568. **Kurtti, T. J. and Brooks, M. A.**, Growth of a microsporidian parasite in cultured cells of tent caterpillars (*Malacosoma*), *Curr. Topics Microbiol. Immunol.*, 55, 204, 1971.
569. **Kurtti, T. J., and Brooks, M. A.**, Propagation of a microsporidan in a moth cell line, in *Invertebrate Tissue Culture, Applications in Medicine, Biology, and Agriculture,* Kurstak, E. and Maramorosch, K., Eds., Academic Press, New York, 1976, 395.
570. **Kurtti, T. J. and Brooks, M. A.**, Propagation of microsporidia in invertebrate cell culture, in Proc. 1st Int. Colloq. Invertebr. Pathol., Kingston, Ontario, 1976, 123.
571. **Tsang, K. R., Brooks, M. A., and Kurtti, T. J.**, Culture conditions regulating the infection of cells by an intracellular microorganism, in *Invertebrate Cell Culture Applications,* Maramorosch, K., Ed., Academic Press, New York, 1982, 125.
571a. **Kurtti, T. J., Tsang, K. R., and Brooks, M. A.**, The spread of infection by the microsporidan, *Nosema disstriae,* in insect cell lines, *J. Protozool.*, 30, 652, 1983.
572. **Ishihara, R.**, Growth of *Nosema bombycis* in primary cell cultures of mammalian and chicken embryos, *J. Invertebr. Pathol.*, 11, 328, 1968.
572a. **Streett, D. A. and Lynn, D. E.**, *Nosema bombycis* replication in a *Manduca sexta* cell line, *J. Parasitol.*, 70, 452, 1984.
572b. **Kawarabata, T. and Ishihara, R.**, Infection and development of *Nosema bombycis* (Microsporida: Protozoa) in a cell line of *Antheraea eucalypti, J. Invertebr. Pathol.*, 44, 52, 1984.
573. **Weidner, E.**, Ultrastructural study of microsporidian invasion into cells, *Z. Parasitenk.*, 40, 227, 1972.
574. **Jaronski, S. T.**, personal communication, 1982.
574a. **Smith, J. E., Barker, R. J., and Lai, P. F.**, Culture of microsporidia from invertebrates in vertebrate cells, *Parasitology,* 85, 427, 1982.
575. **Grobov, O. F. and Zuman, B. V.**, Cultivation of *Nosema apis* Zander in the culture of tissue of honey bee, *Parazitologiya,* 6, 176, 1972.
576. **Trager, W.**, The hatching of spores of *Nosema bombycis* Nägeli and the partial development of the organism in tissue cultures, *J. Parasitol.*, 23, 226, 1937.
577. **Ishihara, R. and Sohi, S. S.**, Infection of ovarian tissue culture of *Bombyx mori* by *Nosema bombycis* spores, *J. Invertebr. Pathol.*, 8, 538, 1966.
578. **Sohi, S. S.**, In vitro cultivation of hemocytes of *Malacosoma disstria* Hübner (Lepidoptera: Lasiocampidae), *Can. J. Zool.*, 49, 1355, 1971.
579. **Sohi, S. S. and Wilson, G. G.**, Persistent infection of *Malacosoma disstria* (Lepidoptera: Lasiocampidae) cell cultures with *Nosema (Glugea) disstriae* (Microsporida: Nosematidae), *Can. J. Zool.*, 54, 336, 1976.
580. **Wilson, G. G. and Sohi, S. S.**, Effect of temperature on healthy and microsporidia-infected continuous cultures of *Malacosoma dissstria* hemocytes, *Can. J. Zool.*, 55, 713, 1977.
581. **Gupta, K. S.**, Cultivation of *Nosema mesnili* Paillot (Microsporidia) *in vitro, Curr. Sci.*, 33, 407, 1964.
582. **Purrini, K. and Zizka, Z.**, More on the life cycle of *Malamoeba scolyti* (Amoebidae: Sarcomastigophora) parasitizing the bark beetle *Dryocoetes autographus* (Scolytidae, Coleoptera), *J. Invertebr. Pathol.*, 42, 96, 1983.
583. **Bailey, L.**, Honey bee pathology, *Annu. Rev. Entomol.*, 13, 191, 1968.
584. **Corbel, J. C.**, Les parasites des Orthoptères, *Ann. Biol.*, 6, 392, 1967.
585. Cited by Vávra in Reference 110.
586. Cited by Levine in Reference 94.
587a. **Pérez, C.**, Le cycle évolutif d'*Adelea mesnili,* coccidian coelomique parasite d'un Lépidoptere, *Arch. Protistenk.*, 2, 1, 1903.
587b. **Steinhaus, E. A.**, A coccidian parasite of *Ephestia kühniella* Zeller and of *Plodia interpunctella* (Hbn.) (Lepidoptera: Phycitidae), *J. Parasitol.*, 33, 29, 1947.
588. **Bhatia, M. L.**, On *Adelina tribolii,* a coccidian parasite of *Tribolium ferrugineum* F., *Parasitology,* 29, 239, 1937.
589. **Sprague, V. and Vávra, J.**, Preface, in *Comparative Pathobiology,* Vol. 1, Bulla, L. A. and Cheng, T. C., Eds., Plenum Press, New York, 1976, IX.
590. **Weiser, J.**, Microsporidia in invertebrates: host-parasite relations at the organismal level, in *Comparative Pathobiology,* Vol. 1, Bulla, L. A. and Cheng, T. C., Eds., Plenum Press, New York, 1976, 163.

591. **Maddox, J. V., Alger, N. E., Ahmad, A., and Aslamkhan, M.**, The susceptibility of some Pakistan mosquitoes to *Nosema algerae* (Microsporidia), *Pakistan J. Zool.*, 1, 19, 1977.
592. **Undeen, A. H.**, *In vivo* germination and host specificity of *Nosema algerae* in mosquitoes, *J. Invertebr. Pathol.*, 27, 343, 1976.
593. **Hazard, E. I.**, The use of microsporidia (Protozoa) for the control of aquatic insect pests, in Impact of the Use of Microorganisms on the Aquatic Environment, **Bourquin, A. W., Ahearn, D. G., and Meyers, S. P.**, Eds., *U.S. Environmental Protection Agency, U.S. Ecological Research Service*, 660/3-75-001, Washington, D.C., 1975, 69.
594. **Undeen, A. H. and Maddox, J. V.**, The infection of nonmosquito hosts by injection with spores of the microsporidian *Nosema algerae*, *J. Invertebr. Pathol.*, 22, 258, 1973.
595. **Undeen, A. H. and Alger, N. E.**, *Nosema algerae:* infection of the white mouse by a mosquito parasite, *Exp. Parasitol.*, 40, 86, 1976.
596. **Alger, N. E., Maddox, J. V., and Shadduck, J. A.**, *Nosema algerae:* infectivity and immune response to normal and nude mice, Mimeographed document, World Health Organization, VBC/80.778, Rome, 1980.
597. **Alger, N. E. and Maddox, J. V.**, unpublished data, 1978.
598. **Hazard, E. I.**, Reported as personal communication by Chapman et al. in Reference 638.
599. Cited in Reference 344.
600. **Van Essen, F. W. and Anthony, D. W.**, Susceptibility of nontarget organisms to *Nosema algerae* (Microsporida: Nosematidae), a parasite of mosquitoes, *J. Invertebr. Pathol.*, 28, 77, 1976.
600a. **Hostounský, Z.**, Action of the microsporidian *Nosema algerae* on hosts of different insect orders (Diptera, Lepidoptera, Coleoptera), *J. Protozool.*, 31, 50A, 1984.
600b. **Hostounský, Z.**, Formation of various morphological types in the spores of *Nosema algerae* after their oral transfer to *Gastrophysa viridula* (Coleoptera), *J. Protozool.*, 29, 504, 1982.
600c. **Hostounský, Z. and Kodys, F.**, Infection potential of *Nosema algerae* for the Colorado potato beetle, *Leptinotarsa decemlineata*, *J. Protozool.*, 29, 504, 1982.
600d. **Hostounský, Z. and Kodys, F.**, Infection potential of *Nosema algerae* to cereal leaf beetles, *Oulema melanopus* and *O. lichenis*, *Abstr. Soc. Protozool.*, 27A, 1985.
601. **Costa, C. A. F. and Bradley, R. E.**, Hyperparasitism of intrasnail stages of *Fasciola hepatica* by a mosquito microsporidian parasite, *J. Invertebr. Pathol.*, 35, 175, 1980.
602. **Lai, P. F. and Canning, E. U.**, Infectivity of a microsporidium of mosquitoes *(Nosema algerae)* to larval stages of *Schistosoma mansoni* in *Biomphalaria glabrata*, *Int. J. Parasitol.*, 10, 293, 1980.
603. **Ignoffo, C. M. and Garcia, C.**, Susceptibility of larvae of the black cutworm to species of entomopathogenic bacteria, fungi, protozoa, and viruses, *J. Econ. Entomol.*, 72, 767, 1979.
604. **Thomson, H. M.**, A list and brief description of the microsporidia infecting insects, *J. Insect. Pathol.*, 2, 346, 1960.
605. **Bird, F. T. and Whalen, M. M.**, A special survey of the natural control of the spruce budworm, *Choristoneura fumiferana* (Clem.) in northern Ontario in 1949, *For. Insect Lab. Annu. Rep. 2*, Sault Ste. Marie, Ontario, Canada, 1949, 7.
606. **Henry, J. E.**, *Nosema locustae:* an alternative method of grasshopper control, in Proc. Advancement Pesticides, State Dept. of Health and Environmental Science, Environ. Sci. Div., Helena, Mont., 1975, 16.
607. **Menapace, D. M., Sackett, R. R., and Wilson, W. T.**, Adult honey bees are not susceptible to infection by *Nosema locustae*, *J. Econ. Entomol.*, 71, 304, 1978.
608. **Shadduck, J. A.**, Maximum challenge safety tests of *Nosema locustae* in rabbits and mice, unpublished report submitted to the U.S. Environmental Protection Agency, Washington, D.C., 1980.
609. **Henry, J. E.**, Tests on the viability of spores in ruminant animals, unpublished report, 1977.
610. **Henry, J. E., Oma, E. A., and Billeb, B.**, Safety of *Nosema locustae:* bioassay of tissues from rats treated with spores, unpublished report, 1973.
611. **McEwen, L. C., Hanson, B. L., Levy, E. R., and Haegele, M. A.**, Wildlife hazards related to new range grasshopper control chemicals and other materials, unpublished progress report, 1977.
612. **Hudson, R. H. and Zinkl, J. G.**, Zenobia basin birds and mammals: microscopic examination of tissues, unpublished report, 1979.
613. **McCracken, R. J.**, Availability of research data for registration of *Nosema locustae*, *Fed. Reg.*, 44, 9609, 1979.
613a. **Germida, J. J., Onofriechuk, E. E., and Ewen, A. B.**, Effect of *Nosema locustae* Canning (Microsporida) and three chemical insecticides on microbial activity in soil, *Can. J. Soil Sci.*, 67, 631, 1987.
614. **Luna, G. C., Henry, J. E., and Ronderos, R. A.**, Infecciones experimentales y naturales con protozoos patogenos en acridios de la Republica Argentina (Insecta, Orthoptera), *Rev. Soc. Entomol. Argentina*, 40, 243, 1981.

615. **Hazleton Laboratories, Inc.,** Primary skin irritation study in rabbits: lactose-virus polyhedra *(A. californica), Hirsutella thompsoni* spores and mycelium, spores of *Nosema locustae*, unpublished report, Vienna, Va., 1972.
616. **Hazleton Laboratories, Inc.,** Acute dermal toxicity-guinea pigs: *A. californica* virus polyhedra; *A. californica* ME virus (released rods); *Nosema locustae* (spores); and *Hirsutella thompsoni* (spores and mycelium), unpublished report, Vienna, Va., 1973.
617. **Hazleton Laboratories, Inc.,** Acute inhalation toxicity-rats: *Hirsutella thompsoni-fungus, Nosema locustae*-microsporidian, and *Trichoplusia ni*-virus *(A. californica)*, unpublished report, Vienna, Va., 1973.
618. **Hazleton Laboratories, Inc.,** 13-Week dietary administration-rats — F, L: *A. californica* ME virus, *Nosema locustae* spores, unpublished report, Vienna, Va., 1973.
619. **Hazleton Laboratories, Inc.,** Subacute oral-rats: *Hirsutella thompsoni* fungus, *Nosema locustae*-microsporidian, and *Trichoplusia ni*-virus *(A. californica)*, unpublished report, Vienna, Va., 1973.
620. **Hazleton Laboratories, Inc.,** Acute LC_{50}-rainbow trout and bluegill sunfish: *Nosema locustae*, unpublished report, Vienna, Va., 1974.
621. **Hudson, R. H. and Zinkl, J. G.,** Results of *Nosema locustae* toxicity tests with pheasants and mallard ducklings, unpublished report, 1979.
622. **Heimpel, A. M.,** Intraperitoneal injections in mice using *Nosema locustae* spores, unpublished report, 1978.
623. **Hill, R. E., Carpino, D. P., and Mayo, Z. B.,** Insect parasites of the European corn borer *Ostrinia nubilalis* in Nebraska from 1948—1976, *Environ. Entomol.*, 7, 249, 1978.
624. **Andreadis, T. G.,** Current status of imported and native parasites of the European corn borer in Connecticut, *J. Econ. Entomol.*, 75, 626, 1982.
625. **Purrini, K. and Skatulla, U.,** Über die natürlichen Krankheiten der Frostspanner *Operophtera brumata* L. und *Erannis defoliaria* Clerck (Lepidoptera, Geometridae) im Spessart, Bayern, *Anz. Schaedlingskd. Pflanz. Umweltschutz*, 52, 20, 1979.
626. **Wilson, G. G.,** Effects of the microsporidia *Nosmea disstriae* and *Pleistophora schubergi* on the survival of the forest tent caterpillar, *Malacosoma disstria* (Lepidoptera: Lasiocampidae), *Can. Entomol.*, 109, 1021, 1977.
626a. **Wilson, G. G.,** Pathogenicity of *Nosema disstriae, Pleistophora schubergi* and *Vairimorpha necatrix* (Microsporidia) to larvae of the forest tent caterpillar, *Malacosoma disstria, Z. Parasitenkd.*, 70, 763, 1984.
627. **Metspalu, L.,** On the effect of microsporidiosis on hibernating pupae of noctuids, *Eesti NSV Tead. Akad. Toim. Biol.*, 25, 13, 1976.
628. **Simchuk, P. A.,** Susceptibility of the codling moth, the pear moth and the plum fruit moth to the microsporidium *Pleistophora schubergi, Biol. Nauki*, 10, 51, 1978.
629. **Jaques, R. P.,** Field efficacy of viruses infectious to the cabbage looper and imported cabbageworm on late cabbage, *J. Econ. Entomol.*, 70, 111, 1977.
630. **Fuxa, J. R. and Brooks, W. M.,** Effects of *Vairimorpha necatrix* in sprays and corn meal on *Heliothis* species in tobacco, soybeans, and sorghum, *J. Econ. Entomol.*, 72, 462, 1979.
631. **Jordan, J. A. and Noblet, R.,** Susceptibility of selected insect pests to two species of microsporidan pathogens, *Nosema heliothidis* and *Vairimorpha necatrix, S. C. Agric. Exp. Stn. Tech. Bull.*, 1069, 1, 1978.
632. **Hamm, J. J. and Lynch, R. E.,** Comparative susceptibility of the granulate cutworm, fall armyworm, and corn earworm to some entomopathogens, *J. Ga. Entomol. Soc.*, 17, 363, 1982.
633. **Mistric, W. J. and Smith, F. D.,** Tobacco budworm: control on flue-cured tobacco with certain microbial pesticides, *J. Econ. Entomol.*, 66, 979, 1973.
634. **Brooks, W. M.,** unpublished data, 1970.
635. **Fuxa, J. R.,** Susceptibility of lepidopterous pests to two types of mortality caused by the microsporidium *Vairimorpha necatrix, J. Econ. Entomol.*, 74, 99, 1981.
636. **Brooks, W. M.,** unpublished observations, 1979.
637. **Anon.,** Data sheet on the biological control agent *Vavraia (Pleistophora) culicis* (Weiser 1946), Mimeographed document, World Health Organization, VBC/80.759, Rome, 1980.
638. **Chapman, H. C., Clark, T. B., and Petersen, J. J.,** Protozoans, nematodes, and viruses of anophelines, *Misc. Publ. Entomol. Soc. Am.*, 7, 134, 1970.
639. **Weiser, J.,** Production of the microsporidian *Plistophora culicis* Weiser in substitute host, *Folia Parasitol.*, 25, 365, 1978.
639a. **Weiser, J.,** Transmission of microsporidia to insects via injection, *Vestn. Cesk. Spol. Zool.*, 42, 311, 1978.
639b. **Sedlacek, J. D., Dintenfass, L. P., Nordin, G. L., and Ajlan, A. A.,** Effects of temperature and dosage on *Vairimorpha* sp. 696 spore morphometrics, spore yield, and tissue specificity in *Heliothis virescens, J. Invertebr. Pathol.*, 46, 320, 1985.
640. **Martignoni, M. M.,** Mass production of insect pathogens, in *Biological Control of Insect Pests and Weeds,* DeBach, P., Ed., Chapman, and Hall, London, 1964, 579.

641. **Ignoffo, C. M.,** Possibilities of mass-producing insect pathogens, in *Insect Pathology and Microbial Control,* van der Laan, P. V., Ed., North-Holland, Amsterdam, 1967, 91.
642. **Ignoffo, C. M. and Hink, W. F.,** Propagation of arthropod pathogens in living systems, in *Microbial Control of Insects and Mites,* Burges, H. D. and Hussey, N. W., Eds., Academic Press, New York, 1971, 541.
643. **Cole, R. J.,** The application of the "triangulation" method to the purification of *Nosema* spores form insect tissues, *J. Invertebr. Pathol.,* 15, 193, 1970.
644. **Undeen, A. H. and Alger, N. E.,** A density gradient method for fractionating microsporidan spores, *J. Invertebr. Pathol.,* 18, 419, 1971.
645a. **Kelly, J. F. and Knell, J. D.,** A simple method for cleaning microsporidian spores, *J. Invertebr. Pathol.,* 33, 252, 1979.
645b. **Undeen, A. H. and Avery, S. W.,** Continuous flow-density gradient centrifugation for purification of microsporidian spores, *J. Invertebr. Pathol.,* 42, 403, 1983.
645c. **Jouvenaz, D. P.,** Percoll: an effective medium for cleaning microsporidian spores, *J. Invertebr. Pathol.,* 37, 319, 1981.
646. **Gast, R. T. and Davich, T. B.,** Boll weevils, in *Insect Colonization and Mass Production,* Smith, C. N., Ed., Academic Press, New York, 1966, 405.
647. **Henry, J. E.,** Microbial control of grasshoppers with *Nosema locustae* Canning, *Misc. Publ. Entomol. Soc. Am.,* 11, 85, 1978.
648. **Pilley, B. M., Canning, E. U., and Hammond, J. C.,** The use of a microinjection procedure for large-scale production of the microsporidian *Nosema eurytremae* in *Pieris brassicae, J. Invertebr. Pathol.,* 32, 355, 1978.
649. **Lai, P. F. and Canning, E. U.,** Some factors affecting spore replication of *Nosema algerae* (Microspora, Nosematidae) in *Pieris brassicae* (Lepidoptera), *J. Invertebr. Pathol.,* 41, 20, 1983.
649a. **Hostounský, Z.,** Production of microsporidia pathogenic to the Colorado potato beetle *(Leptinotarsa decemlineata)* in alternate hosts, *J. Invertebr. Pathol.,* 44, 166, 1984.
650. **Hostounský, Z. and Weiser, J.,** Production of spores of *Nosema plodiae* Kellen et Lindegren in *Mamestra brassicae* L. after different infective dosage, I, *Vestn. Cesk. Spol. Zool.,* 36, 97, 1972.
651. **Fowler, J. L. and Reeves, E.,** In vivo propagation of a microsporidian pathogenic to insects, *J. Invertebr. Pathol.,* 25, 349, 1975.
652. **Wilson, G. G. and Kaupp, W. J.,** Application of a microsporidia, *Nosema fumiferanae* against spruce budworm on Manitoulin Island, 1975, *Environ. Can. For. Serv. Inf. Rep.,* 1P-X-11, Ottawa, Ontario, Canada, 1975.
653. **Wilson, G. G. and Kaupp, W. J.,** Application of *Nosema fumiferanae* and *Pleistophora schubergi* (Microsporida) against the spruce budworm in Ontario, 1976, *Environ. Can. For. Serv. Inf. Rep.,* 1P-X-15, Ottawa, Ontario, Canada, 1976.
654. **Wilson, G. G.,** A method for mass producing spores of the microsporidian *Nosema fumiferanae* in its host, the spruce budworm, *Choristoneura fumierana* (Lepidoptera: Tortricidae), *Can. Entomol.,* 108, 383, 1976.
655. **Wilson, G. G. and Kaupp, W. J.,** Application of *Nosema disstriae* and *Pleistophora schubergi* (Microsporida) against the forest tent caterpillar in Ontario, 1977, *Environ. Can. For. Serv. Inf. Rep.,* FPM-X-4, 1977.
656. **Lewis, L. C. and Lynch, R. E.,** Foliar application of *Nosema pyrausta* for suppression of populations of European corn borer, *Entomophaga,* 23, 83, 1978.
657. **Hostounský, Z.,** The utilization of a natural sedimentation and Brownian movement in a concentration and separation of microsporidian spores from insect tissue, *J. Invertebr. Pathol.,* 38, 431, 1981.
658. **Nilova, G. N.,** Vlijanie mikrosporidioza na nekotorye fiziologiceskie funkeii ozimoj sovki *(Agrotis segetum* Schiff.), *Izv. Akda. Nauk Tadzzh. SSR Biol. Otd.,* 3, 66, 1967.
659. **Weiser, J. and Hostounský, Z.,** Production of spores of *Nosema plodiae* Kellen et Lindegren in *Mamestra brassicae* L. after different infective dosage II. Comparison with *Nosema heterosporum* Kellen et Lindegren, *Vestn. Cesk. Spol. Zool.,* 37, 234, 1973.
660. **Sprenkel, R. K.,** Studies on *Nosema plodiae* (Sporozoa, Microsporida) and *Mattesia dispora* (Sporozoa, Gregarina) Pathogens of the Indian Meal Moth, *Plodia interpunctella* (Lepidoptera, Pyralidae) in Illinois, Ph.D. thesis, University of Illinois, Urbana, 1973.
661. **Schwalbe, C. P., Burkholder, W. E., and Boush, G. M.,** *Mattesia trogodermae* infection rates as influenced by mode of transmission, dosage and host species, *J. Stored Prod. Res.,* 10, 161, 1974.
662. **Guttman, H. N. and Wallace, F. G.,** Nutrition and physiology of the Trypanosomatidae, in *Biochemistry and Physiology of the Protozoa,* Vol. 3, Hutner, S., Ed., Academic Press, New York, 1964, 459.
663. **Hutner, S. H., Bacchi, C. J., and Baker, H.,** Nutrition of the Kinetoplastida, in *Biology of the Kinetoplastida,* Vol. 2, Lumsden, W. H. R. and Evans, D. A., Eds., Academic Press, New York, 1979, 653.
664. **Elliot, A. M., Ed.,** *Biology of Tetrahymena,* Van Nostrand Reinhold, New York, 1973.
665. **Kidder, G. W. and Dewey, V. C.,** The biochemistry of ciliates in pure culture, in *Biochemistry and Physiology of Protozoa,* Vol. 1, Lwoff, A., Ed., Academic Press, New York, 1951, 323.

666. **Kramer, J. P.,** *Herpetomonas muscarum* (Leidy) in the haemocoel of larval *Musca domestica* L., *Entomol. News,* 72, 165, 1961.
667. **Henry, J. E.,** Strategies for using pathogenic microorganisms to control noxious insects in the pasture and rangeland ecosystems, in Microbial Control of Insect Pests: Future Strategies in Pest Management Systems, **Allen, G. E., Ignoffo, C. M., and Jaques, R. P., Eds.,** Workshop, Univ. Florida, Gainesville, 1979, 195.
668. **Sneller, V. P.,** Inhibition of *Dirofilaria immitis* in gregarine-infected *Aedes aegypti:* preliminary observations, *J. Invertebr. Pathol.,* 34, 62, 1979.
669. **Thompson, A. C. and McLaughlin, R. E.,** Comparison of the lipids and fatty acids of *Mattesia grandis* and the fat body of the host, *Anthonomus grandis, J. Invertebr. Pathol.,* 30, 108, 1977.
670. **Schwalbe, C. P., Boush, G. M., and Burkholder, W. E.,** Physical and physiological characteristics of *Trogoderma glabrum* infected with the schizogregarine pathogen *Mattesia trogodermae, J. Invertebr. Pathol.,* 22, 153, 1973.
671. **Shapas, T. J., Burkholder, W. E., and Boush, G. M.,** Population suppression of *Trogoderma glabrum* by using pheromone luring for protozoan pathogen dissemination, *J. Econ. Entomol.,* 70, 469, 1977.
672. **Burkholder, W. E.,** Manipulation of insect pests of stored products, in *Chemical Control of Insect Behavior: Theory and Application,* Shorey, H. H. and McKelvey, J. J., Eds., John Wiley & Sons, New York, 1977, 345.
673. **Burkholder, W. E.,** Application of pheromones for manipulating insect pests of stored products, Proc. Symp. Insect Pheromones and Their Applications, Ministry of Agriculture and Forestry, Tokyo, 1977.
674. **Burkholder, W. E. and Shapas, T. J.,** Use of entomopathogens with pheromones and attractants in pest management systems for stored-product insects, in Microbial Control of Insect Pests: Future Strategies in Pest Management Systems, Allen, G. E., Ignoffo, C. M., and Jaques, R. P., Eds., Workshop, Univ. Florida, Gainesville, 1979, 236.
675. **Canning, E. U.,** Insect control with protozoa, in *Biological Control in Crop Production, 5. Beltsville Symposia in Agricultural Research,* Papavizas, G. C., Ed., Allanheld Osmun, New Jersey, 1981, 201.
676. **Haq, N., Reisen, W. K., and Aslamkhan, M.,** The effects of *Nosema algerae* on the horizontal life table attributes of *Anopheles stephensi* under laboratory conditions, *J. Invertebr. Pathol.,* 37, 236, 1981.
677. **Savage, K. E. and Lowe, R. E.,** Studies of *Anopheles quadrimaculatus* infected with a *Nosema* sp., in *Proc. IVth Int. Colloq. Insect Pathol.,* 1971, 272.
678. **Anthony, D. W., Savage, K. E., and Weidhaas, D. E.,** Nosematosis: Its effect on *Anopheles albimanus* Wiedemann, and a population model of its relation to malaria transmission, *Proc. Helminthol. Soc. Wash.,* 39, 428, 1972.
679. **Anthony, D. W., Lotzkar, M. D., and Avery, S. W.,** Fecundity and longevity of *Anopheles albimanus* exposed at each larval instar to spores of *Nosema algerae, Mosq. News,* 38, 116, 1978.
680. **Hulls, R. H.,** The adverse effect of a microsporidan on sporogony and infectivity of *Plasmodium berghei, Trans. R. Soc. Trop. Med. Hyg.,* 65, 421, 1971.
681. **Savage, K. E., Lowe, R. E., Hazard, E. I., and Lofgren, C. S.,** Studies of the transmission of *Plasmodium gallinaceum* by *Anopheles quadrimaculatus* infected with a *Nosema* sp., *Bull. WHO,* 45, 845, 1971.
682. **Ward, R. A. and Savage, K. E.,** Effects of microsporidian parasites upon anopheline mosquitoes and malarial infection, *Proc. Helminthol. Soc. Wash.,* 39, 434, 1972.
683. **Gajanana, A., Tewari, S. C., Reuben, R., and Rajagopalan, P. K.,** Partial suppression of malaria parasites in *Aedes aegypti* and *Anopheles stephensi* doubly infected with *Nosema algerae* and *Plasmodium, Indian J. Med. Res.,* 70, 417, 1979.
684. **Thomson, H. M.,** The effect of a microsporidian parasite on the development, reproduction, and mortality of the spruce budworm, *Choristoneura fumiferana* (Clem.), *Can. J. Zool.,* 36, 499, 1958.
685. **Wilson, G. G.,** Effects of *Nosema fumiferanae* (Microsporida) on rearing stock of spruce budworm, *Choristoneura fumiferana* (Lepidoptera: Tortricidae), *Proc. Entomol. Soc. Ontario,* 111, 115, 1980.
685a. **Wilson, G. G.,** A dosing technique and the effects of sub-lethal doses of *Nosema fumiferanae* (Microsporida) on its host the spruce budworm, *Choristoneura fumiferana, Parasitology,* 87, 371, 1983.
685b. **Wilson, G. G.,** The transmission and effects of *Nosema fumiferanae* and *Pleistophora schubergi* (Microsporida) on *Choristoneura fumiferana* (Lepidoptera: Tortricidae), *Proc. Entomol. Soc. Ontario,* 115, 71, 1984.
686. **Wilson, G. G.,** The effects of feeding microsporidian *(Nosema fumiferanae)* spores to naturally infected spruce budworm *(Chorsitoneura fumiferana), Can. J. Zool.,* 55, 249, 1977.
687. **Wilson, G. G.,** Effects of larval age at inoculation, and dosage of microsporidian *(Nosema fumiferanae)* spores, on mortality of spruce budworm *(Choristoneura fumiferana), Can. J. Zool.,* 52, 993, 1974.
688. **Wilson, G. G.,** Microsporidian infection in spruce budworm, *(Choristoneura fumiferana)* 1 and 2 years after application, *Environ. Can. For. Serv. Bimonthly Res. Notes,* 34, 16, 1978.
689. **Henry, J. E.,** Protozoan and Viral Pathogens of Grasshoppers, Ph.D. thesis, Montana State University, Bozeman, 1969.

690. **Ewen, A. B. and Mukerji, M. K.**, Evaluation of *Nosema locustae* (Microsporida) as a control agent of grasshopper populations in Saskatchewan, *J. Invertebr. Pathol.*, 35, 295, 1980.
691. **Decker, G.**, Microbial insecticides — and their future, *Agric. Chem.*, 15, 30, 1960.
692. **Van Denburgh, R. S. and Burbutis, P. P.**, The host-parasite relationship of the European corn borer, *Ostrinia nubilalis,* and the protozoan, *Perezia pyraustae*, in Delaware, *J. Econ. Entomol.*, 55, 65, 1962.
693. **Brindley, T. A. and Dicke, F. F.**, Significant developments in European corn borer research, *Annu. Rev. Entomol.*, 8, 155, 1963.
694. **Lewis, L. C., Mutchmor, J. A., and Lynch, R. E.**, Effect of *Perezia pyraustae* on oxygen consumption by the European corn borer, *Ostrinia nubilalis, J. Insect Physiol.*, 17, 2457, 1971.
695. **Lewis, L. C. and Lynch, R. E.**, Influence on the European corn borer of *Nosema pyrausta* and resistance in maize to leaf feeding, *Environ. Entomol.*, 5, 139, 1976.
696. **Lynch, R. E. and Lewis, L. C.**, Influence on the European corn borer of *Nosema pyrausta* and resistance in maize to sheath-collar feeding, *Environ. Entomol.*, 5, 143, 1976.
697. **Lewis, L. C., Cossentine, J. E., and Gunnarson, R. D.**, Impact of two microsporidia, *Nosema pyrausta* and *Vairimorpha necatrix*, in *Nosema pyrausta* infected European corn borer *(Ostrinia nubilalis)* larvae, *Can. J. Zool.*, 61, 915, 1983.
698. **Wilson, G. G.**, Effects of *Pleistophora schubergi* (Microsporida) on the spruce budworm, *Choristoneura fumiferana* (Lepidoptera: Tortricidae), *Can. Entomol.*, 114, 81, 1982.
699. **Simchuk, P. A.**, The effect of microsporidians of *Pleistophora carpocapsae* and *P. schubergi* on the growth, development and mortality of caterpillars of *Malacosoma neustria* silkworm, *Parazitologiya*, 14, 158, 1980.
700. **Wilson, G. G.**, Susceptibility of the larch sawfly to *Pleistophora schubergi* (Microsporida), *Can. For. Serv. Res. Notes*, 1, 1, 1981.
701. **Chu, W. H. and Jaques, R. P.**, Pathologie d'une microsporidiose de l'arpenteuse du chou, *Trichoplusia ni* (Lep.: Noctuidae), par *Vairimorpha necatrix*, *Entomophaga*, 24, 229, 1979.
702. **Lewis, L. C., Gunnarson, R. D., and Cossentine, J. E.**, Pathogenicity of *Vairimorpha necatrix* (Microsporida: Nosematidae) against *Ostrinia nubilalis* (Lepidoptera: Pyralidae), *Can. Entomol.*, 114, 599, 1982.
703. **Brooks, W. M.**, unpublished observations, 1980.
704. **Grundler, J. A.**, Laboratory and Field Evaluations of *Vairimorpha necatrix* (Kramer) (Protozoa: Microsporida) for Biological Control of *Agrotis ipsilon* (Hufnagel) (Lepidoptera: Noctuidae), M.S. thesis, University of Missouri, Columbia, 1981.
704a. **Grundler, J. A., Hostetter, D. L., and Keaster, A. J.**, Laboratory evaluation of *Vairimorpha necatrix* (Microspora: Microsporidia) as a control agent for the black cutworm (Lepidoptera: Noctuidae), *Environ. Entomol.*, 16, 1228, 1987.
705. **Reynolds, D. G.**, Laboratory studies of the microsporidian *Plistophora culicis* (Weiser) infecting *Culex pipiens fatigans* Wied., *Bull. Entomol. Res.*, 60, 339, 1970.
706. **Reynolds, D. G.**, Experimental introduction of a microsporidian into a wild population of *Culex pipiens fatigans* Wied., *Bull. WHO*, 46, 807, 1972.
707. **Grassmick, R. A. and Rowley, W. A.**, Larval mortality of *Culex tarsalis* and *Aedes aegypti* when reared with different concentrations of *Tetrahymena pyriformis*, *J. Invertebr. Pathol.*, 22, 86, 1973.
708. **Miller, L. K., Lingg, A. J., and Bulla, L. A.**, Bacteria, viral, and fungal insecticides, *Science*, 219, 715, 1983.
709. **Kirschbaum, J. B.**, Potential implication of genetic engineering and other biotechnologies to insect control, *Annu. Rev. Entomol.*, 30, 51, 1985.
710. **Lüthy, P. and Arif, B. M.**, Designing microorganisms for insect control, *Bioessays*, 2, 22, 1985.
711. **Kawanishi, C. Y., Huang, Y. S., and Pounds, K.**, unpublished data, 1986.
712. **Hamm, J. J. and Hare, W. W.**, Applications of entomopathogens in irrigation water for control of fall armyworms and corn earworms (Lepidoptera: Noctuidae) on corn, *J. Econ. Entomol.*, 75, 1074, 1982.

ENTOMOGENOUS FUNGI

Clayton W. McCoy, Robert A. Samson, and Drion G. Boucias

INTRODUCTION

The fungi comprise a large, heterogeneous, and ubiquitous group of heterotrophic organisms living as saprophytes or parasites, or associated with other organisms as symbionts. They are devoid of chlorophyll but resemble plants by generally having definite cell walls. They are characterized by a distinctive, filamentous, multinucleate vegetative structure known as the mycelium and uni- or multicellular bodies called spores or conidia that detach from the reproductive structures of the parent and give rise to new individuals. Two general types of reproduction are recognized; teleomorphic (sexual) and anamorphic (asexual or imperfect). Anamorphic reproduction, sometimes called somatic or vegetative, does not involve the union of nuclei or sex organs. Teleomorphic reproduction, on the other hand, is characterized by a union of two nuclei.

The majority of known fungi, whether normally parasitic or not, are capable of living on dead organic material. Therefore, fungi can be facultative or obligate parasites of different living organisms such as plants, vertebrates, invertebrates, and other microorganisms. It is well to note that these organisms can be plant or animal "pathogens" and the disease symptoms or pathologies caused by these fungi are generally termed "mycoses". Mycoses caused by parasitic fungi of insects and related arthropods sometimes are severe enough to eliminate a host population in a given habitat. Such outbreaks, termed epizootics, are generally attributable to a combination of circumstances that favor the fungal disease. Paramount among these are physical factors of the environment, particularly moisture and a dense host population. This unique feature of microclimate dependency by fungi can be explained by their mode of infection. Rather than killing their host by toxigenic action following oral ingestion, as is the case for many parasitic microorganisms, fungi usually invade their host directly through the integument via a germinating spore. Our review will be limited to those fungi which have a parasitic or pathogenic association with arthropods, particularly the true insects (Hexapoda) and mites (Acarina); these are termed "entomogenous" or "entomopathogenic" fungi.

At first glance, consideration of the entomogenous fungi might appear to be a rather restricted topic. However, extensive studies of entomogenous fungi have been done. As an example, Muller-Kogler[1] cited approximately 1200 publications in his book on entomogenous fungi. During the past 15 years, with renewed interest in entomological biological control and integrated pest management, more research has been directed to the role of naturally occurring entomogenous fungi in the regulation or suppression of populations of insect pests. In addition, entomogenous fungi are receiving more attention as artificially introduced microbial insecticides and also are being identified as key natural mortality factors in the ecology of many important invertebrate pests. It should be pointed out, however, that entomogenous fungi hold a relatively modest position, when compared to bacteria, as microbial control agents today.

Many scientists have been directly contacted for current information for this review because many publications devoted to entomogenous fungi are published in journals of limited international distribution. Even then, it has been impossible to study all the literature on this subject and no doubt omissions will result.

Our purpose is to review the taxonomy, host specificity, mode of parasitism, and environmental stability of entomogenous fungi with specific references to their use as microbial insecticides to control both aquatic and terrestrial pests.

The works of Ainsworth et al.,[2,3] Muller and Loeffler,[4] and Alexopoulos and Mims[5] are recommended as a general taxonomic review of fungi. The textbook edited by Batra[6] should be referred to for information on insect-fungus symbiosis, mutualism, and commensalism. For a more general review of various biological aspects of entomogenous fungi, the books by Evlakhova,[7] Koval,[8] and Burges[9] also should be consulted. Diagnostic manuals by Poinar and Thomas[10] and Weiser[11] give keys and illustrations of entomogenous fungi. Steinhaus'[12] classical *Advanced Treatise of Insect Pathology*, Muller-Kogler's[1] textbook, *Pilzkrankheiten bei Insekten*, and review articles by Madelin[13] and Ferron[14,15] also are valuable sources on the general subject of entomogenous fungi. General reviews on the use of fungi for insect control can be found in Burges and Hussey,[16] Burges,[9] Bell,[17] Roberts and Humber,[18] and Hall and Papierok.[19] *Ainsworth & Bisby's Dictionary of the Fungi* should be consulted for a more complete definition of mycological terms.[20]

HISTORICAL BACKGROUND

Recognition of fungi attacking insects can be traced to early civilizations of the Far East. Two millennia ago, the Chinese identified species of *Cordyceps* and *Isaria* from cadavers of silkworm and cicada, respectively. Precious stone effigies of these insects were placed in the mouths of their dead in an attempt to instill some degree of immortality that was associated with fungal mummified insects.[21] Caterpillars infected with *Cordyceps* have been used in ancient and modern Chinese and Indonesian folk pharmacopoeia to treat a large number of ailments such as opium addiction. They have also been used as a general tonic, as food, and as an aphrodisiac.[22] It is reasonable to assume that as sericulture became a major industry of the Orient, fungal diseases of silkworm increased in incidence and general importance. Many students of sericulture believe that the muscardine fungus, probably *Beauveria bassiana*, was recognized about 1000 A.D. in China.[23]

In Europe, entomogenous fungi attacking honeybee, silkworm, and a few other insects were alluded to by poets and naturalists around the 18th century. First to mention the disease we know as muscardine in sericulture circles of Europe was Antonio Vallisnieri around 1710.[23] Although disease, as an abnormal condition in insects, was in all probability first critically observed in the honeybee and silkworm, the first published record was associated with the larva of a noctuid which harbored the characteristic vegetative growth of a *Cordyceps*.[23] This disease was reported by de Reaumur in 1726 as a malady of the "Chinese plant worm". One should refer to Cooke's work of 1892[24] for an interesting history of this fungus. Another early record of an entomogenous fungus was by Christian Paulinus, at the beginning of the 18th century, when he wrote that "certain trees in the island of Sombrero in the East Indies had large worms *(Cordyceps* — infected insects) attached to them underground in place of roots".[23] Insects parasitized by *Cordyceps* fungi were frequently known as "vegetable wasps" and "plant worms".[24]

Some early naturalists such as Fries, Persoon, and Ditmar recorded the occurrence of disease in insects that they observed during taxonomic studies. DeGeer probably published the first description of an *Entomophthora* (= *Empusa*) infection in flies.[25] In 1826, Kirby and Spence observed associations between insects and fungi and included these in their publication.[26]

The illustrious work of Agostino Bassi in 1834 was a historical landmark for insect pathology. Bassi was the first to experimentally demonstrate that a microorganism (the fungus, *Beauveria bassiana*) could be the cause of an infectious disease (of silkworms).[23] Not only did Bassi postulate a germ theory of disease of animals, but also implied that microbial agents might be used to destroy harmful insects. Furthermore, he correctly ascertained that warm, humid, environmental conditions facilitated the growth and development of fungi.

Following the paramount work of Bassi, other species of fungi were rapidly reported as parasitic on insects. Thereafter, numerous researchers contributed to the basic understanding of entomogenous fungi. Noteworthy among these early scientists were Thaxter, Giard, Gray, Robin, and Cook. In more recent times, the English and American mycologists, Petch and Mains, extensively published on entomogenous fungi.

In 1874, the American entomologist, J. L. LeConte, and the famous French microbiologist, Louis Pasteur, addressed for the first time the use of microorganisms to control insect pests.[23,27] Shortly thereafter, the first significant experimental tests using large quantities of artificially produced spores of *Metarhizium anisopliae* (Metchnikoff) Sorokin, to control the wheat cockchafer, *Anisoplia austriaca* Herbst, and the sugarbeet curculio, *Cleonus punctiventris* Germ, were performed by Metchnikoff in 1879 in Russia.[28] Metchnikoff developed a method for growing fungal spores on sterilized beer mash and was the first to realize the importance of mass production of entomogenous fungi by artificial means. Later, Krassilstschik following the lead of Metchnikoff, organized the first small production plant for *Metarhizium*.[29] The work was subsequently abated due to inconsistent results and a lack of basic information on epizootiological factors. Meanwhile, observations of a white parasitic fungus, *Beauveria bassiana* (= *globulifera*) of the chinch bug, *Blissus leucopterus* (Say), stimulated a major control campaign in the midwestern U.S.[30] In 1888, Lugger tried to initiate an outbreak of *Beauveria* on chinch bugs by scattering diseased insects in localized areas in Minnesota. At the same time, Snow began experimentation in Kansas.[30a] Virtually, the whole state of Kansas was treated with the fungus, during the period from 1891—92, and observers reported favorable results. However, a subsequent appraisal of the program showed no appreciable increase in infected chinch bugs following artificial distribution of spores.

After the turn of the century, considerable attention was focused on the role of fungi in the control of citrus whitefly with *Aschersonia aleyrodis* Webber in Florida.[31,32] Most of this work involved the movement of naturally occurring fungus from tree to tree or the statewide release of fungus artificially grown in culture vessels. Establishment of this fungus resulted in a biological control of whiteflies that still is successful today.[33]

During the last half century, papers, monographs, and reviews on the taxonomy, biology, and use of entomogenous fungi to control insect pests have steadily increased in number, diversity, and quality.[34] However, programs using fungi have not been successfully incorporated into the pest control recommendations for crops in North America. This is somewhat disappointing since there are many examples of the successful use of fungi to control pests (more than 28 species or groups of insects) over the last 40 years.[34]

In 1962, Nutrilite Products Inc., Lakeview, Calif., mass produced experimental quantities of a dust formulation of *Beauveria bassiana* for distribution to researchers in an attempt to reawaken U.S. interest in the use of this fungus to control pest insects. However, the viability of commercial formulations was questionable: attempts toward registration of the fungus as Biotrol FBB® were unsuccessful.[35] In the last decade, industry, particularly Abbott Laboratories (Chicago, Ill.), has taken a more active interest in the commercialization of entomogenous fungi in agriculture in the U.S. Their work with *Hirsutella thompsonii* resulted in the granting of the first Experimental Use Permit registration and labeling under the trade name, Mycar,[36] for this fungus in the U.S. Likewise, Tate and Lyle, a firm in Great Britain, obtained governmental approval for the marketing of *Verticillium lecanii* under the label Vertalec and Mycotal.[37] Although this action does not assure immediate widespread utilization of any fungus, the availability of a standardized industrial product fills a significant void that had previously prevented the implementation of many field programs. Heretofore, small quantities of the fungus produced by scientists limited research to small scale trials that many times resulted in more experimental variability than a researcher could physically and mentally address in an effective scientific manner.

During the last 20 years, considerable work on the industrial pilot production and use of fungi, primarily *Beauveria bassiana*, for insect control has been done by workers in eastern Europe and the U.S.S.R.[38] In 1977, permanent production facilities in the U.S.S.R. produced 22 metric tons for the control of the Colorado potato beetle.

TAXONOMY OF POTENTIAL CANDIDATES

Traditionally, the classification of entomogenous fungi paralleled the system established for all fungi. Morphology (characteristics of the mycelium and types of fructification) has been the most fundamental taxonomic criterion used to determine natural phylogenetic relationships. In many respects taxonomic characterization, confined strictly to morphology, has created many problems in fungal classification. Furthermore, variations in preparatory techniques such as choice of mounting medium, source and age of fungi in relation to host, culture medium, and physical factors such as temperature, light, etc., have been particularly crucial in systematically treating fungi. Light microscopy has been commonly used in taxonomic studies with entomogenous fungi. However, electron microscopy has been widely used in recent years to study ultrastructure of spores,[39] conidia,[40] and conidiogenesis.[41,42]

Although the classification of entomogenous fungi is still based on morphological criteria, more recent biological factors, reflecting specialization with respect to host, have changed the direction of taxonomic research. A broader definition of a species based on physiological, biochemical, and genetic differences has revealed the necessity for more precise characterization. Biochemical, X-ray, pyrolysis, gas chromatographic, and electrophoretic methods have been used to separate closely related taxa that appear morphologically alike.[43-45] Biological characteristics such as virulence, host specificity, and mass culture are basic to the successful development of fungi as microbial control agents. Thus, identification or characterization of strains and defining key attributes should accelerate the future development and use of virulent pathotypes.

There are approximately 750 species of fungi found throughout the five major taxa comprising the division Eumycota. Eumycota have been identified as being entomogenous or having confirmed or suspected entomogenous species.[46] The major taxa according to Ainsworth et al.[2,3] are the Mastigiomycotina (zygosporic fungi), the Ascomycotina (sac fungi), the Basidiomycotina (club fungi), and the Deuteromycotina, better known as Fungi Imperfecti (Table 1). In some recent classifications the Mastigiomycotina have been placed in the Zygomycotina. The Deuteromycotina, consisting of an artificially arranged grouping of various fungi, imply no phylogenetic relationships. These fungi have been placed into form genera on the basis of morphological features. The taxon was developed for those fungi recognized as being distinct but without a known teleomorphic stage of development. When teleomorphic stages are eventually identified for species of the Deuteromycotina, most end up in the Ascomycotina.

Most insect groups have little association with fungi and there appears to be no uniformity of distribution of fungal genera with the parasitic habit. This biological diversity suggests that infectivity for insects has been acquired several times during the course of fungal evolution. The most important entomogenous fungi produce asexual propagules at some point in their life cycle and most publications deal with spore or conidium producing species within the Entomophthorales and Hyphomycetes. Based on origin, the asexual propagules of the Entomophthorales are called spores and those for the Hyphomycetes are called conidia.

Mastigiomycotina

The Mastigiomycotina represent a primitive fungal group which is almost wholly aquatic and is known collectively as water molds or aquatic phycomycetes. They show the closest phylogenetic resemblances to protozoa, in that they produce zoospores and in many of the

Table 1
LIST OF THE MAJOR TAXA WHICH CONTAIN ENTOMOGENOUS SPECIES (GENERA CONTAINING ECTOPARASITIC SPECIES ARE SHOWN IN PARENTHESES)[a]

Mastigiomycotina

  Chytridiomycetes-Blastocladiales
    Coelomomyces
    Coelomycidium
    Myiophagus

  Oomycetes-Lagenidiales
    Lagenidium

Zygomycotina

  Zygomycetes-Entomophthorales
    Conidiobolus
    Entomophthora
    Massospora
    Neozygites
    Zoophthora
    Erynia
    Tabanomyces

Ascomycotina
  Plectomycetes-Ascosphaerales
    Ascosphaera

  Pyrenomycetes-Clavicipitales
    Cordyceps
    Torrubiella

  Pyrenomycetes-Hypocreales
    Nectria
    Hypocrella
    Calonectria
  Laboulbeniomycetes-Laboulbeniales
    Laboulbeniales

  Loculoascomycetes-Myriangiales
    Myriangium

  Loculoascomycetes-Pleosporales
    Podonectria

Deuteromycotina[b]
Hyphmycetes-Moniliales
  (Acariniola)
  Acremonium
  Acrodontium
  Akanthomyces
  (Amphoramorpha)
  Antennopsis
  Aspergillus
  Beauveria
  (Chantransiopsis)
  (Coreomycetopsis)
  Culicinomyces
  Engyodontium
  (Endosporella)
  Desmidiospora
  Fusarium
  Funicularis
  Gibellula
  Granulomanus
  Hirsutella
  Hymenostilbe
  Isaria
  (Mattirolella)
  Metarhizium
  (Muiaria)
  (Muiogone)
  Nomuraea
  Paecilomyces
  Paraisaria
  Pleurodesmospora
  Polycephalomyces
  Pseudogibellula
  Sorosporella
  Sporothrix
  Stilbella
  Syngliocladium
  Synnematium
  (Termitaria)
  (Termitariopsis)
  Tetracrium
  (Thaxteriola)
  Tilachlidium
  Tolypocladium
  (Trichothecium)
  Verticillium
  Coelomycetes-
    Sphaeropsidles
  Aschersonia
  Tetranacrium
  Mycelia Sterilia
  (Aposporella)
  Aegerita
  (Hormisciodeus)
  (Hormiscium)

Table 1 (continued)
LIST OF THE MAJOR TAXA WHICH CONTAIN ENTOMOGENOUS SPECIES (GENERA CONTAINING ECTOPARASITIC SPECIES ARE SHOWN IN PARENTHESES)[a]

Basidiomycotina
Phragmobasidiomycetes-
Septobasidiales
Septobasisium
Uredinella
Filobasidiella

[a] Classification follows Ainsworth et al.[2,3]
[b] Many species in the genera listed are anamorphs of ascomycetous insect parasites.

simpler forms (e.g., chytrids), the vegetative structures are not mycelial. The principal insect parasites are found in the Chytridiomycetes and Oomycetes (Table 1). The genera, *Coelomomyces* (Blastocladiales) and *Lagenidium* (Lagenidiales) comprise aquatic fungi, which are obligate parasites in mosquitoes. The genus *Coelomomyces* consists of about 40 described species.[47] The genus is widespread and has been found in all continents except Antarctica.[48] The fungus is unique in that many species require a copepod as an obligate alternate host to complete their life cycle[47,49] (Figure 1). Successful experimental hybridizations of two closely related species, *Coelomomyces dodgei* and *C. punctatus* have been completed using strains with differential pigmentation in the gametophytic phase.[50-52]

The genus *Lagenidium* contains *L. giganteum*, a facultative parasite of mosquito larvae with worldwide distribution and wide host range for mosquitoes (Figure 2). *L. giganteum* can grow vegetatively either as a parasite of mosquito larvae or as a saprophyte where it apparently prefers a littoral habitat.[53] Its life cycle is typical of the genus *Lagenidium* with both sexual (dormant oospore) and asexual (zoospore) reproduction. The zoospore is the infectious agent. The mosquito larva is killed when the zoospore invades the hemocoel and forms an extensive mycelium throughout the body.[54]

A new chytrid-like genus, *Coelomycidium*, has recently been described from the larvae of blackflies.[55,56] *C. simulii* Debaisieux attacks the larvae by initially locating in the fat body. Spherical thalli fill the body cavity and subsequently form single uninuclear zoospores. Zoospores leave the body and attack a new host.

Several species of aquatic Chytridiomycetes, e.g., *Catenaria* (Blastocladiales) and *Aphanomycopsis* and *Atkinsiella* (Saprolegniales) are reported by Martin[57-60] as water mold parasites of eggs of midges (Diptera: Chironomidae) in Virginia. Recently, *Couchia circumplexa*[61] was described as a new pathogen, causing high mortality in egg masses of *Chironomus attenuatus*, *Tendipes decorus*, and *Pentaneura carnea*. This mold also belongs to the Saprolegniaceae and produces dimorphic zoospores and appressorial complexes aiding in penetration of the egg chorion.

The genus, *Myiophagus* (Chytridiales), accommodates the single species *M. ucrainicus* (Wize) Sparrow which was originally described from larvae of *Cleonus* and *Anisoplia* from the Soviet Union.[13] Chytrid diseases of scale insects caused by *Myiophagus* species have been reported from numerous species of armored scales occurring on citrus in Florida and other areas of the Caribbean region. *M. ucrainicus* is rarely encountered in citrus groves today.[655] Changing horticultural practices and successful biological control programs against armored scales using parasites apparently have reduced the incidence of *Myiophagus*. An excellent account of the life cycle of *M. ucrainicus* in females of the armored scale *Lepidosaphes beckii* (Newman), has been published by Karling.[62]

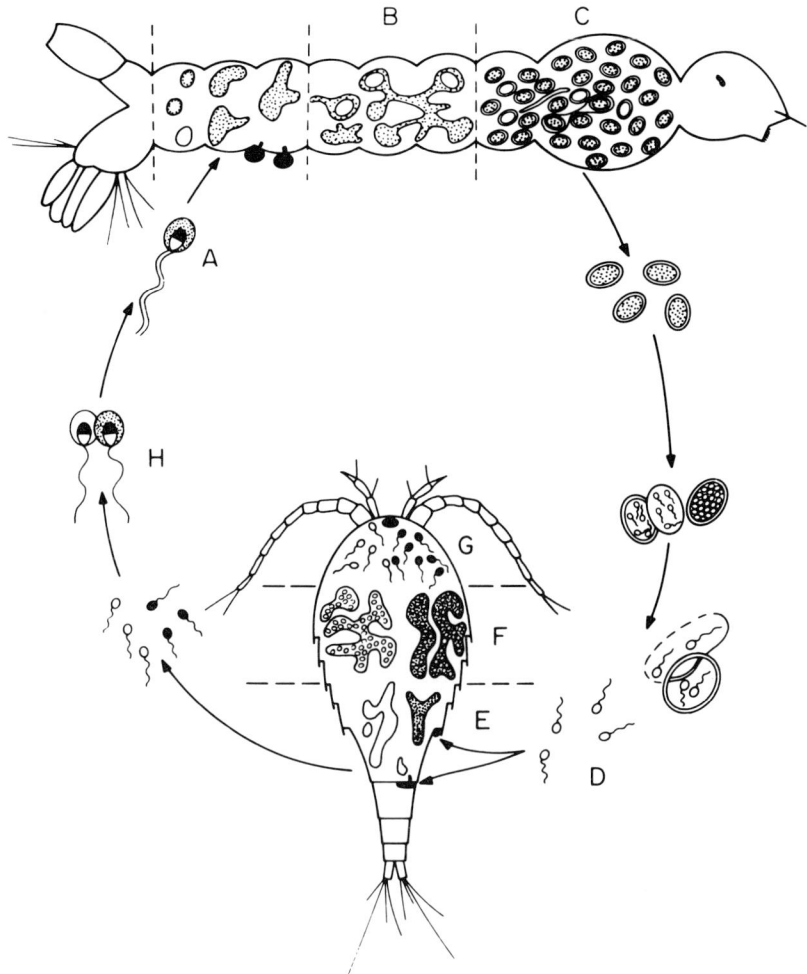

FIGURE 1. Generalized life cycle for *Coelomomyces*. (A) Biflagellate zygote infects larva by encysting, preferentially on cuticle at an intersegmental membrane, and then invading hemocoel via a germ tube. Subsequently, hyphae (B and C) proliferate throughout the hemocoel and fat body forming sporangia at their tips. After liberation from the larva, sporangia release meiospores (D) which infect a copepod by encysting on an intersegmental membrane and invading the hemocoel via a germ tube. Each meiospore forms a gametophyte (E) which eventually becomes a gametangium (F) and ruptures, releasing gametes (G) of a single mating type. The gametes escape from the carcass of the dead copepod and opposite mating-types fuse (H) forming a mosquito-infective zygote. (Courtesy of Dr. Brian A. Federici, University of California, Riverside.)

Zygomycotina

The Zygomycetes are characterized by spores formed endogenously in sporangia or sporagiola, coenocytic (nonseptate) mycelia, and by sexual reproduction through the formation of zygospores. Generally, Zygomycetes produce infections which involve the formation of well-defined hyphal bodies that eventually develop into spores, sporangia, and resting spores.

Entomogenous Zygomycetes are located principally in the order Entomophthorales. The genus *Sporodiniella* (Mucorales), however, with the single species *S. umbellata*, causes epizootics particularly on membracids.[63] With the sole exception of the genus *Massospora*, members of the Entomophthorales are distinguished from all other members of the Zygomycotina by the presence of forcibly discharged spores.

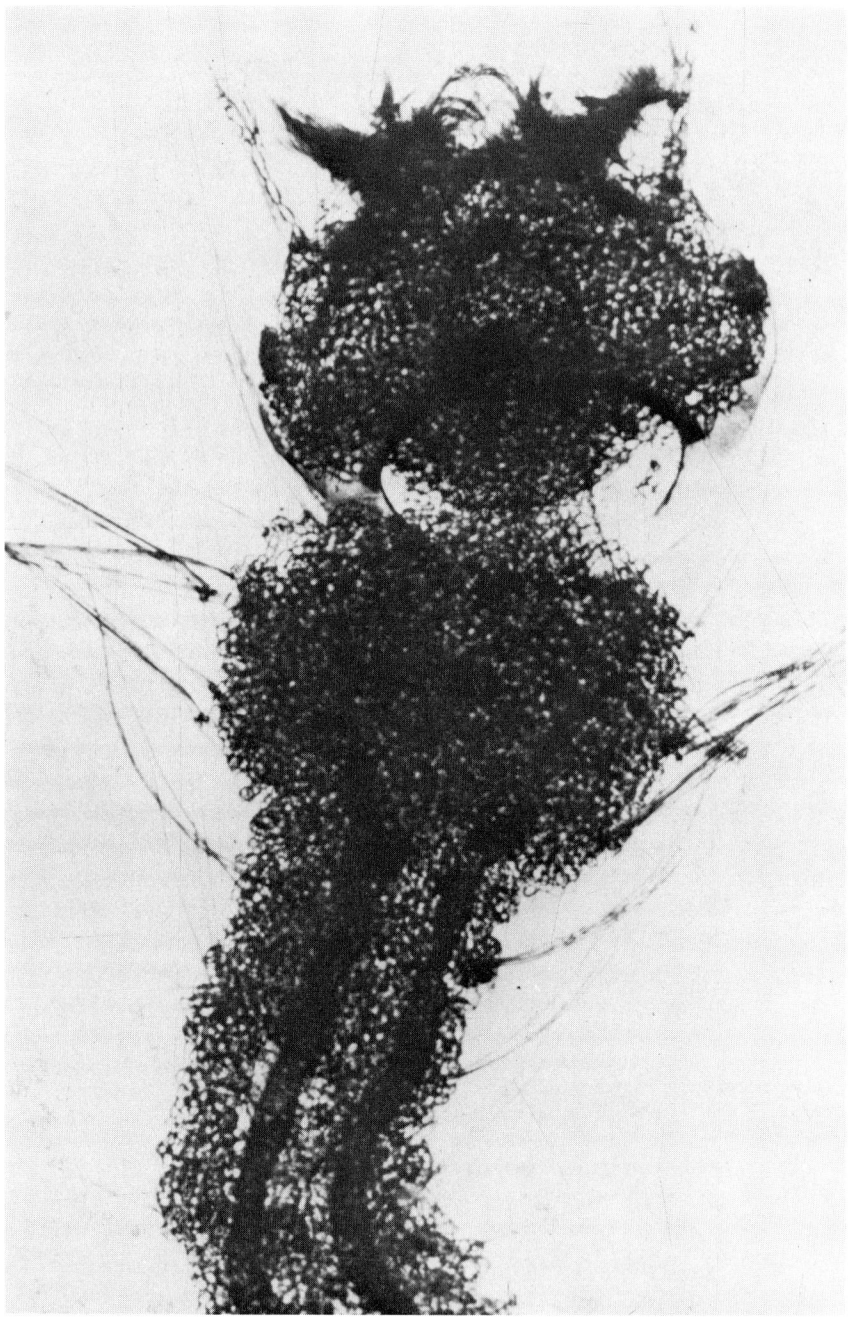

FIGURE 2. *Lagenidium giganteum* in *Culex pipens quinquefasciatus* larvae showing masses of segmented hyphae in the body cavity (magnification × 720). (Courtesy of Dr. Thomas McInnis, Clemson University, Clemson, South Carolina.)

Entomophthora has been considered the largest and most well-known genus of the Entomophthoraceae.[64] The genus has worldwide distribution attacking many species of Homoptera, Hemiptera, Diptera, and Lepidoptera as well as numerous Acari and millipeds (Figure 3): Species of Entomophthorales parasitizing aphids have attracted much attention. Nomenclatorial problems, with this genus, have existed for nearly 20 years. MacLeod[65] recognized

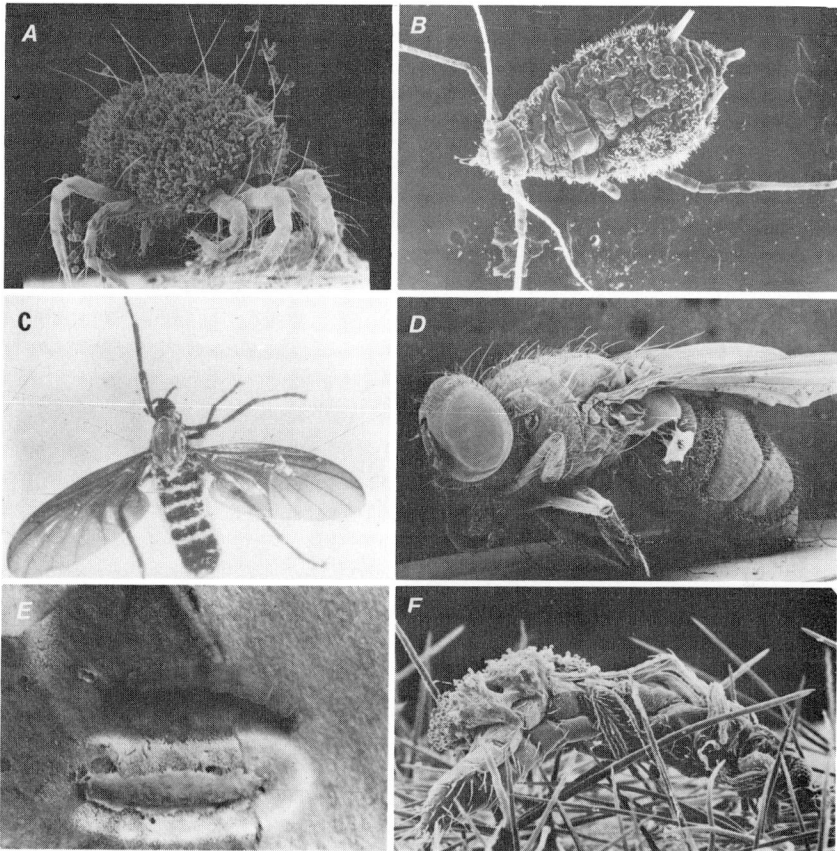

FIGURE 3. Micrographs of different species of arthropods infected with Entomophthoraceae. (A) *Tetranychus urticae* infected with *Entomophthora* sp. (magnification × 70). (B) Unknown species of aphid infected with *E. fresenii* (magnification × 50). (C) *Plecia nearctica* infected with *Conidiobulus coronatus* (magnification × 15). (D) *Musca domestica* infected with *E. muscae* (magnification × 50). (E) *Plutella maculipennis* infected with *Entomophthora* sp. (magnification × 15). (F) *Thrips tabaci* infected with *E. thripidum* (magnification × 85). (Photographs A and B—D courtesy of Drs. G. R. Carner, Clemson University, Clemson, South Carolina and R. A. Samson, C.B.S., Baarn, Netherlands.)

that species of this genus probably represented more than one genus. Since 1970, MacLeod, Muller-Kogler, and Wilding[64,66,67] have revised the Entomophthorales and questioned the validity of precedent genera on the basis of conidial attachment and morphology. Evlakhova[7] and Gustafsson,[68] in 1974 and 1965, proposed 12 and 8 groups respectively based on morphological features. Zimmerman[69] reviewed all the species of this important group in 1978.

Taxonomic controversy resulted following Batko's[70,71] drastic division of the genus with the creation of many new genera (Figure 4). Batko's grouping of the species of *Entomophthora* at the generic rather than into subgenera or infra-generic taxons was questioned by Waterhouse and Brady.[72] They felt that the separation of the genus *Entomophthora* into several genera was premature because of the lack of information on hyphal bodies, pseudocystidia, nuclear contents, and description of the secondary spores. Waterhouse and Brody[72] developed a key to the *Entomophthora* species *sensu lato*, using features of the sporophore branching, spore shape, and presence or absence of rhizoids.

Remaudiere and Keller[73] have presented a revised classification of the Entomophthoraceae

BATKO	REMAUDIÈRE & KELLER	BEN ZE'EV & KENNETH	HUMBER
Conidiobolus	*Conidiobolus*	*Conidiobolus*	*Conidiobolus*
Entomophaga	--	*Entomophaga*	*Entomophaga*
Entomophthora	*Entomophthora*	*Entomophthora*	*Entomophthora*
Culicicola	--	--	--
Triplosporium	*Neozygites*	*Triplosporium*	*Triplosporium*
Zoophthora subg. *Zoophthora* subg. *Pandora* subg. *Erynia* subg. *Furia*	*Zoophthora* *Erynia*	*Erynia* subg. *Zoophthora* subg. *Neopandora* subg. *Erynia*	*Erynia*
Strongwellsea	--	*Strongwellsea*	*Strongwellsea*
--	--	*Meristacrum*	*Meristacrum*

FIGURE 4. Recent reclassification of the genera of the Entomophthoraceae as proposed by Remaudiere and Keller (1980).

based mainly on morphological characters of the primary and secondary spores (Figure 4). In their scheme, *Entomophthora* now includes the previously described genera *Empusa*, *Myiophyton*, and *Culicicola*. The genus is primarily characterized by campanulate spores and now contains only six species. *Conidiobolus* has been redefined and now comprises about 40 species. The genus *Entomophaga* Batko is synonymized with *Conidiobolus*. Chemotaxonomy, particularly fatty acid composition and isozyme analyses,[74,75] has proven valuable in elucidating infrageneric relationships among those species available in axenic culture. King and Humber[76] suggested the following criteria for separation: (1) the shape of primary conidiophores and primary conidia; (2) whether an outer wall layer separates readily from the primary conidium; (3) the presence or absence and nature of any multiplicative conidia and/or villose spores; (4) the shape, color, surface decorations, and mode of formation of zygospores or azygospores; (5) ease of culture on common mycological media; and (6) presence or absence and characteristics of rhizoids and/or cystidia. In view of these findings, *Entomophaga coronata* has been placed in the genus *Conidiobolus* or in the subgenus *Delacroixia*.[77,78]

Remaudiere and Keller[73] reintroduced the genus *Neozygites* for ten species and discussed in detail the genus, *N. aphidis* (= *N. fresenii*), a common pathogen on aphids. For nomenclatorial reasons, the genus *Triplosporium* was rejected and considered to be a synonym of *Neozygites*. The genus *Strongwellsea* was placed in synonymy with *Erynia* and the genus *Zoophthora* reintroduced.[79] *Erynia* has about 30 species, while *Zoophthora* Batko contains 8 species. Important nomenclatorial changes included the naming of *E. neoaphidis* (= *Entomophthora aphidis* sensu Thaxter[79]) and the synonymy of *Entomophthora virulenta* and *E. thaxteriana* with *Conidiobolus obscurus*.[80,81]

Humber[81] discussed certain taxonomic criteria used for the classification of the Entomophthorales and reviewed the use of generic names by Batko and Remaudiere. The number of nuclei in the primary spores, the branching of sporophores, and the mode of discharge of the primary spores were regarded as major characters suitable for delimiting genera. The validity of *Strongwellsea* and *Entomophaga* was upheld. Species of *Conidiobolus* not placed in the genus *Entomophaga* were placed in the family Ancylistaceae. Humber[82] and Ben Ze'ev and Kenneth[83,84] have emended the genus *Erynia* and propose a different classification than the one of Remaudiere and Keller (Figure 4). They rejected the genus *Zoophthora* but subdivided *Erynia* into the four original subgenera as proposed by Batko. In addition, the genus *Neozygites* will be elevated to the family Neozygitaceae and include the genera *Neozygites* and *Thaxterosporium* (= Triplosporium).[656] There is still need for further clarifications of genera in the Entomophthorales despite the number of recent publications. General acceptance of one of the proposed classifications is not expected in the near future.

The genus *Conidiobolus* contains mainly saprobic species; however, a few pathogenic species have been reported from insects, mites, Myriapoda, Nematoda, and Mammifera (King).[85] Several species of *Conidiobolus*, originally described from plant detritus, have also been recovered from insects. *C. coronatus*, a ubiquitous saprobe and occasional entomopathogen, also infects humans and other mammals.[86,87] Human mycoses of this species have been reported from the tropics around the world and have been isolated from horses in southern Texas.[88] With the exception of *C. incongrus*, none of the remaining 25 species of *Conidiobolus* is known to attack mammals.

Strongwellsea is a little known and unusual genus of Entomophthorales that parasitizes adult flies particularly the adult cabbage root fly, *Hylemya brassicae* Bouche.[89] Neither the lifespan nor behavior of infected flies is affected, however, flies may be sterile. Infected flies have a gapping hole in the ventral abdomen through which internally formed spores are dispersed. Inside the hole, a large cavity is lined by a hymenium of sterile cells and sporophores. A layer of multi-nucleate, coenocytic, vegetative hyphae is found between the hymenium and the hemocoele of the host. A single uninucleate spore, analogous to a monosporic sporangiole, is formed on each sporophore and is forcibly discharged toward a light source. The primary spores can form a germ tube that is 2 mm long or may form a single secondary spore that is forcibly discharged toward light. The nucleus, one of the largest among all fungi, may be larger than 20×5 µm.

The genus *Massospora* represents 11 species of entomogenous fungi, all specific to periodical cicadas.[90] *Massospora* grows only in the abdominal cavity of adults. The abdomen swells until several segments of the exoskeleton fall away exposing a mass of conidia. Active, infected adult cicadas transmit the disease to healthy cicadas. Resting spores, produced by *Massospora*, ensure the long-term survival of the fungus and also are disseminated by living cicadas. A generalized life cycle and epizootiology for *M. levispora* was published by Soper et al.[91,92]

Tabanomyces, proposed as a new genus, was based on one species (*Coelomomyces milkoi* Dudka and Koval) isolated from larvae of tabanid flies in the Soviet Union.[93] The zygospores germinate to form a linear four-celled sporophore with each cell producing a spherical spore ejected as in *Conidiobolus* (thus confirming a similarity to other Entomophthorales). Humber[81] recognized the similarity of *Tabanomyces* with the nematophagous genus *Meristacrum*. He combined the genera as *M. milkoi* and proposed inclusion of the genus in the Ancylistaceae.

Research concerning intraspecific characterization of the different isolates of strains of *Entomophthora* has received increased attention. Remaudiere et al.[94] recognized different strains within the *E. sphaerosperma* group. Latge and de Bievre[95] confirmed conspecificity for several strains of *E. obscura* with physiological information on lipid composition. Enzyme electrophoresis has recently been used to determine intraspecific genetic variability in laboratory strains of *Entomophthora*.[96]

Before terminating discussion of the Zygomycotina, brief mention should be made of the class Trichomycetes. A number of families in this class, which morphologically suggest an affinity to the Zygomycetes, are entomogenous. However, evidence of true parasitism is lacking and they are now generally considered to be endocommensals, i.e., inhabiting the hindgut or external cuticle of insects, crustaceans, and diplopods.[97,98]

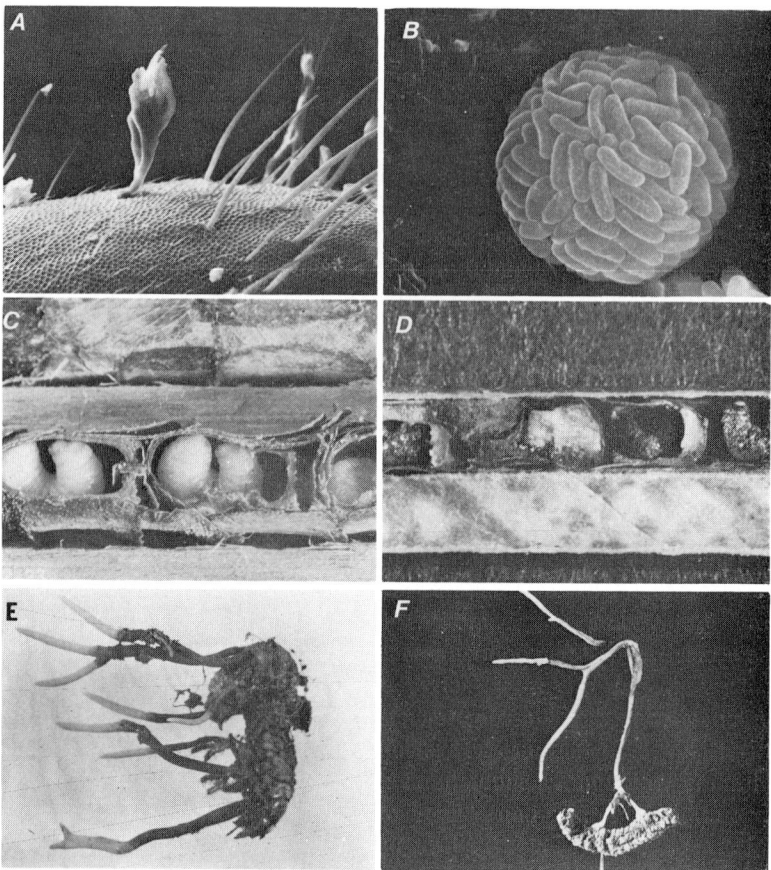

FIGURE 5. Representative genera of the Ascomycotina with entomogenous species. (A) Micrograph of a staphylinid beetle cuticle supporting a highly specialized ectoparasitic Laboulbeniales (magnification × 150). (B) Spore packet from a cyst of *Ascosphaera aggregata* (magnification × 2500). (C—D) Healthy and diseased larvae of *Megachile rotundata* infected with *A. aggregata*. (E—F) Soil inhabiting scarabeid larvae infected with *Cordyceps* sp. (Photographs A and F and B—D courtesy of Drs. R. A. Samson, C.B.S., Baarn, The Netherlands and L. P. Kish, University of Idaho, Moscow, Idaho, respectively.)

Ascomycotina

The Ascomycotina represent a large taxon in which insect parasitism only is found in a few diverse genera (Figure 5). Apart from approximately 1300 species and varieties of highly specialized ectoparasitic Laboulbeniales,[99] which will not be discussed in this review, the entomogenous genera within the Ascomycotina are endoparasitic with the vegetative phase developing within the body of the host. Sexual development for these fungi are characterized by the formation of an ascus and ascospores. Entomogenous fungi have been identified from the Plectomycetes, Pyrenomycetes, and Loculoascomycetes (Table 1).

The Plectomycetes, characterized as producing closed ascocarps in which globose evanescent asci are borne, contain a few unique entomogenous species within the genus *Ascosphaera* (= *Pericystis* Betts). Species within this genus have been reported to cause chalkbrood disease which affects the larvae of honeybees, leaf-cutting bees, and other pollinator species.[100-105] Skou[102] in a revision of the Ascosphaerales placed *A. alvei* in a new genus *(Bettsia)* and concluded it was solely saprophytic on pollen. He also has reported an anamorph of *B. alvei* as *Chrysosporium farinaecola*. Other new species, found in association with (leaf-cutting) bees but of which no pathogenicity was proven, have been reported.[106,107] The infective unit is probably ascospores, formed in spore balls in dark-colored cysts, that appear as tiny black specks on mummified bee larvae (Figure 5 A to D). All pathogenic species cause a swelling, probably due to development of spore cysts beneath the larval integument, that easily bursts thus giving the dead larvae a ragged appearance. Prophylactic and fungicidal agents are being developed in the western U.S., Canada, and Europe where chalkbrood disease is a serious problem of domestic and wild bees.[105]

Another plectomycetous genus, *Microascus*, contains one species *M. exsertus*,[108] which caused a disease of the leaf-cutting bee *Megachile willinghbiella* in Denmark.

The Pyrenomycetes bear unitunicate, club-shaped, or cylindrical asci, in virtually closed ascocarps with an ostiole. The ascocarp is generally a globose or flask-shaped perithecium. Entomogenous Pyrenomycetes have perithecia with virtually all entomogenous species found in the orders Hypocreales or Clavicipitales (Table 1).

Within the Clavicipitales, two morphologically similar genera are known to be entomogenous. *Cordyceps* and *Torrubiella* (Figure 6) represent the telemorph stage of a number of deuteromycetous genera. The perithecia of these genera are either embedded in a stromatal tissue or are external. However, a sharp delimitation of both genera is difficult since these generic characteristics may appear in the same species, e.g., *C. tuberculata*. Host distribution and parasitism for these genera will be discussed with their corresponding deuteromycetous forms. More than 250 species of *Cordyceps* have been found on species of Diptera, Hymenoptera, Coleoptera, Lepidoptera, Hemiptera, Isoptera, and spiders.[21,109-113]

On insects, the appearance of a *Cordyceps* may be quite striking and it is not surprising that they were the first entomogenous fungi reported by man. *Cordyceps* spp. are rather host specific and generally favor moist habitats, infecting all developmental stages of insects that occur in moist soil, decaying logs, bark, and on foliage. At times, the stroma of *Cordyceps* may be parasitized by other fungi, notably *Sphaeronemella* sp.

The aerial portion or stromata of *Cordyceps* are basically compact bundles of longitudinal parallel or somewhat interwoven hyphae. It forms from a compact mycelial mass or sclerotium on or within the host. The perithecia develop upon the fertile upper part of the stroma in various ways. In most species, the asci are long and narrow, generally cylindrical, but narrowing toward the bases and thickened at the apices. The ascospores are hyaline multiseptate and in most species are filiform, i.e., nearly as long as the asci. At maturity, they can break into one-celled round cylindric or fusiod segments. Several deuteromycetous genera are anamorphs of *Cordyceps* and *Torrubiella*. Genera, found to be associated with Ascomycetes from recent taxonomic research of entomogenous fungi from tropical rain forests, are listed in Table 2.

The large and heterogeneous order Hypocreales contain fungi which have a bright colored ostiolate perithecium with relatively soft or waxy walls.[114,115] Stroma generally appear as brightly colored fleshy structures in the *Hypocrella*. In this genus, the species are parasitic on scale insects, Coccidae, and whiteflies. Many of the species were first described, based on conidia, as species of *Aschersonia*.

The genus *Nectria*, which represents a number of phytopathogenic species, also contains four weakly pathogenic species of scale insects (Figure 7J. *N. flammea*, *N. aurantiicola*, *N. diploae*, and *N. coccidophaga*,[116] with an anamorph referable to the genus *Fusarium*,[117] have been reported from species of armored scale insects inhabiting citrus.[31]

The little known genus *Cordycepioideus*,[118,119] which is probably pathogenic to termites, also belongs to the Hypocreales. The asci in both species of the genus are clavate with a long stipe and contain two or eight, ellipsoidal, black, or purple gray ascospores.

The Loculoascomycetes, which are characterized by having bitunicate asci borne in specialized stromatic locules, contain a few weakly pathogenic species in the orders Myriangiales (locules uniascal) and Pleosporales (locules polyascal). Entomogenous species of *Myrian-*

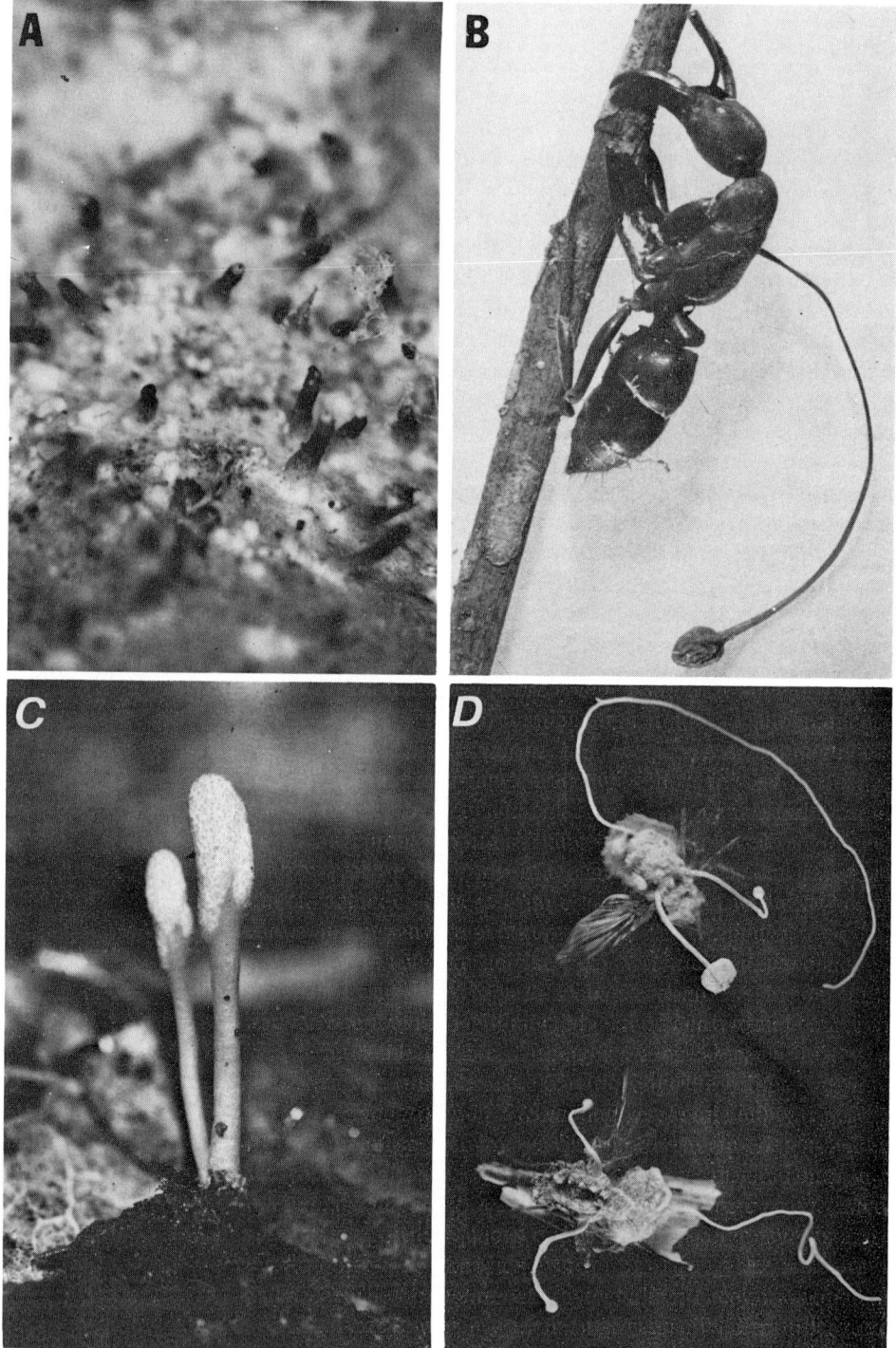

FIGURE 6. Various stromata in *Cordyceps* and *Torrubiella*. (A) Solitary and distinct perithecia of *Torrubiella rubra* on coccids (magnification × 15). (B) *Cordyceps australis* on a *Paltothyreus tarsatus* ant (magnification × 5). (C) Two stromata of a *Cordyceps* species on a lepidoptera larvae completely buried in litter (magnification × 2). (D) *Cordyceps dipterigena* on Diptera, showing the stromata of the *Cordyceps* teleomorph and synnemata of the *Hymenostilbe* anamorph (magnification × 3).

Table 2
DEUTEROMYCETOUS GENERA REPRESENTING ANAMORPHS OF THE ASCOMYCETES *CORDYCEPS* AND *TORRUBIELLA*

Akanthomyces	*Nomuraea*
Gibellula	*Paecilomyces*
Granulomanus	*Paraisaria*
Hirsutella (= *Synnematium*)	*Stilbella*
Hymenostilbe	*Tilachilidiopsis*
Metarhizium	*Verticillium*

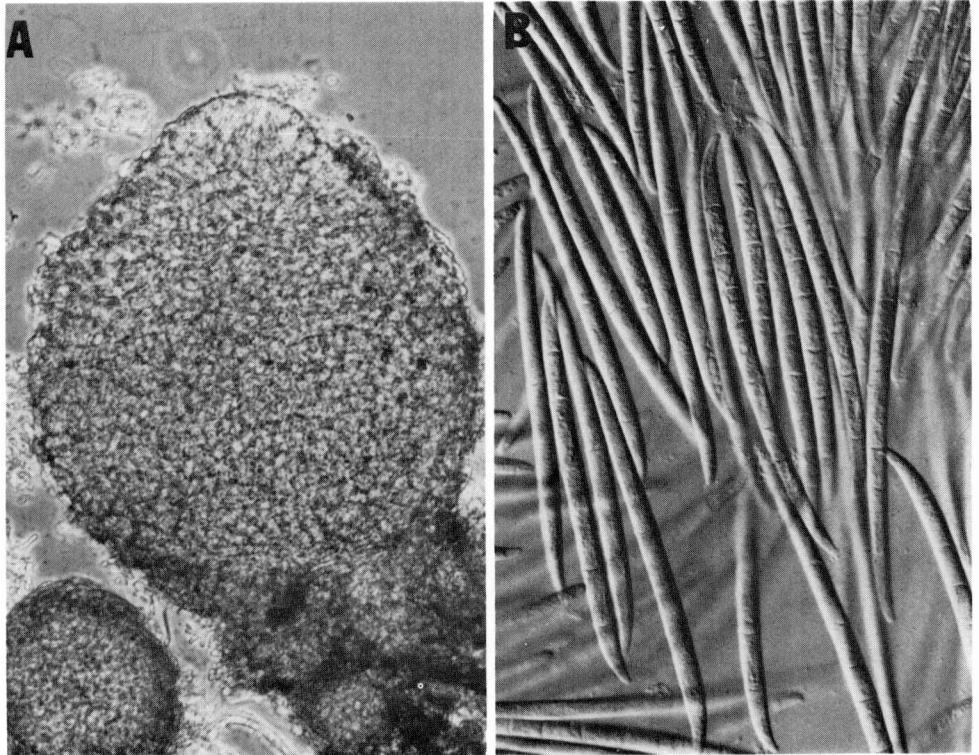

FIGURE 7. *Nectria flammea*, an ascomycetous parasite on scales (A) perithecium (magnification × 240). (B) Septate conidia of the *Fusarium* anamorph (magnification × 900).

gium are parasites of scale insects in the tropics.[31] The mycelium forms a black cushion-shaped mat covering one or more scales; no conidial form is known. A systematic key to the Myriangiales, pathogenic on Coccoidea, recently was published.[7]

Within the Pleosporales, eight species of the genus *Podonectria* infect scales.[120] Some species, e.g., *P. coccicola* and *P. echinata*, have a *Tetracrium* and *Tretanacrium* anamorph. Although these scale fungi have been reported to be weakly pathogenic, some question the accuracy of these reports.

Basidiomycotina

The Basidiomycotina, which produce basidiospores, on the outside of a specialized spore producing basidium cell, contain only a few entomogenous genera. *Septobasidium* and a

Table 3
LIST OF THE IMPERFECT GENERA CONTAINING ECTOPARASITIC SPECIES

Acariniola	*Hormiscium*
Aegeritella	*Laboulbeniopsis*
Amphoramorpha	*Mattirolella*
Antennopsis	*Muiaria*
Aposporella	*Muiogone*
Chantransiopsis	*Termitaria*
Coreomycetopsis	*Termitariopsis*
Endosporella	*Thaxteriola*
Entomocosma	*Trichothecium*
Hormisciodeus	

similar genus *Uredinella*[121] are found in the Septobasidiales. They are characterized by transversely septate basidia and are associated with scale insects. Recently, Coles and Talbot identified the anamorph, *Harpographium coryneliodies*, *S. clelandii*,[122] a parasite of female coccids. In their study of entomogenous fungi from the Galapagos Islands, Evans and Samson[123] found an anamorph of *S. pilosum* on *Ischnapsis* nymphs (Diaspididae), called *Aegerita webberi*. Dykstra,[124] studying ultrastructural features of the genus *Septobasidium*, found morphological similarity with the rust fungi.

Septobasidium and *Uredinella* differ on the basis of the type structure producing the epibasidium and the parasitic-symbiotic relationship that exists with their host. This complex relationship begins with the basidiospores germinating on the insect and sending their hyphae into the host body. As the insect settles itself on its plant host, the fungi grow and form a hyphal mat over the insect. *Septobasidium* parasitizes a whole colony of scale insects whereas *Uredinella* parasitizes single insects. The insects subsequently live under the hyphal mat. It is significant that neither the fungus nor the insect occurs without the other. Generally, parasitized insects are not killed, but they are rendered sterile.

In the genus *Filobasidiella* one species, *F. depauperata* (= *F. arachnophila*),[125] is described that lacks the typical features of a basidiomycete but is morphologically very similar to other species of the genus. *F. depauperata* often occurs together with *Verticillium lecanii* and is reported from spiders, aphids, and scales.

Deuteromycotina

The *Deuteromycotina* (Fungi Imperfecti), a conglomerate of form genera, contain the majority of entomogenous fungi with ectoparasitic and endoparasitic modes of infection. Entomopathogenic genera have been listed in the form orders Hyphomycetes, Coelomycetes, and Mycelia Sterilia (Table 1). The reader is referred to Roberts and Humber[18] for a recent review on entomogenous Deuteromycetes and taxonomic details and a key to the common genera is presented by Samson.[126] Ectoparasitic genera are listed in Table 3. Interested researchers should refer to Madelin[13] and Ainsworth et al.[2,3] for a more detailed account of the fungi. Many ectoparasitic fungi are associated with termites. Sands[127] summarized the pathological effects of a number of these Deuteromycetes, while Blackwell and Kimbrough[128] keyed out some known termite infesting ectoparasites.

Hyphomycetes are mycelial fungi that reproduce by conidia generally formed on free or aggregated conidiophores arising from the substratum surface. Coelomycetes are fungi that produce by means of conidia borne in pycnidia, acervuli, or sporodochia. Mycelia Sterilia have no known reproductive structures.

The classification of the Fungi Imperfecti has evolved from the Saccardian system (ca. 1886) which was based on superficial morphological features.[2,3] Most recent taxonomic studies rely on classical parameters combined with ontogenetic information on conidiogenesis. Older literature contains taxonomic errors at all levels of classification of the entomogenous Deuteromycotina. However, recent monographs (on many key genera) have resolved many of these taxonomic errors. Those species having potential as microbial control agents will be treated in detail in other parts of this review.

A number of hyphomycetes have conidiophores frequently united into loosely or tightly formed synnemata (Figure 8). Synnemata formation appears to be related to the phototrophic response of the host in its natural habitat and they are found on insects buried in the soil or hidden in trunks and tree bark.[126] Synnematous genera are usually associated with ascomycetous fungi such as *Cordyceps, Torrubiella,* and *Nectria*. Many synnematous fungi may lose this typical feature when grown on artificial media. On the other hand, some mononematous species, e.g., *Hirsutella thompsonii*, exhibit an unexpected synnematal development in pure culture (Figure 8C).

The genus, *Akanthomyces* is characterized by having distinct cylindrical or clavate synnemata with phialides arranged along the synnema in a hymenium-like manner. This genus was revised by Mains,[129] and Samson and Evans[130] and described four species that attack various arthropods in Ghana. In Africa, *A. gracilis* has mainly been found on ants but also infects beetles and caterpillars as well as froghoppers. The fungus grows and sporulates on artificial media. Species of Isarioides of the genus *Paecilomyces*, mostly described as *Isaria*, develop synnemata similar to *Akanthomyces* but are characterized by having phialides with distinct necks that are divergent in loose verticillate patterns. Samson[131] revised the genus *Paecilomyces* and erected 31 species of which 14 are entomogenous. Species of *Paecilomyces* and *Nomuraea*, occurring on various arthropods in Ghana, were reported by Samson and Evans[132] and three new entomogenous species were described from Canada and China by Bissett[133] and Liang.[134] *P. farinosus* is the most common species attacking a wide range of hosts such as the pine shoot moth, sawflies, and the Colorado potato beetle. Another common species is *P. tenuipes* (on lepidopteran larvae) and *P. fumoso-rosea* (on adult damsalflies and other insects).[135,136] Aoki reported mixed infection of gypsy moth by *P. canadensis* (= *P. lavanicus*) and *Entomophthora aulicae*.[137] Most species of *Paecilomyces* will grow on artificial media.

The genus *Gibellula* also has conidiophores united in distinct synnemata; however, it usually appears to be a specific parasite of spiders.[138] In temperate and tropical areas, spider cadavers are frequently found attached to the underside of leaves by this fungus. All four entomogenous species within *Gibellula* have been identified from Ghana.[138] The genus *Gibellula* seems to be rather common in South American tropical rain forests and several new species are reported.[139] Species of *Gibellula* are anamorphs of *Torrubiella*. *Granulomanus*[140] is usually found as a pleomorphic form of *Gibellula* spp. This form is characterized by polyblastic conidiogenous cells producing needle-like conidia. In some species, the *Granulomanus* form can be very pronounced and apparently plays an important role in the ecology and distribution of the fungus. *G. formicarum* is different from all *Gibellula* species because it produces conidia singly on sympodial conidiogenous cells and subsequently it was included in the new genus *Pseudogibellula*. *P. formicarum* has a wide host range and will readily grow in pure culture.

Recently, Baker and Zaim[141] (in Zaim et al.) described the hyphomycete, *Funicularis triseriatus* from the tree hole breeding mosquito, *Aedes triseriatus*. The fungus is characterized by funiculose colonies and thallic conidia arranged sympodially on short conidiophore branches. This fungus is rare and no data of its parasitic nature have been obtained.

The genus *Hirsutella*,[142] often associated with *Cordyceps* spp.,[143] is typified by phialides which are swollen at the base, mostly solitary, and producing conidia in mucus singly or in heads. Mains revised the genus identifying only nine synnematous species from insects in the order Homoptera, Orthoptera, Coleoptera, Lepidioptera, Diptera, and Hymenoptera

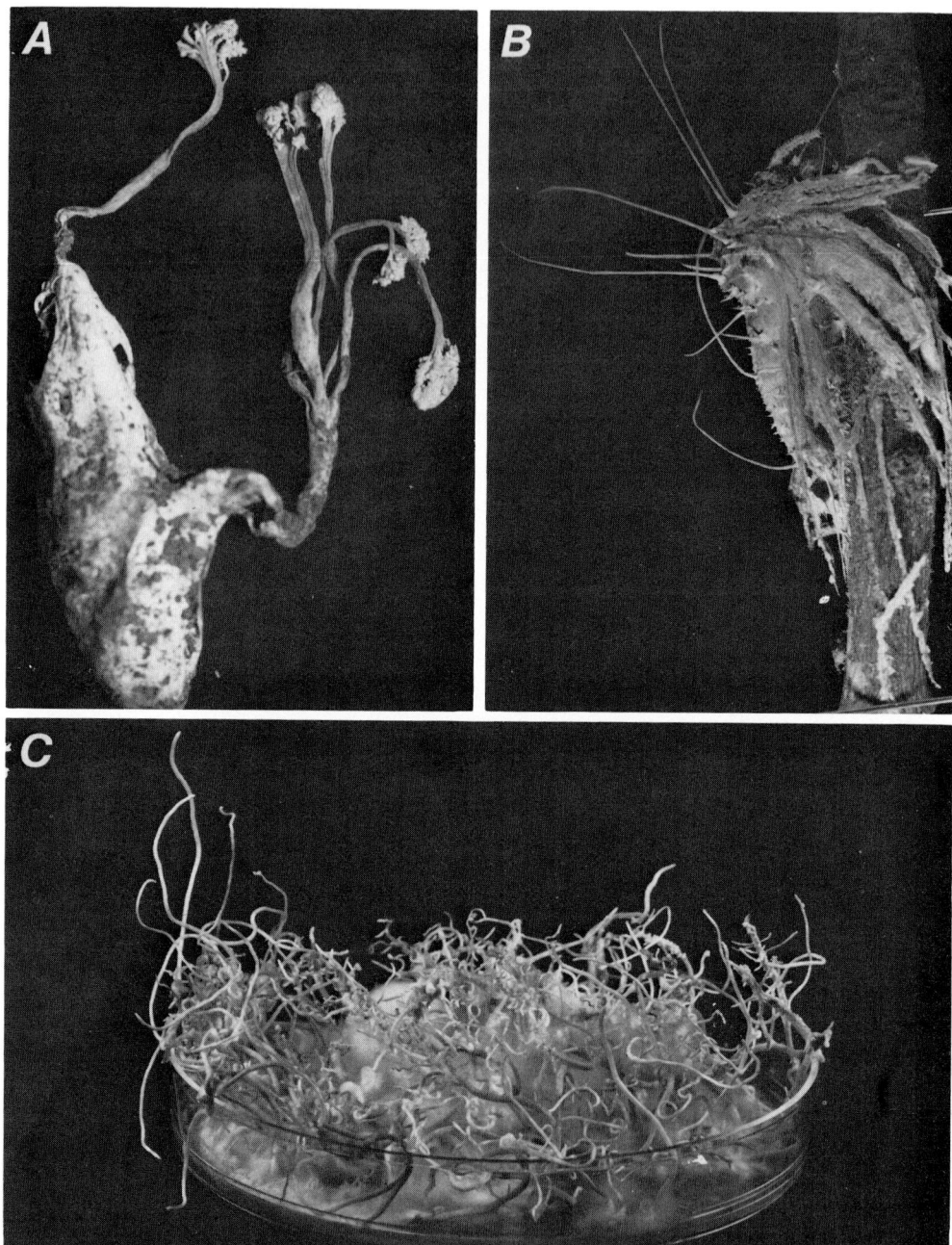

FIGURE 8. Some examples of synnematal development in the Deuteromycotina (A) *Paecilomyces tenuipes* on a lepidoptera larva, showing distinct synnematal growth. Note spore blooms at the apex (magnification × 2.5). (B) Adult moth infected with distinct cylindrical synnemata of *Akanthomyces aculeata* (magnification × 1). (C) Synnematal development of *Hirsutella thompsonii* var. *synnematosa* on oatmeal agar.

but failed to consider species without synnemata. Pacioni[144] reviewed some of the common species previously called *Isaria*. Recently, Minter and Brady[145] divided the genus *Hirsutella* based on the presence or absence of synnemata. They described and illustrated ten mononematous species including four new species. The teleomorph for *H. versicolor* was described as *Torrubiella pruinosa* (Petch) Minter and Brady, yet its placement in this genus is still questioned. The host range for the mononematous species is mainly restricted to Homoptera or mites (Figure 9). However one species, *H. rhossiliensis*, is reported as pathogenic to nematodes. Sturhan and Schneider[146] described this fungus as *H. heteroderae* from *Heterodera humuli*. Recent research, however, showed that this fungus is conspecific with *H. rhossiliensis*, first described from a soil sample, but later found to be pathogenic to several species of nematodes, e.g., *Criconemella xenoplax*.[147]

One of the mononematous species is a potential microbial control agent namely, *Hirsutella thompsonii*. This species is pathogenic to the citrus rust mite, *Phyllocoptruta oleivora*, and other phytophagous mites.[148] *H. thompsonii* is somewhat different from the other species of this genus in that it produces spherical rough-walled conidia. Samson et al.[149] described and illustrated three morphologically distinct groups of the pleomorphic species, *H. thompsonii*. The variety *vinacea* is characterized by vinaceous colonies on agar media and a different host, *Acalitus vaccinii* (Keifer). The variety *synnematosa* occurs on *Eriophyes* spp. and related genera in the tropics. In artificial culture, they produce cream-colored cylindrical synnemata bearing two kinds of phialides with conidia in chains. The variety *thompsonii* is characterized by gray-green colonies on artificial medium and the morphological characteristic defined for the original species. A polyblastic conidial state has been observed for *H. thompsonii* when grown on artificial media.

Two mutants with unique conidiogenesis have been isolated from a wild pathotype of *H. thompsonii*.[150] The mutants differ from the wild type in virulence, rate of conidiogenesis, and number of conidiogenous cells.

Boucias and McCoy[151] recently found distinct enzymatic differences using vertical acrylamide gel electrophoresis for 15 geographical isolates of *Hirsutella thompsonii*. Biochemical differences were correlated with defined varietal groups. Enzymes with multiple electromorphs commonly occurred indicating considerable molecular heterogeneity within the species. Nonregulatory and substrate-dependent enzymes demonstrated more heterogenicity than regulatory enzymes.

The genus *Synnematium*,[142] is similar to the genus *Hirsutella*, and was distinguished from it by its slender synnemata, elongate slender phialides narrow at the base, conidia in heads and covered with mucus, and typical sclerotia. The genus was revised by Mains who recognized two species reported to attack Homoptera and Coleoptera. Recently, Evans and Samson[143] showed that both *Hirsutella* and *Synnematium* conidiophores occur together on the same ants as the anamorphs of *Cordyceps* species (e.g., *C. kniphofioides* on *Cephalotus atratus*) and that a systematic separation is not possible. Therefore, the genus *Synnematium* was synonymized with *Hirsutella* and all species were transferred to this genus. *Hirsutella* (= *Synnematium*) *jonesii* was isolated from naturally infected adults of *Promecotheca papuana* in New Guinea[152] and cultured in vitro on malt extract agar and autoclaved eggs. Beetles were successfully infected by applying an aqueous suspension of conidia or a dust prepared from dried sclerotia. Natural control of the purple scale, *Lepidosaphes beckii* by *H. jonesii* was reported in New Caledonia.[153] Although *H. jonesii* causes significant mortality, biological characteristics of the host life cycle suggest that further ecological research is necessary before its potential as a microbial control agent can be fully assessed.

A new genus *Engyodontium* with two species, *E. album* and *E. parvisporum*, was recently described by de Hoog.[140] This genus, characterized by dense clusters of short-branched conidiophores and fertile cells on undifferentiated hyphae, was reported attacking scale insects and the whitefly. The little known genus *Hymenostilbe*, also frequently associated with *Cordyceps*, is an entomogenous hyphomycete with cylindrical synnemata covered by a hymenium-like layer of conidiogenous cells. It is distinguished from *Akanthomyces* by production of solitary conidia on sympodal conidiogenous cells, rather than conidia in chains on phialides. Samson and Evans[154] described nine species in a recent revision of *Hymenostilbe*, but many other undescribed species exist in new material collected in South America.

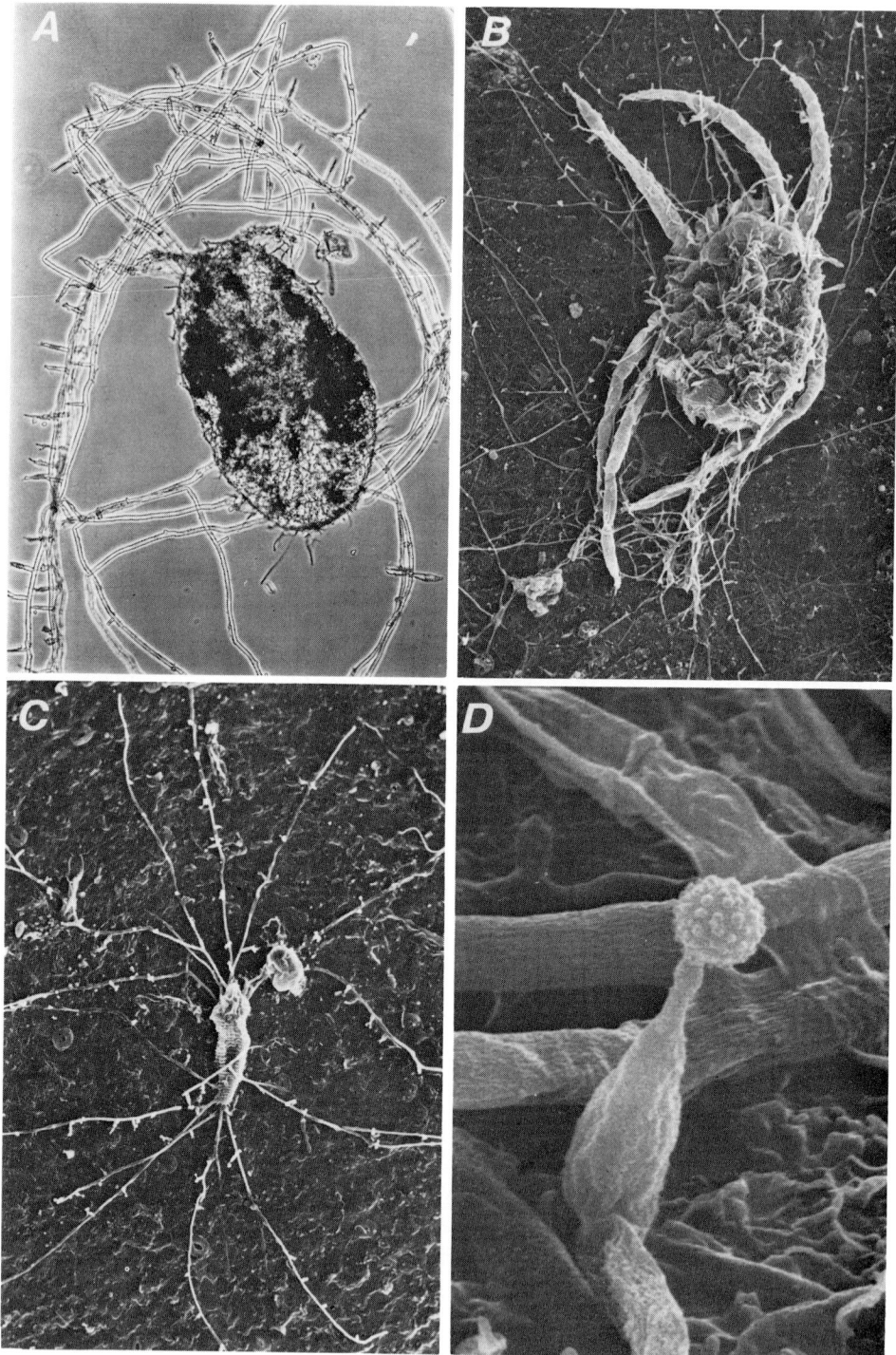

FIGURE 9. Representative mononematous species of the genus *Hirsutella*. (A) Aerial hyphae on phialides of *Hirsutella besseyi* on nymphs of *Ischnapsis* sp. (magnification × 130). (B—D) *Hirsutella thompsonii* (B) on Texas citrus mite (magnification × 150), and (C) on *Phyllocoptruta oleivora* (magnification × 150). (D) Typical phialide and verrucose, globose conidium of *H. thompsonii* (magnification × 5000).

Tropical and subtropical species have been found on Diptera, Myrmicine ants, Homoptera Heteroptera, and spiders. *Hymenostilbe* species have not been cultured on agar media despite numerous attempts to do so.

The genus *Tilachlidium* as circumscribed by Mains[142] and Gams[155] is characterized by having simple or branched synnemata bearing phialides that produce conidia in white, glutinous globules. Most species are saprophytes on various substrata; however *T. larvarum* has been found parasitizing species of Lepidoptera.

The ill-defined genus *Isaria* previously was used to accommodate many insect fungi with synnemata. Recent taxonomic research demonstrated, however, that most species belong in other genera, i.e., *Paecilomyces, Beauveria*, and *Metarhizium*. Samson and Brady[156] transferred *I. dubia*, a rather common pathogen of *Hepialus* spp., to a new genus *Paraisaria*. This genus has white, loose synnemata (consisting of verticillately branched conidiophores with phialides bearing several necks) and conidia produced in heads. *P. dubia* is the anamorph of *Cordyceps gracilis*. *Paraisaria* resembles the genus *Syngliocladium* and associated *Sorosporella*. *Sorosporella uvella* was reported to kill noctuid moth larvae (in the U.S.) by Speare.[157]

The genus *Beauveria* (Figure 10) has been monographed by MacLeod[158] and de Hoog.[159] de Hoog[159] recognized two species, *B. bassiana* and *B. brongniartii*, that attack all stages of insects of all groups; lungs of wild rodents; and nasal passages of horses, man, and giant tortoises. *B. bassiana* also occurs in the soil as a ubiquitous saprophyte. *B. bassiana*, known since 1835 as the cause of the white muscardine disease of silkworms, has been considered an important candidate for development into a microbial insecticide. Conidiogenous cells in *Beauveria* are denticulate with an elongated rhachis generally arising in clusters from subtending cells (Figure 10D). The species *B. bassiana* has subglobose to globose conidia with conidiogenous structures forming dense clusters. The closely related important entomogenous species, *B. brongniartii* (= *tenella*), has ellipsoid conidia. Studies with conidiogenous structure of different isolates of *B. bassiana* using immunoelectrophoresis and enzyme activity analysis have been conducted by Fargues et al.[160] Two strains from different hosts produced partially identical electrophoretic diagrams and similar enzyme activity indicating the same serotype. Serological techniques have also been used to characterize strains of *B. bassiana* in the U.S.S.R.[161] Recently, Evans and Samson[162] added two new species to the genus, *B. amorpha* on Coleoptera in Brazil and *B. velata* on lepidopterous larvae from Ecuador. Both taxa can be readily grown in culture.

The genus *Acrodontium* was formed by de Hoog[159] to accommodate seven species similar to *Beauveria*. These species are characterized by a proliferating tip of conidiogenous cells and a straight denticulate rhachis giving rise to apiculate conidia. The species, *A. crateriformis*, has been frequently observed on spiders and aphids.[126] The genus *Nomuraea* has been resurrected by Kish et al.[163] and described in detail by Samson[131] to accommodate two entomogenous species, *N. rileyi* (= *Spicaria prasina* = *Spicaria rileyi*) and *N. atypicola* (= *Isaria atypicola*). *N. rileyi* is pathogenic to a number of economically important lepidopterous pests[164] (Figure 11A, B) and *N. atypicola* is a subtropical to tropical species attacking mainly spiders.[132] Pathotypes of *N. rileyi* from different hosts have distinct biochemical differences.[165] The genus *Nomuraea* is characterized by having irregular to verticillate conidiophores and phialides with short inconspicuous necks. Infected hosts are covered with a dense white mat of hyphae that, upon conidiogenesis, turns pale-green or purple. *N. rileyi* and *N. atypicola* can be separated readily on the basis of conidial morphology and color.[126] *N. atypicola* is the anamorph of *Cordyceps cylindrica* as confirmed by recent cultural studies.[166]

The genus *Metarhizium* was recently monographed by Tulloch[167] who recognized two species, *M. anisopliae* and *M. flavoviride*. The two species are separated on the basis of conidial morphology: *M. anisopliae* was divided into two varieties, variety *anisopliae* (pos-

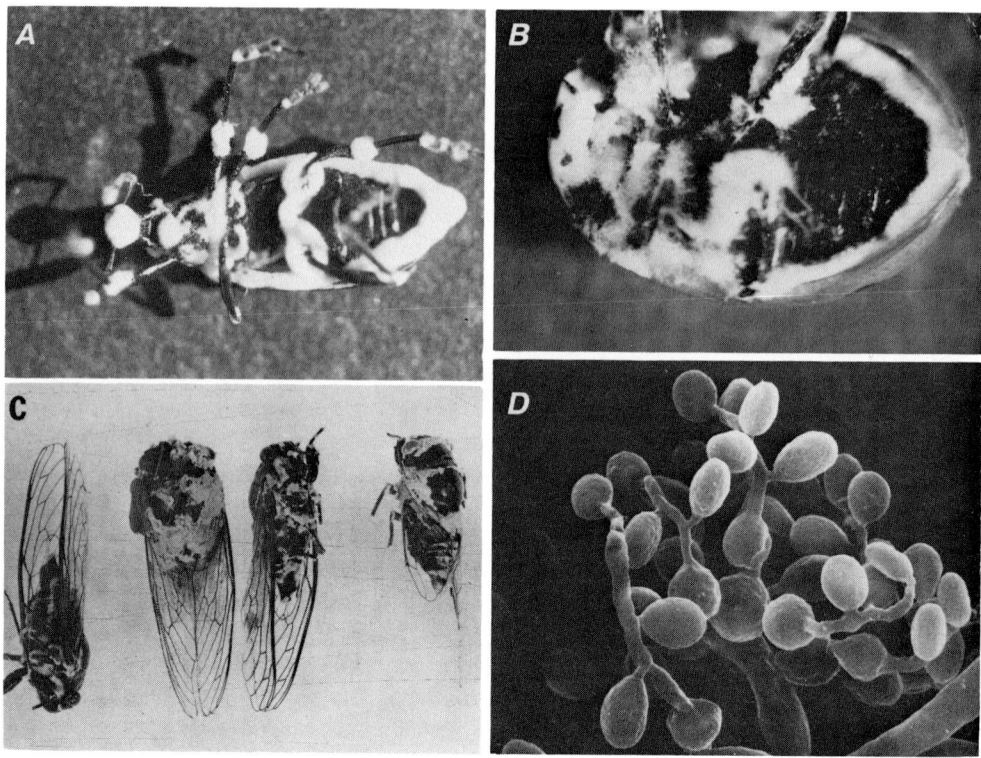

FIGURE 10. Various examples of *Beauveria bassiana*, the causal agent of the white muscardine disease. (A) Adult *Diaprepes abbreviatus*. (B) An unknown ladybird beetle. (C) Adult cicadids. (D) Conidiogenous structures of *B. bassiana* depicting the typical denticulate conidia on an elongated zig-zag-shaped rhachis (×6700).

sessing short conidia, 3.5 to 9.0 μm and attacking species within Orthoptera, Coleoptera, Lepidoptera, Hemiptera, Hymenoptera, and Arachnida) and variety *major* (possessing long conidia, 9.0 to 18.0 μm). The genus is characterized by having dry catenulate conidia arising from densely packed conidiophores borne over the surface of the host (Figure 12). Fargues et al.[168] was able to distinguish different strains of *M. anisopliae* using biochemical techniques.

The genus *Verticillium* is characterized as producing conidia on awl-shaped phialides borne in verticils or whorls on the conidiophore. The generic concept was emended by Gams[155] to accommodate the entomogenous species of the genera *Cephalosporium* and *Acrostalagmus*. These fungi have been placed in the section *Prostrata* that is characterized by a velvet or cotton-like development of the aerial mycelium. The mycelium may occasionally contain mesotonous to acrotonous whorls of phialides. In contrast, Balazy[169] believed that Gams' taxonomic work was too general and further questioned its inclusion in *Prostrata*. The taxonomic controversy also involved synonymy of a number of entomogenous species within the *V. lecanii* complex. About ten species within this diverse genus attack a number of different insects. *V. lecanii*, however, is considered the most important candidate as a microbial control agent of aphids and scales[170] (Figure 11C, D). The teleomorph connection of *V. lecanii* with the ascomycete *Torrubiella confragosa* was established based on specimens and cultures of Coccidae from the Galapagos Islands.[125]

The genus *Acremonium* was monographed by Gams,[155] who placed it in synonymy with *Cephalosporium*. This genus contains two entomogenous species (*A. larvarum* and *A. zeylanicum*) both mainly attacking species of caterpillars. The genus is characterized by conidia borne in mucus balls at the tips of solitary, awl-shaped phialides.

The genus *Aspergillus* contains a number of entomogenous species including *A. flavus*,

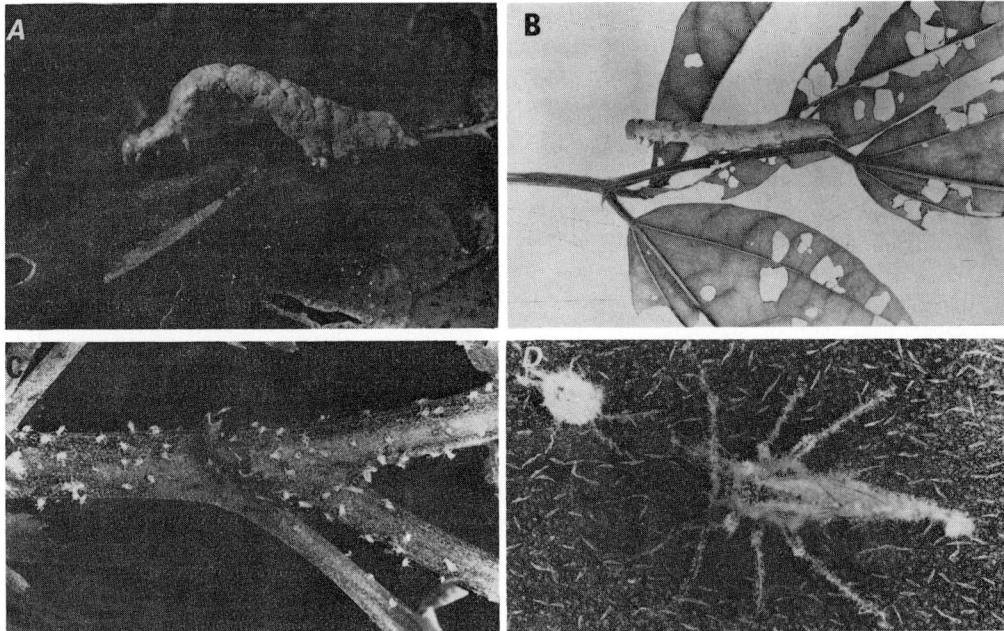

FIGURE 11. Entomogenous fungi of the (A—B) Deuteromycotina. The cabbage looper, *Trichoplusia ni*, and an unidentified lepidopterous larvae infected with *Nomuraea rileyi* and (C—D) aggregation and single aphid (close-up) infected with *Verticillium lecanii*. (Photographs A and C—D courtesy of Mr. James V. Bell, USDA, Stoneville, Mississippi and Dr. Richard A. Hall, Glasshouse Crops Research Institute, Littlehampton, England, respectively.)

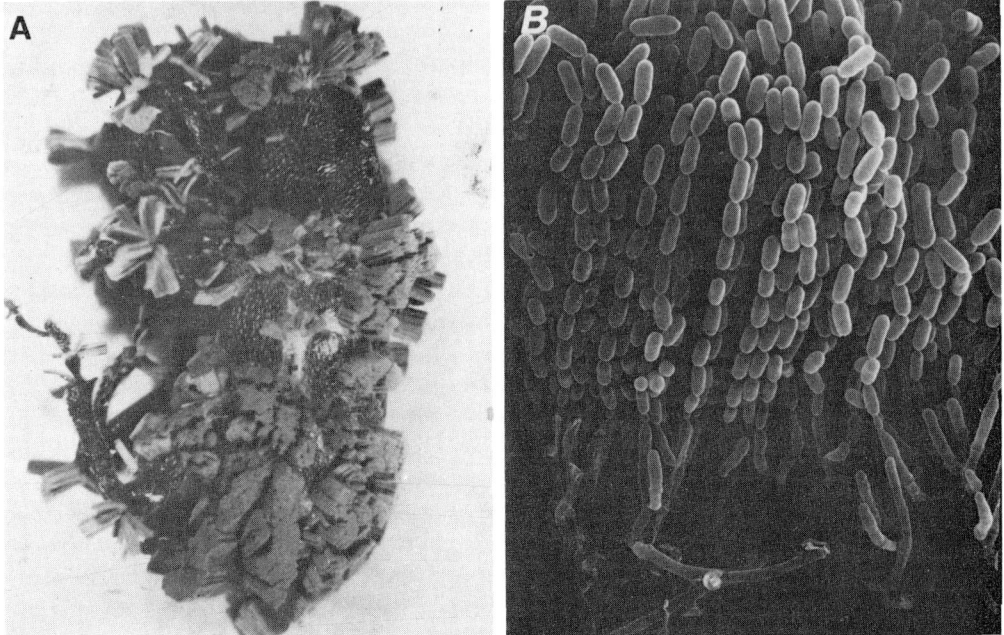

FIGURE 12. *Metarhizium anisopliae*, the causal agent of the green muscardine disease on (A) *Hylobius pales*. (B) Note the pallisade-like conidial masses scanning electron micrograph of the phialides and typical compact conidial chains ($\times 1100$). (Photograph A courtesy of Dr. H. Schabel, University of Wisconsin, Stevens Point, Wisconsin.)

A. parasiticus, and *A. ochraceus* that attack grasshoppers, bees, ants, and lepidopterous larvae.[124] *Aspergillus* has been observed to grow in nature on dead substrates indicating that a saprophytic existence may be its dominant life habit. This genus is characterized by having phialides borne in whorls on swollen apices (vesicles) of erect conidiophores.

Some species of the genus *Fusarium*, known to parasitize insects or other arthropods (e.g., American lobster[171]), enter the host through wounds.[117] As previously mentioned, the anamorphs are associated with *Nectria*. *F. solani* (Mart.) Sacc. has been reported as a weak pathogen of beetles and other invertebrates.[172] The genus is characterized by two types of phialides producing septate, i.e., curved, banana-shaped macroconidia and smaller microconidia. *F. coccophilum* is an important pathogen on coccids and scales. This species often produce synnemata on the hosts. Besides the anamorph, the *Nectria* teleomorph is often found (Figure 7).

The genus *Culicinomyces* is described from one species described as *C. clavisporus* from an isolate from a laboratory larval colony of *Anopheles quadrimaculatus* in the U.S.[173] In addition to mosquito larvae, Sweeney[174] found a similar isolate from larvae of Chironomidae and Ceratopogonidae susceptible to infection in Australia. The Australian isolate proved to be identical with the American species.[175] The fungus normally gains entry into host larvae through the walls of the digestive tract via germinated conidia having entered the gut through the oral opening.[176] The genus is characterized by having flask-shaped phialides and clavate conidia in slimy heads (Figure 13).

The genus *Tolypocladium* was described by Gams[177] for a group of mainly soilborne fungi, resembling the genus *Beauveria*. Federici et al.[178] and Soares[179] found *T. cylindrosporum* causing as much as 90% mortality in larvae of the tree hole mosquito, *Aedes sierrensis*, in California. This fungus can be grown on artificial media where it forms blastospores in shake liquid cultures and conidia on semisolid substrates. The same fungus was also reported from *A. australis* in New Zealand. Samson and Soares[180] described *T. cylindrosporum* a pathogen isolated from mosquitoes and a new species, *T. extinguens*, found infecting the glowworm, *Arachnocampa luminosa*.

Several species of the usually saprophytic genus *Stilbella* are known to parasitize arthropods.[181] *Stilbella* is characterized by distinct synnemata, terminating in swollen head, bearing phialides and conidia in slime. A very common species is *S. buquetti*[182] which infects weevils and ants. This species is often associated with the common *Cordyceps australis*. The genus *Polycephalomyces*, of which several taxa are reported, is similar to *Stilbella*. It differs from *Polycephalomyces* in that it has regular branched synnemata.

The monotypic genus, *Desmidiospora*,[183] has been reported from ants. This genus typically has large lobed conidia as well as a *Hirsutella* anamorph. Evans and Samson[184] consider *Desmidiospora* as a potential anamorph of *Cordyceps, unilateralis*, a common fungus on *Camponotus* ants.

de Hoog[185] described four entomogenous species in the genera *Sporothrix*. Some species are apparently common on leafhoppers in South America. Mains[181] identified three entomogenous species from the genus *Tilachlidiopsis*. The nomenclatorial status of this genus is, however, unclear and must await further taxonomic study.[182]

Two new genera recently have been proposed. *Pleurodesmospora*[186] was established for *Gonatorrhodiella coccorum* an entomogenous hyphomycete of scale insects, whiteflies, mites, aphids, and leafhoppers. The conidia in this fungus are produced in chains by reduced, peg-like phialides. The genus *Clathroconium*[187] is a peculiar fungus reported from spiders. It is characterized by yellow-brown clathroid conidia, consisting of filaments forming a hollow network. Only one species, *C. arachnicola*, has been found from Ghana.

Within the Coelomycetes, two genera have been identified as being entomogenous. The genus, *Aschersonia* (Figure 14), is a well-known pathogen of scale insects and whiteflies. The taxonomy of the genus was treated by Petch[114] and Mains,[188,189] who recognized several

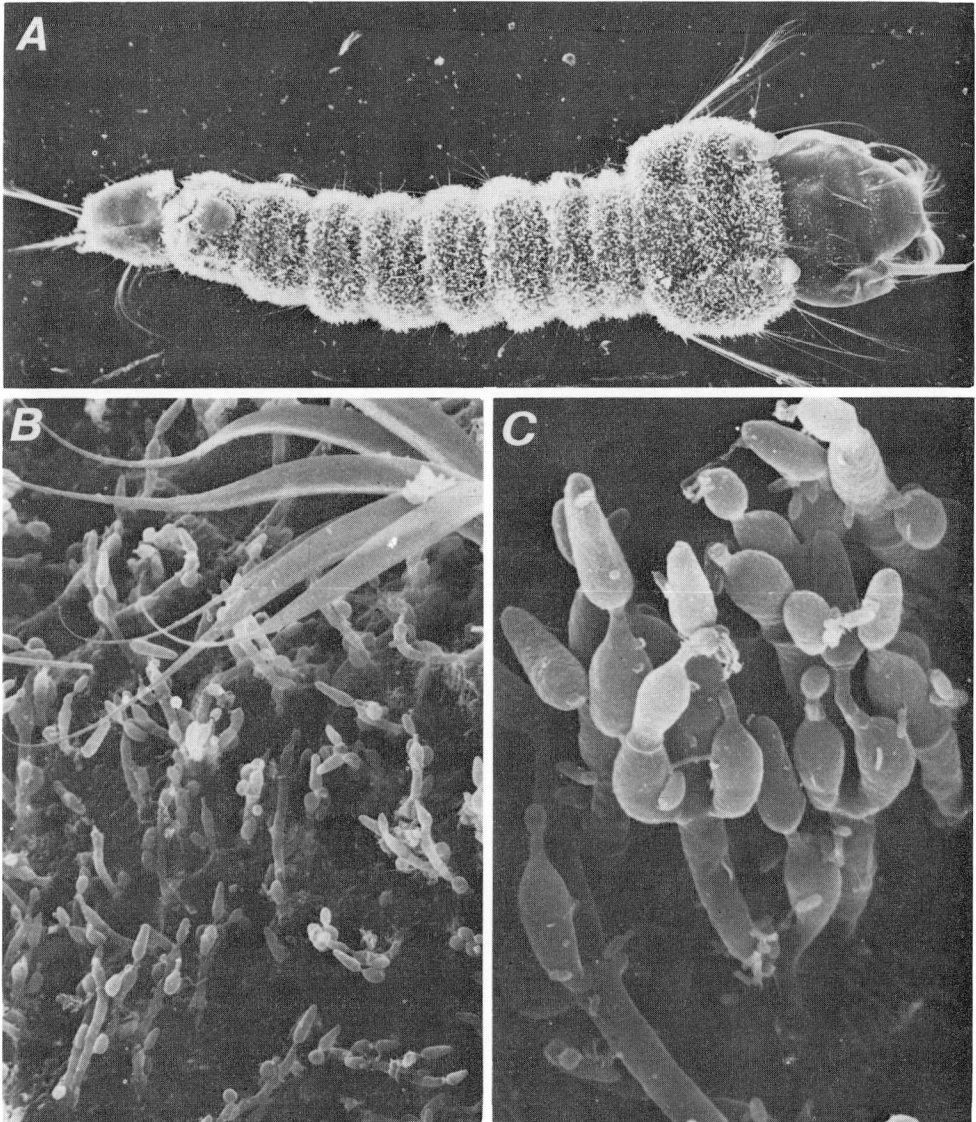

FIGURE 13. Scanning electron micrographs of *Culicinomyces clavisporus*, an entomogenous fungus of (A) mosquito larvae completely covered by hyphyae and conidiophopres (magnification × 40). (B) Part of larval body depicting the sporulating structures (magnification × 800). (C) Conidiophores with flash-shaped phialides and clavate conidia of *Culicinomyces clavisporus* (magnification × 3500).

species occurring on whiteflies in North America. A recent systematic key of the genus *Aschersonia* was prepared by Protsenko.[190] Species of *Aschersonia* have small simple stromata consisting of densely interwoven thick-walled hyphae in which the phialides and slimy conidia are produced in pycnidia that are locules without differentiated walls[191] (Figure 14C, D). They are anamorphs of species of *Hypocrella* where the perithecial stages are known. In most species, perithecia are rarely produced. Some species, e.g., *A. aleyrodis*, *A. placenta*, and *A. cubensis* are very common in the subtropics. The use of these fungi as possible bioinsecticides has been investigated.[121]

The pycnidial genus *Tetranacrium* is mostly associated with species of the ascomycete genus *Podonectria*.

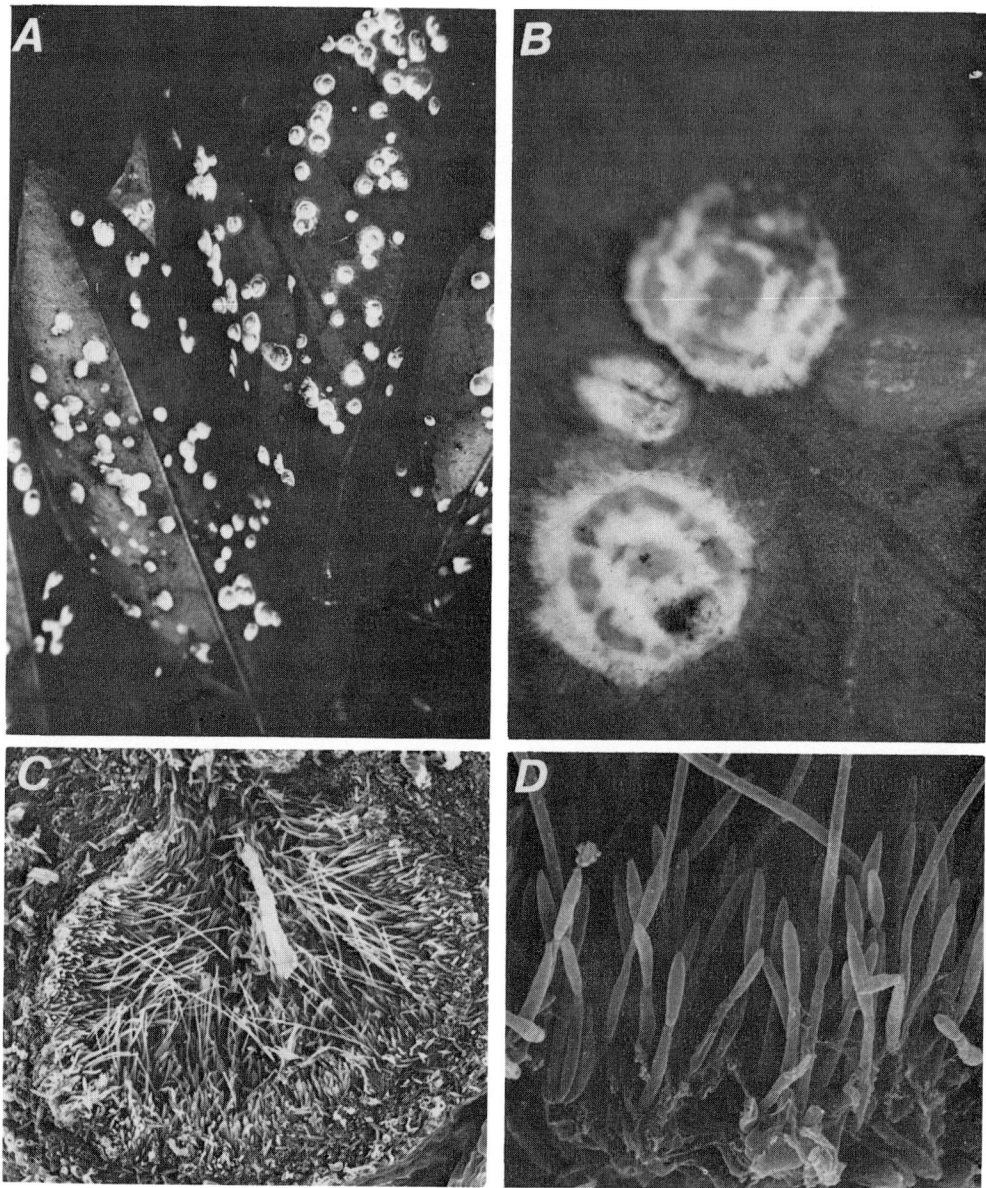

FIGURE 14. *Aschersonia aleyrodis*, a common pathogen on whiteflies. (A) Citrus leaves covered with infected larvae of *Dialeurodes citri*. (B) Larvae completely parasitized with white mycelium showing sporulating structures indicated by conidial slime (magnification × 15). (C) Scanning electron micrograph of a pycnidium, containing conidiogenous cells and paraphysis. (D) Detail of a pycnidium with phialides and conidia (magnification × 2300).

Within the Mycelia Sterilia, only the genus *Aegerita* is endoparasitic. Fawcett[192] reported *A. webberi* as a parasite on *Aleyrodes citri*, the citrus whitefly. *A. webberi* forms dark "sporodochia" in which no true conidia are produced, but these sporodochia are easily dislodged and act as propagules. Petch[193] considered *A. webberi* as the anamorph of the basidiomycete *Septobasidium pilosum*, but his findings do not agree with our observations on specimens from Florida.

INVASION OF THE HOST

Mode of Parasitism

Approximately 1500 species of the Laboulbeniales and a few genera within the Deuteromycotina contain species that are ectoparasites (Table 1).[194] These entomogenous fungi have a thallus that is mainly superficial and when it penetrates the host integument, its function is strictly haustorial. For example, the termite parasite, *Termitaria snyderi*, has a haustoria which penetrates the cuticle of the host and forms lobed structures beneath.[195] Generally, these parasites may impair host activities and/or cause severe debilitation, but do not cause early death.

Entomogenous fungi with an endoparasitic mode of parasitism include all species that are considered important candidates as microbial control agents. These fungi usually cause early death by nutritional deficiency, invasion and/or digestion of tissue, and/or by the release of toxins. Unlike a few species of fungi which are only capable of invasion through a wound in the host cuticle (i.e., *Fusarium, Mucor,* and *Penicillium*), endoparasitic entomogenous species usually penetrate the host's integument and sometimes invade the host via the gut,[176,196,197] buccal cavity,[198] or through spiracles.[199] Impairment of respiration, due to blockage of the air tube at the perispiracular valve, has also been reported in mosquito larvae.[200,201]

Infection Cycle for Representative Fungi

Endoparasitic infections are usually initiated by spores or conidia (i.e., Zygomycotina, Deuteromycotina) though in the Mastigiomycotina and Ascomycotina zoospores, planonts or ascospores, respectively, are responsible (Figure 15). One exception is the whitefly parasite, *Aegerita webberi*, where propagules consisting of aggregations of inflated cells are the infective unit.[192] Infective propagules vary widely in shape, size, and place of origin; some are dry (*Beauveria, Metarhizium*) while others are covered with a sticky mucus (i.e., *Hirsutella* spp.). In the Entomophthoraceae, resting spores (zygospores, azygospores) and primary conidia produce adhesive spores which are infectious to host insects.[202-204] Sclerotia of *Synnematium jonesii* also have been reported to be infectious.[152]

Deuteromycetes

The mode of infection of the various entomopathogenic hyphomycetes are quite similar. Conidia, the infective propagules of these pathogens, are usually produced in copious amounts on externally borne conidiophores. These propagules, passively disseminated by abiotic and biotic factors, contact the susceptible host and initiate the infection sequence. A typical infection cycle of a hyphomycetes is as follows: conidial attachment, germination, germ tube penetration, vegetative growth, and then conidiogenesis.

The epicuticle of the host integument, a composite structure containing wax and several lipoprotein layers,[205] is the site of initial fungus-host interaction. In a recent review, Fargues[206] proposed that adhesion of conidia to the cuticle involves both an adsorption stage and a consolidation stage. Initially, pregerminated conidia will adsorb to the cuticular surface via nonspecific, passive interactions involving charged groups (Figure 16). The specific chemicals involved in this adsorption stage have not been determined. Conidia of many Deuteromycetes (*Beauveria bassiana, Nomuraea rileyi, Metarhizium anisopliae*) possess a dry, hydrophobic surface which presumably interacts with the outer surface of the host cuticle. Grula et al.[207] reported the presence of hemagglutinin activity associated with the outer surface (pellicle) of *B. bassiana* conidia. Conidial samples pretreated with carbohydrates that competitively inhibit hemagglutination activity (glucose, glucosamine, N-acetylglucosamine) had less affinity to host cuticle than untreated conidial samples. Certain Deuteromycetes, i.e., *Hirsutella* and *Culicinomyces*, produce wet conidia coated with a mucilagenous outer layer which is believed to play an important role in conidial attachment.[149,208]

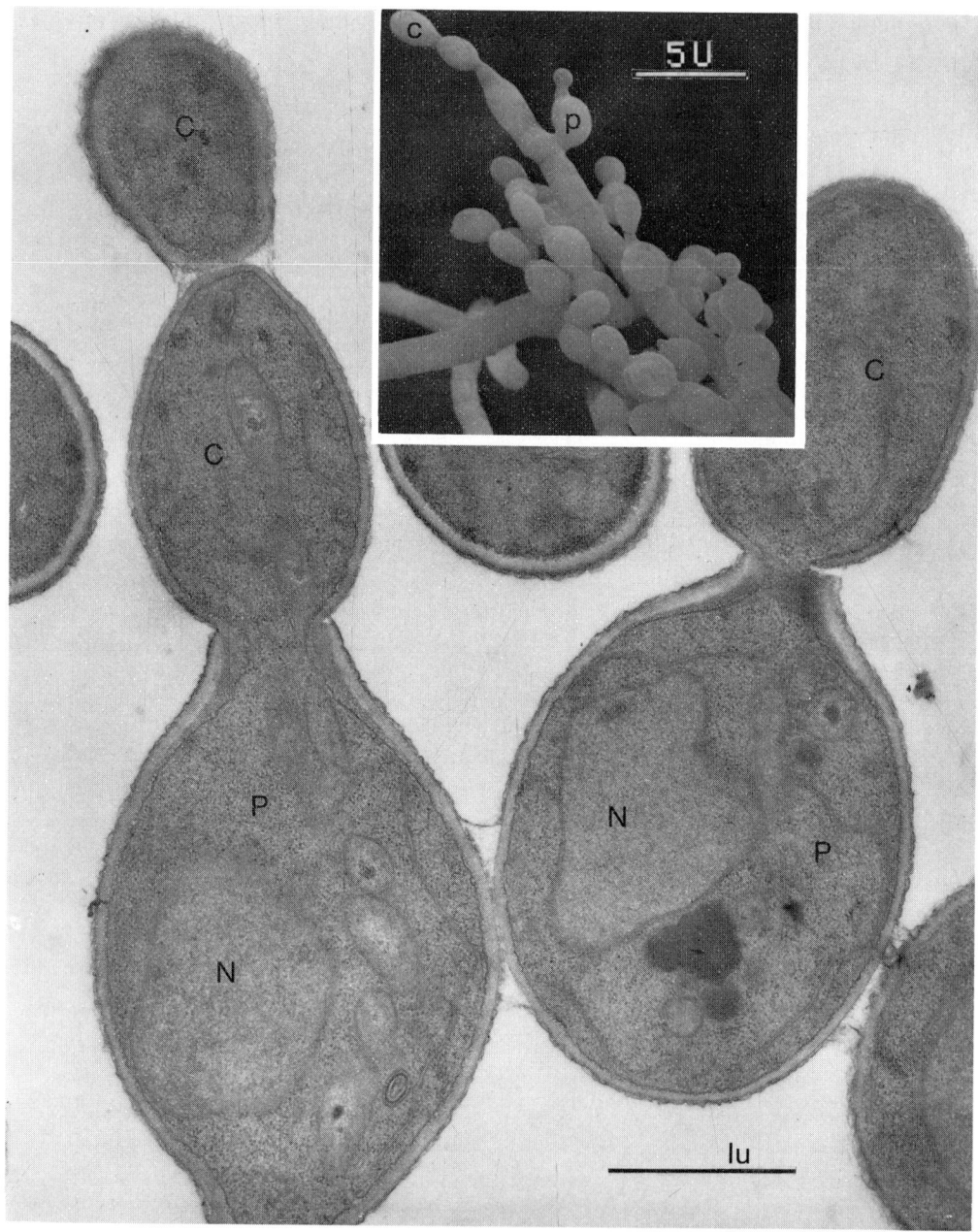

FIGURE 15. Scanning and transmission electron micrographs of conidiogenesis of *Nomuraea rileyi*. Note the large phialide cell giving rise to a linear array of mature conidia. C = conidium, P = phialide, N = nucleus.

The second phase of the conidial adhesion process is the consolidation stage. It has been proposed that chemical cues associated with the cuticle may passively stimulate adsorbed pregerminated conidia to become "actively" attached to the cuticle by inducing either certain enzymatic processes and/or the secretion of mucilagenous substances.[206] Early work by Wallengren and Johansson[209] reported that *Metarhizium anisopliae* conidia will react with the epicuticular surface before germination has commenced. More recently, Michel[210] has reported that *Beauveria bassiana* pregerminated conidia produce esterase, lipase, and *N*-

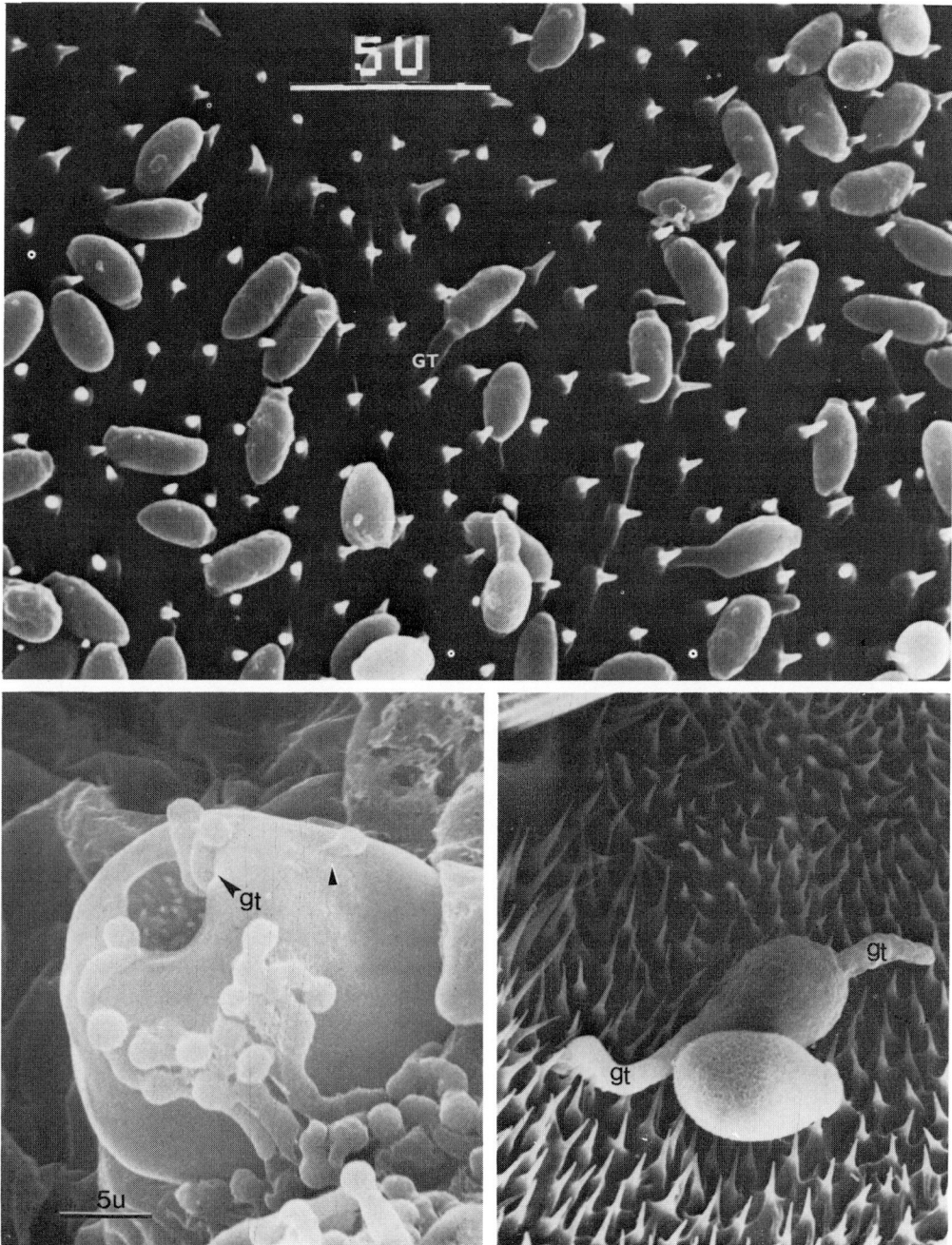

FIGURE 16. Scanning electron micrograph of conidial germination on the epicuticular surface of host insect larvae. Note production of germ tube (GT) and its orientation to the cuticle surface. Upper *Nomuraea rileyi* on *Anticarsia gemmatalis*; lower left *Beauveria bassiana* on *Heliothis zea*; lower right *Erynia blunckii* on *Plutella xylostella* L.

acetylglycosaminidase activities when deposited on the cuticle of the wax moth, *Galleria mellonella*. Such enzymes may be important in the adhesion process. The tenacity by which conidia adhere to the host cuticle has been recently investigated by Boucias and Pendland.[657] Conidia of *Nomuraea rileyi*, applied to larvae of the velvetbean caterpillar, *Anticarsia gemmatalis*, remain firmly attached to host cuticle following a 2 to 3 hr treatment in boiling

1% sodium dodecyl sulfate. Examination of the resulting ghost-cuticle revealed the presence of an undefined fibrilla sheath material at the conidial-cuticle interface. Similar treatment of larvae contaminated with conidia of a nonentomopathogenic fungus resulted in the complete removal of conidia from the cuticular surface.

Cuticular components, in addition to being important determinants for conidial adhesion, also play a nutritional role in the stimulation and subsequent growth of the conidial germ tube. The importance of such components will vary according to the specific pathogen being studied. Fungi, such as *Metarhizium anisopliae*, that germinate on a water-agar substrate,[211] typically contain large numbers of lipoprotein inclusions and lipid droplets.[212] These substances are metabolized during germ tube formation. However, cuticular components, while not requisite for germ tube formation, have enhanced the germination rate of conidia of *M. anisopliae*.[213,214] *Beauveria bassiana* requires both a carbon and nitrogen source for stimulation and growth of the germ tube.[215-217] Compounds such as *N*-acetylglucosamine, chitin, starch, and long chain fatty acids stimulated germination whereas either an inorganic or organic nitrogen source is required for germ tube growth. Conidia of *Nomuraea rileyi*, not containing extensive nutrient reserves, require either a complex medium such as yeast extract or a crude extract of host cuticle for germ tube development.[164] A simple carbon and nitrogen source tested independently or in combinations did not stimulate conidial germination.[218] Fractionation of the crude extract on a silicic acid column and subsequent in vitro assay of resulting fractions revealed that the sterol, diacylglycerol, and polar lipids produced germination rates equivalent to that achieved with the original crude extract.[218]

Upon germination, the germ tube may penetrate directly[219-221] or grow over the surface of the epicuticle before penetrating.[222] The tropisms controlling these processes are poorly understood. Certain hyphomycetes produce a specialized appressorial cell at the germ tube-epicuticle interface.[212,213,223] The appressorium has been reported to be coated with an amorphous mucilaginous material that sticks it to the epicuticle. A similar material, an extracellular fibrillar sheath, is deposited at the germ tube-penetration site of *Nomuraea rileyi*. *N. rileyi* does not form an appressorium (Figure 17).[224] This sheath material, although readily stained with ruthenium red, did not bind to commercial lectin conjugates, thus indicating the presence of polysaccharides with possible 1-3 linkages.[225]

Successful penetration of entomogenous fungi through host cuticle, a requisite for infection, depends upon both the inherent capabilities of the penetrating germ tube and the physiological state of the host. Penetration is believed to involve both mechanical and enzymatic activities of the germ tube.[226,227] A germ tube must penetrate both the epicuticle, a layer characterized by a cross-linked network of lipids and protein, and the procuticle, a layer of chitin and protein. The procuticles can be further differentiated into sclerotized exocuticle and nonsclerotized epicuticle layers. Histological studies[212,221,228-230] have shown that penetration involves lysis of the cuticular layers. Chemical studies by Samsinakova et al.[231] showed that digestion of insect cuticle required a protease-chitinase combination. Similarly, Smith et al.[232] found that digestion of detergent-produced "insect ghosts" required a protease-chitinase treatment. However, these studies utilized commercial enzyme preparations and not enzymes extracted from germ tubes. Many entomopathogenic hyphomycetes produce extracellular proteases, chitinases, and/or lipases.[229,233-239] At present, the relative significance these enzymes play in the penetration process is unclear. In many cases, the assayed extracts were derived from vegetative mycelia and not from the developing germ tube. Furthermore, no clear correlation has been made between enzymatic activity and pathogenicity[240] and the lytic enzymes, chitinase, proteases, and lipases are common to many nonentomopathogenic fungi. The specific enzymes or enzyme combinations responsible for penetration of the germ tube have yet to be defined or characterized.

As indicated earlier, both the inherent properties of the fungus and the physiological state of the host's cuticle will influence the penetration process. In most cases, germ tubes will

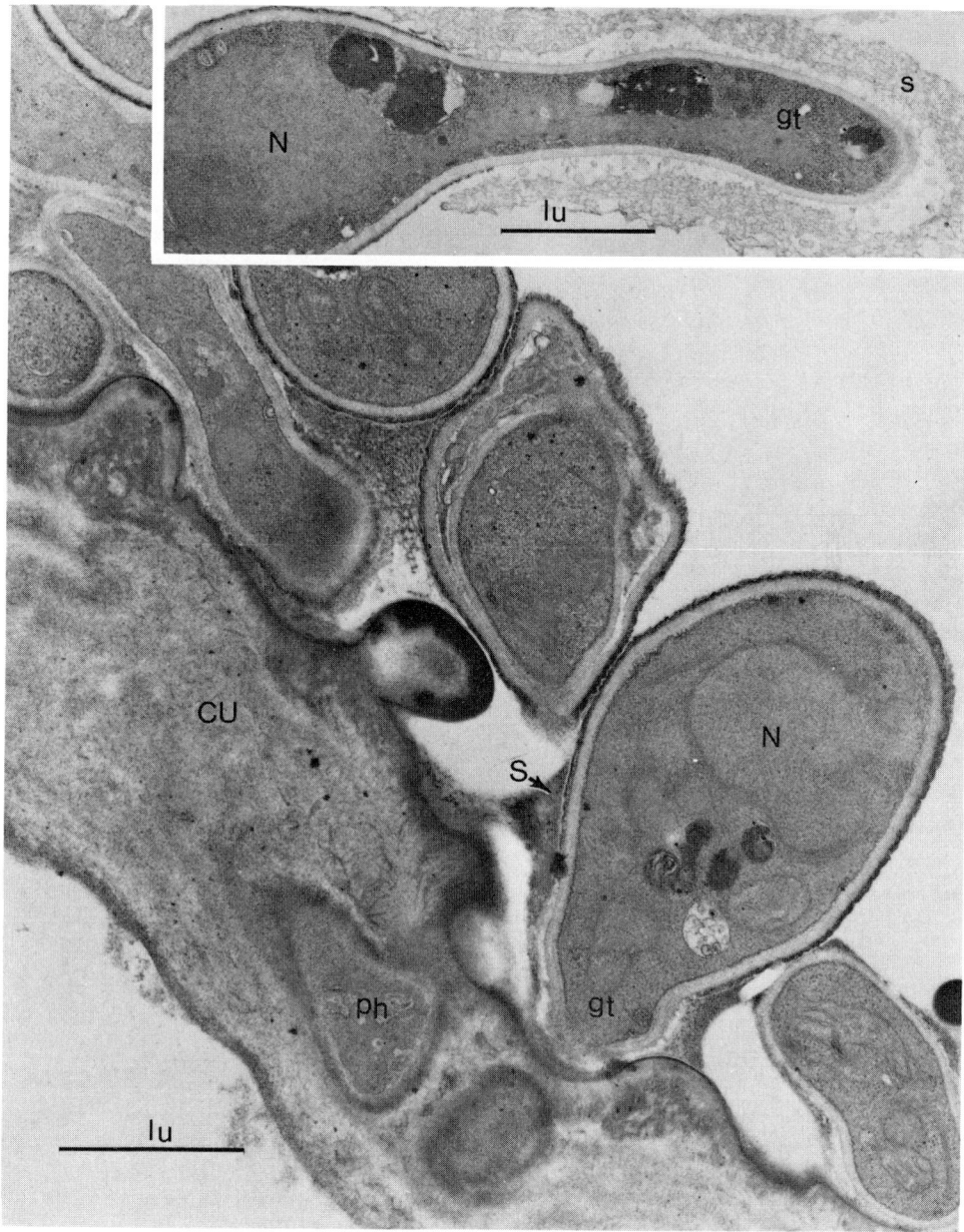

FIGURE 17. Transmission electron micrograph of *Nomuraea rileyi* penetrating the epicuticle of host *Anticarsia gemmatalis*. Note production of sheath material(s) by germ tube (gt). Cu = cuticle, s = sheath, N = nucleus, ph = *penetration hyphe*.

penetrate the nonsclerotized membranous zones quicker and more directly than the sclerotized cuticle. Germ tubes contacting heavily sclerotized regions may respond by producing an appressorial-like structure or by growing over the cuticle surface until it reaches a suitable penetration site.[222] Histological studies have shown that penetration-hyphae, after transversing the epicuticle, may grow laterally and mechanically displace the procuticular laminae.[221,223,241] With *Metarhizium anisopliae*, a specialized penetration plate may form beneath the epicuticle. Various structures, produced from this plate, will transverse the laminae in

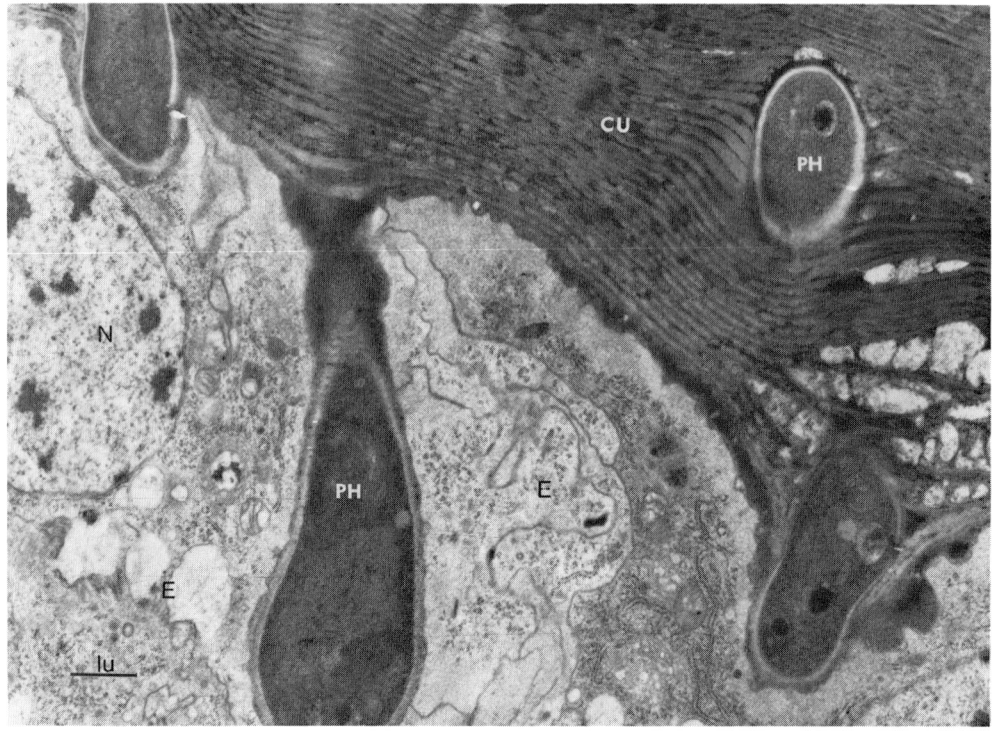

FIGURE 18. Penetration of the endocuticle (U) and the epithelial layer (E) by *Nomuraea rileyi* penetration hyphae (PH). U = endocuticle, Cu = cuticle, N = nucleus.

a stepwise fashion.[241] The extensive growth of penetrating hyphae of many hyphomycetes strongly suggests an active digestion and absorption of cuticular components. The time required for germ tube penetration through the host integument is variable. As examples, several days are required for *M. anisopliae* to penetrate the heavily sclerotized cuticle of wireworms.[241] and Kawakami and Mikuni[242] found that it takes 16 to 40 hr for *Beauveria bassiana*, *Spicaria prasina* (= *Nomureae rileyi*) , and *Paecilomyces farinosa* to penetrate the integument of silkworm, *Bombyx mori* larvae. Similarly, *Nomuraea rileyi* can penetrate the velvetbean caterpillar, *Anticarsia gemmatalis*, within 24 hr (Figure 18).[221]

Colonization of host tissues after penetration of the germ tube may involve production of elongated hyphae and/or yeast-like hyphal bodies (blastospores). The specific host tissues infected and the kinetics of the infection process vary according to the pathogen-host system (Figure 19). Certain hyphomycetes induce death before extensive invasion of host tissues. This type of fungal-induced mortality has been attributed to the production of toxic substances.[243] As an example, *Metarhizium anisopliae* produces an array of toxic components including prodestruxin, destruxins A + B, destruxins C + D, proteases, and cytochalasins.[244-249] The effectiveness of these secondary toxic metabolites in killing the host is influenced by the host itself. Hanel[230,250] reported that *M. anisopliae* kill termites within 2 days postinoculation at which time only very few hyphal bodies could be detected in the hemolymph. Penetration of the majority of the host's tissues occurred after death. Alternatively, Zacharuk[251] reported that the hypodermis, fat body, malpighian tubules, midgut, ventral abdominal ganglion, and muscle tissues of elaterid larvae were invaded by *M. anisopliae* prior to death. He suggested the presence of toxins based on the observation of membrane integrity of cell organelles and dehydration of tissue cells. Recently, it has been suggested that these secondary metabolites act as immunosuppressants, inhibiting the cellular and/or

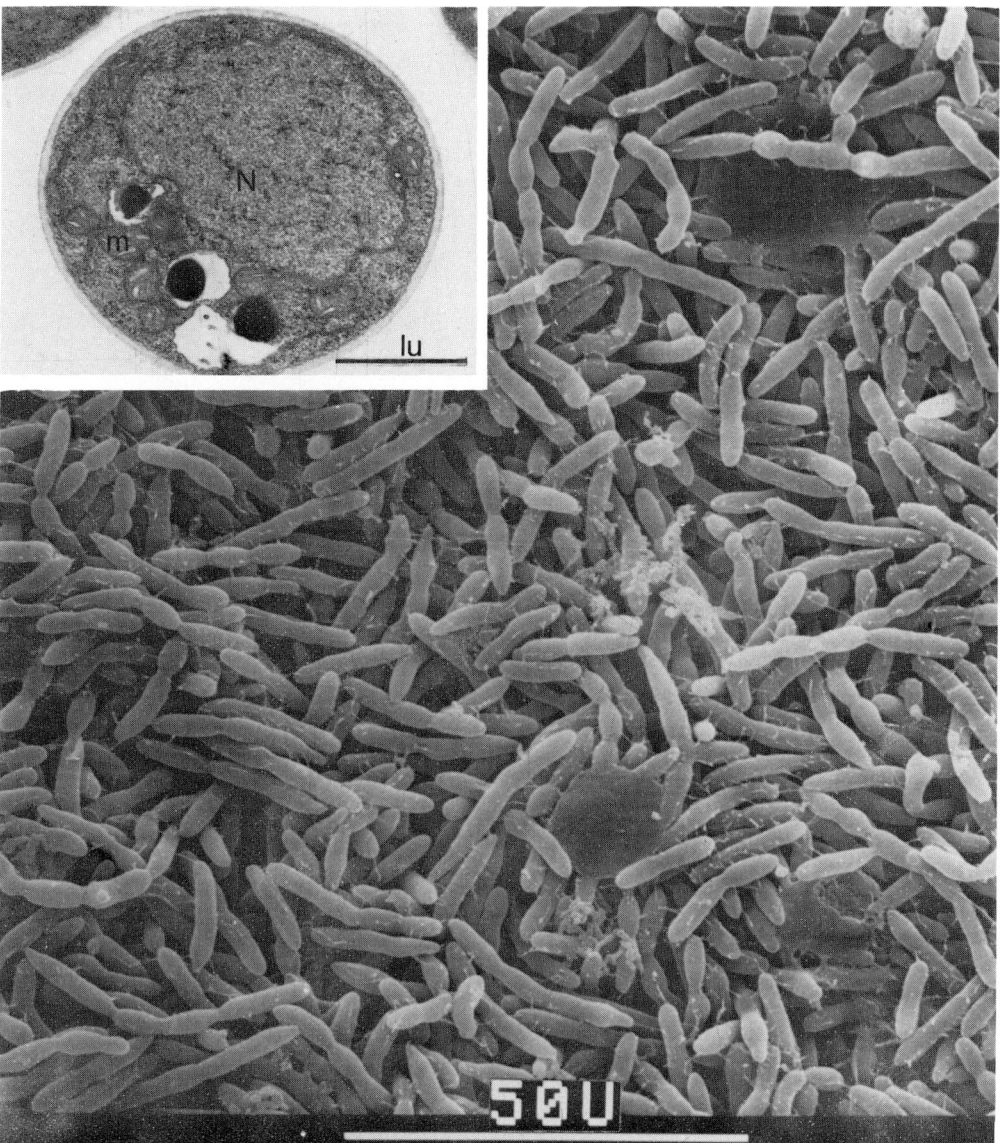

FIGURE 19. Scanning and transmission electron micrographs of vegetative hyphal bodies of *Nomuraea rileyi* found in the blood.

humoral host defense response.[252] *Beauveria bassiana* also produces toxic substances,[253-255] but evidence that these toxic compounds are involved in in vivo pathogenicity is still lacking. For example, Champlin and Grula[256] using an autobiographic assay were unable to detect the toxic beauvericin in *B. bassiana* infected corn earworm larvae, *Heliothis zea*, as compounds isolated from in vitro cultures of *B. bassiana* were insecticidal. The cyclic depsipeptide bassianolide induces atony when injected (1 to 6 μg/larvae) into larvae of both *Bombyx mori* or *H. zea*.[256-258] In addition to toxins, several researchers have proposed that secondary compounds produced by entomogenous fungi (i.e., beauvericin, oosporein) are antibiotics[256-258] that prevent bacterial putrefaction and thus permit fungal mummification of host tissue. Certain hyphomycetes, i.e., *Nomuraea rileyi*, will undergo extensive vegetational multiplication in the host prior to death. In such cases, physiological starvation and not mycotoxins is probably the cause of the host's death.

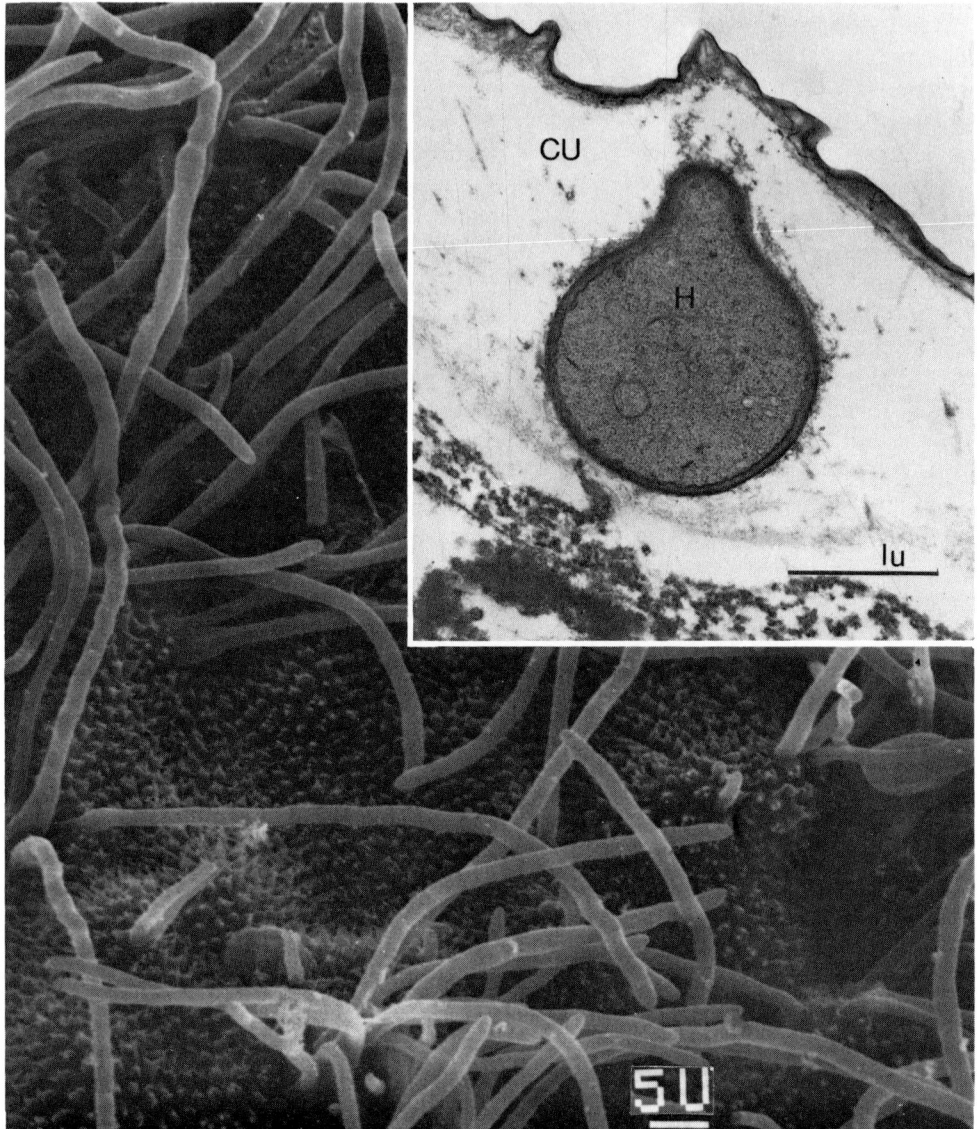

FIGURE 20. Emergence of *Nomuraea rileyi* hyphae from the cuticle of a diseased *Anticarsia gemmatalis* larva. The emergent hyphae will form a conidial producing phialide cell. Vegetative growth of *N. rileyi* has digested cuticular lamella (insert). Cu = cuticle, H = hyphae.

Insects, before succumbing to fungal infection, exhibit various symptoms including restlessness, cessation of feeding, and loss of coordination. Diseased insects commonly move to high places, i.e., on vegetation, or if subterranean, rise to the soil surface. Following death, the fungal hyphae continue to grow usually resulting in mummification. As the host nutrients are depleted, the internal fungal cells (hyphal bodies) may differentiate into elongate hyphae which extend outward from the body cavity forming a mycelial mat of conidiophores over the surface of the integument.[244,259] In many cases, fungal outgrowths will occur initially and most predominantly at the intersegmental regions of the host (Figure 20). The procuticle and internal tissues are usually totally degraded at the onset of the conidiophore stage. Under proper environmental conditions, conidiophores mature giving rise to numerous conidia.

Under adverse environmental conditions (e.g., low humidity), hyphae may produce various structures believed to enhance environmental stability.[251,260] Ultrastructural studies of mummified *Nomuraea rileyi* infected tissues placed in low humidity produced three morphologically distinct structures.[260] Intrahyphal hyphae, characterized by a double cell wall, were commonly produced in the external aerial hyphae. Thick-walled chlamydospores, also observed in the extracuticular fungal mat, were usually produced as apical swellings of thick-walled hyphae. Both chlamydospores and thick-walled hyphae remained viable for at least 12 weeks. Rehydration of these structures generally resulted in germination. The third type of resistant structure was resting bodies which lacked extensive wall thickening. These structures, possessing large lipid reserves, were located throughout the body cavity. After prolonged periods of desiccation and upon rehydration these intralarval resting bodies regenerated new mycelia. Zacharuk[251] also reported the presence of chlamydospores of *Metarhizium anisopliae*, morphologically similar to the resting bodies of *N. rileyi*,[260] within disintegrating elaterid larvae.

Entomophthorales

Entomophthorales species are unique in that they actively discharge conidia as a means of dispersal.[261] Discharged primary and/or secondary conidia produce penetration germ tubes that initiate fungal infection. Thick-walled resting spores (azygospores, zygospores), produced by many species of *Entomophthorales*, are not infective but may produce one or more infective conidia.

The sequential development or life cycle of *Entomophthorales* species has been studied in several different insects,[262-267] but only recently have the invasive and developmental processes been examined in detail.[225-229] In many *Entomophthorales*, the primary conidia, upon contact with susceptible host insects, adhere to the cuticle and produce a germ tube that penetrates the host's cuticle.[265,268] Discharged primary conidia that do not contact a host may then germinate and produce secondary conidia which after discharge and attachment initiate an infection. Alternatively, the discharged primary conidia of certain *Entomophthorales* species, i.e., *Neozygites fresenii* or *Entomophthora floridana* germinate and produce sticky "anadhesive" secondary conidia (capillispores) which are borne on narrow capillary condiophores.[265,269-271] In general, the attachment of primary and secondary conidia is nonspecific and occurs over the entire cuticular surface of the host. In certain cases, a disk-like "adhesive pad", present on adhesive secondary conidia, may enhance conidia attachment to the host.[270,271] Discharged conidia of *E. muscae*, upon contacting the host, flatten and form an adhesive pad at the conidia-epicuticle interface.[267] Similarly, the outer "mucilaginous" layer, containing carbohydrate binding proteins, of *Conidiobolus obscurus* conidia will flatten upon impact with the cuticular surface.[272,273]

Germination of conidia can be influenced by cuticular components. Water soluble and/or lipidic extracts of aphid cuticle stimulated formation of germ tubes by "aggressive" strains of *Conidiobolus obscurus*. Nonpathogenic strains, however, were not stimulated by these extracts, e.g., they formed and discharged secondary conidia instead of producing germ tubes.[274] Kerwin[275] determined that germination of *Entomophthora culicis* primary-discharge conidia was influenced by the presence of fatty acids and that this response was affected by concentration, chain length, and degree of unsaturation of the fatty acids. Oleic acid (C18:1), in combination with chitosan, dilute Emerson's YPS media, or yeast extract (0.1% w/v) induced germ tube formation whereas, both linolenic (C18:2) and linolenic (C18:3) were toxic. Fully saturated fatty acids having a chain length of C12:0 to C22:0 usually induced primary-discharge conidia to produce secondary conidia. Shorter chain saturated fatty acids (≤ C10:0) were toxic to conidia.

In most cases, a germ tube will directly orient to and penetrate the host's cuticle. However, heavily sclerotized regions (such as the head capsule) have been reported to be more resistant to penetration than thinner sclerite and interscleral membrane regions.[265-267] Penetration of the cuticle is believed to be mediated by both enzymatic and mechanical processes.[262-267] Frequently, zones of melanization have been observed at the penetration sites. Both lipase and protease activities, detected in a variety of *Entomophthorales* species, have been implicated in the penetration process.[263,276,277] According to Lambiase and Yendol,[264] germ tubes, after direct penetration of the epicuticle, will bifurcate and expand as digit-like projections, gradually extending inward cleaving the lamellae of endocuticle, and eventually entering the hemocoel.

Invading germ tubes will produce segmented hyphal bodies or filamentous hyphae which will rapidly colonize the hemocoel and subsequently invade fat body and nervous tissue.[265] The amoeboid-like hyphal bodies, produced in the hemocoel by various *Entomophthorales* species, have been shown to be osmotically sensitive protoplasts.[278,279] Ultrastructural examination of these cells revealed the presence of an invaginated plasma membrane that is covered by a thin fibrous coat instead of a formal cell wall. The reasons for the production of protoplasts by certain *Entomophthorales* species is not well understood. Butt et al.[279] proposed that protoplasts feeding via pinocytosis (invaginations of plasma membrane) would be able to multiply more efficiently than walled cells. Alternatively, Dunphy and Nolan[280,281] have demonstrated that protoplasts of *Entomophthora egressa*, unlike hyphae, are capable of avoiding encapsulation by host granulocytes. Although regeneration of protoplasts to the walled hyphal form has been demonstrated under in vitro conditions,[282] its mechanism(s) are still unclear. Several *Entomophthorales* species, i.e., *Conidiobolus obscurus* and *Zoophthora radicans*, do not produce protoplasts but do replicate in the host hemocoel as walled, hyphal bodies and filamentous hyphae.

The time required for *Entomophthorales* species to kill an insect is dependent upon the host it infects. Brobyn and Wilding[265] reported that *Entomophthora thaxteriana* killed *Acyrthosiphon pisum* within 2 to 3 days whereas *E. planchoniana* killed *Aphis fabae* within 4 to 6 days. Several researchers have reported the in vitro production of insecticidal mycotoxins by vegetative cells of *Entomophthorales* species. Freeze-dried filtrates of *Conidiobolus coronata*, a species capable of killing termites within 20 hr post-exposure,[262] were toxic when injected into both *Galleria mellonella* and *Musca domestica*.[283,284] Insecticidal metabolites, identified as azoxybenzoids, have been extracted from in vitro cultures of *E. virulenta*.[285] More recently, Dunphy and Nolan[286] detected a mycotoxin(s) from spent medium of protoplast cultures of *E. egressa*. These toxins, believed to be proteinaceous in nature, caused paralysis when injected into *Choristonuera fumiferana* larvae. The exact role of these toxins in the in vivo development of the pathogen is not known. In many instances, *Entomophthorales*-induced death occurs only after the fungus has colonized the host's tissues, suggesting a minor role for toxin(s).[265]

In the asexual cycle, vegetative, fungal cells after ramifying through host tissues, may differentiate into rhizoids, pseudocystida, or conidiophores.[265] Rhizoid structures, which serve to anchor the mummified host to the substrate, are produced by some isolates of *Entomophthorales*. These structures, arising from enlarged hyphal body cells, will exit the ventral surface of the dead insect. Upon contact with the substrate (generally a leaf surface), the rhizoid apices will differentiate into digitately branched holdfasts. In some instances, as in *Entomophthora planchoniana*, a viscous fluid is secreted from the rhizoids which glues the host to the substrate.[265] The absence of a rhizoid, however, does not mean that the infected insect will not be attached to a substrate. For example, *Musca domestica* adults infected with *E. muscae*, attach to the substrate via their proboscis. Pseudocystida, developing from enlarged vacuolate, hyphal body cell, will emerge from dorsal and lateral regions of infected insects. These structures, produced by several *Entomophthorales* species (*Erynia neoaphidis*, *E. aphidis*), are believed to assist emergence of conidiophores.

Conidiophores, the principal external structures, develop from elongating hyphal bodies.

Conidiophores generally emerge through the dorsal and lateral region of the infected insect producing a felt-like hymenium over the cuticle surface. Often clusters of conidiophores have been observed to emerge through the cuticle. Terminally borne conidia will be produced via septum formation. Mature conidia absorb water, that increases internal osmotic pressure, and results in their discharge. In many cases, the discharged conidia form a halo around the infected insects providing a useful diagnostic characteristic of an *Entomophthorales* infection.

In addition to the above-mentioned structures, many species of *Entomophthora* produce asexual chlamydospores and azygospores. Chlamydospores are thick-walled vegetative cells, produced during unfavorable dry periods, that readily germinate when placed under favorable conditions. Azygospores, reported to arise from multinucleate hyphal bodies[287] and/or from primary conidia,[288] possess a much thicker cell wall than chlamydospores and are considered to be the "overwintering" stage. Azygospores from certain *Entomophthora* species require a maturation period before germination can occur. This maturation period is thought to synchronize the fungal life cycle with that of its host.[289]

The "sexual" cycle of *Entomophthorales* is induced by a fusion of two hyphal body cells. A bud forms at the point of conjugation and the protoplasm of the fused cells flows into the bud resulting in a distinct ampulla. Whether or not nuclear fusion and subsequent genetic recombination occurs during the maturation of zygospores is not clear. Like the azygospores, zygospores of many *Entomophthora* species are multinucleate[290] and possess a thick trilaminar cell wall. Thick-walled zygospores, like the azygospore, are resistant and are important for survival in temperate climatic zones. Zygospores under the proper environmental conditions will produce a germ tube which may produce infectious primary-discharge conidia or secondary anadhesive spores (germ capillispores).[269,290,291] Differentiation of vegetative, hyphal bodies into the externally borne conidiophore or the spore form (azygospore, zygospore) is influenced by an array of intrinsic and extrinsic factors.[292,293] The physiological age of the host is thought to be an important intrinsic factor. Newman and Carner[292] found that *E. gammae* produced mostly conidia in young *Pseudoplusia includens* larvae and spores in mature larvae.[292] Extrinsic, ambient temperature can determine whether the conidial or spore form of *E. sphaerosperma* is produced in infected brown plant hoppers.[294]

Coelomomyces

Until recently, the mode of infection for the zoosporic, obligate, pathogen of dipterous larvae, *Coelomomyces*, was unknown. The establishment of the infective unit of *C. psorophorae* as a biflagellate zygote, a fusion product of gametes produced in the alternate copepod hosts,[49,295] led to the detailed studies of Travland.[296,297] Infection of mosquito larvae by this heteroecic fungus is initiated by the encystment of biflagellate zygotes on the cuticle. Zebold et al.,[298] utilizing a 1% methylene blue solution to selectively stain fungal cysts, demonstrated that zygote attachment was concentrated at the intersegmental membranes, head capsule, and the base of the anal papilliae. Very few cysts were observed on the abdomen or thorax of mosquito larvae. Chemical cues have implicated potential determinants for cyst attachment. Ultrastructural studies by Travland[296] revealed that the fungal cysts of *C. psorophorae* possess adhesion vesicles that secrete an amorphous glycoprotein that glues the cysts to the cuticular surface. Chemical studies on mosquito larval exuviae, a substrate receptive to encystment, indicated that cuticular carbohydrates are involved in the interaction of *C. psorophorae* zygotes with *Culiseta inornata*.[299]

Morphologically, germination of the cyst is initiated as a lateral outgrowth which expands into a bulbous appressorium.[49] An amorphous electron-dense material is deposited at the cuticle-appressorium interface. The narrow penetration tube, originating from the appressorial base, directly penetrates the host cuticle. Protoplasts of the cyst, traversing the penetration tube, are injected into host epidermal cells[49] or directly into the hemocoel.[300,301] They are now described as spheroidal cells or hyphagens.[302] The spheroidal cells grow as a diploid thallus (sporophyte) which eventually produce thousands of thick-walled, multinucleate, oval sporangia at their tips.[303] Mosquito larvae, after their death, release more sporangia into the water.

Sporangial germination, a process preceded by meiosis, results in the release of hundreds of haploid meiospores that selectively encyst and penetrate the copepod host. This fungus will proliferate as a haploid gametophyte within the copepod body cavity. The coenocytic, wall-less gametophyte becomes a gametangium at maturity releasing hundreds of uniflagellate gametes of a particular mating type.[295] The number and type of meiospores infecting a particular copepod will determine the mating type(s) of the progeny gamete population. Gametes of opposite mating types will fuse forming the biflagellate zygote which will encyst on a susceptible mosquito larva. Federici,[50] taking advantage of the color difference between gametangia of male (orange) and female (amber) mating types, produced laboratory hybrids of *Coelomomyces dodgei* and *C. punctatus*. The hybrid biflagellate zygote, capable of infecting mosquito larvae (sporophyte phase), did not produce a functional gametophyte in the copepod host.[304]

Lagenidium giganteum

Lagenidium giganteum, a facultative pathogen of mosquito larvae has a life cycle with both an asexual and sexual phase.[305-307] Host infection is initiated by laterally, inflagellate, zoospores which encyst on the cuticle of the cephalic region of larvae. Germ tubes, emerging from the cysts, penetrate the cuticle. Penetration is often detected by a melanin deposition in the area of a "bore hole". Penetration hyphae, upon reaching the larval hemocoel, branch to form nonseptate hyphae that in turn grow and ramify in the body cavity resulting in death. Vegetative growth of *L. giganteum* (results in decreased levels of endogenous free amino acid, total protein, and free sugar reserves) is believed to kill larvae via "physiological starvation".[308,309] Cessation of growth of the nonseptate hyphae (in host larva) will induce septum formation. The resulting hyphal segments will form either an asexual sporangium, an oogonium, or antheridium. The asexual sporangium, the most prevalant form produced by hyphal segments, swells into an enlarged spherical structure. Mature sporangia produce discharge tubes which then grow out through the host cuticle. Extracellular proteases, having trypsin-like, weak elastase, and collagenase activities have been detected in in vitro cultures of *Lagenidium*[310] and have been implicated as playing a role in initial germ tube penetration, digestion of tissue, and exiting of discharge tube. Zoospore protoplasm migrates through the discharge tube and undergoes differentiation into mature biflagellate zoospores within a terminal membrane-bound vesicle. This vesicle ruptures, releasing zoospores that then either infect mosquito larvae or develop saprophytically on organic detritus. The sexual cycle involves the differentiation of hyphal segments into either male antheridia cell or female oogonia. Conjugation of different mating types results in the formation of thick-walled oospores. This stage, resistant to adverse environmental conditions, is responsible for the persistence of *Lagenidium* during winter or summer droughts. The oospore, when placed under favorable conditions, germinates and grows saprophytically and/or produces infectious zoospores.

Toxins

Some entomogenous fungi appear to kill their host via a toxic action rather than invasion of the host tissue. The importance of fungal toxins, however, extends beyond their presence or absence in the hemocoel of an insect to their production in vitro and potential as microbial insecticides. At present, however, there are no fungal metabolites or toxins produced by entomogenous fungi under commercial development as insecticides.[311]

Toxin production in both nonentomogenous and entomogenous fungi is widespread and has been reviewed extensively by Roberts.[311] Insects may be exposed to toxic fungal metabolites in three ways: (1) ingestion, (2) cuticular contact, or (3) within the hemocoel. Those toxins ingested or absorbed through the cuticle wall have potential as pest control agents.

Several toxic compounds have been identified and/or isolated from culture filtrates or mycelium of *Beauveria, Metarhizium, Nomuraea, Aspergillus, Verticillium, Paecilomyces, Isaria, Fusarium, Cordyceps,* and *Entomophthora*.[311] One compound from *B. bassiana*, the depsipeptide beauvericin active against mosquito larvae,[312] has received the most attention. Two insecticidal substances, destruxins A and B, were isolated from culture filtrates of *M. anisopliae*.[313] These destruxins are toxic to mosquito larvae when added to culture water and are toxic to mice when injected.[314] Of the 14 possible destruxins,[315] only destruxin A has been isolated from *M. anisopliae* var. *major*.[316] Fungi belonging to the genus *Aspergillus* are frequently associated with toxin production. Burnside[317] found an aflatoxin of *A. flavus* toxic to honeybees. *A. ochraceus* produces an insecticidal metabolite (aspochracin) that is toxic to silkworm, fall webworm, and mice.[318] Toxins from *A. oryzae, A. flavus,* and *A. flavus* var. *columnaris* are active against immature milkweed bugs, houseflies, and a number of mosquito species.[319]

Various high molecular weight fungal enzymes secreted in significant quantities into culture media and into the host body, are toxins. Proteolytic and lipolytic enzymes have been identified from *Metarhizium anisopliae, Beauveria bassiana, Aspergillus flavus,* and *Entomophthora* spp.[311] Proteases of *B. bassiana* produced in both stationary and submerged culture were toxic to *Galleria mellonella*.[320]

HOST SPECIFICITY AND VIRULENCE

Invertebrate Host Specificity

The specificity of entomogenous fungi has been defined as "the expression of reciprocal adaptations and affinities between a pathogenic organism and the entirety of its host species".[321] Understanding the specificity of a fungal pathogen is an essential aspect in choosing the best candidate for use as a microbial control agent. A variety of extrinsic and intrinsic factors may determine and/or influence the specificity of fungal entomopathogens. In nature, certain extrinsic factors may preclude the occurrence of the necessary pathogen-host interaction required for successful infection. For example, the spatio-temporal coincidence between the host and pathogen has been reported to account for the specificity of various *Entomophthora* species for aphid hosts.[321]

A review of the current literature on entomopathogenic fungi does not provide a basis to discern the general principle(s) of specificity. Fungal species and/or fungal pathotypes that are morphologically very similar may possess different host ranges. Fungi, reported to have a broad spectrum of activity, in many cases, are comprised of a variety of pathotypes. For example, *Beauveria bassiana* and *Metarhizium anisopliae* are pathogens reported to be capable of infecting over 100 different insect species belonging to a variety of insect orders. Bioassay data on selected *B. bassiana* and *M. anisopliae* isolates, however, demonstrate a high degree of host specificity.[322,323] Similarly, host-range pathotypes of *Nomuraea rileyi*, a fungus isolated from a variety of lepidopteran hosts, also have been identified.[324-326] In contrast, Hall[170] found that *Verticillium lecanii* isolated from different sources (insect hosts, rust and mildew fungi, contact lens) were all pathogenic for the aphid *Macrosiphoniella sanborni*. *Hirsutella thompsonii* and its related species appear to be specific only to mites inhabiting various foliar substrates.[149] Species of *Aschersonia* infectious to whiteflies and a few scale insects appear to have similar host spectra.[188,189]

Successful infection by a particular fungus may be influenced by the physiological state of the host system. Many entomopathogenic fungi will infect only a particular life stage, i.e., egg, larval, pupal, or adult. Dose mortality data for many insects challenged with a fungal pathogen indicate the presence of an age-maturation immune response. In these cases, young larval instars are susceptible to very low dosages whereas mature larvae are resistant to relatively high dosages of the fungus.[326,327] Infraspecific differences in susceptibility among biotypes of the pea aphid *Acyrthosiphon pisum* to two strains of *Conidiobolus obscurus* were initially reported by Papierok and Wilding.[328] Recently a biotype of *A. pisum* (from field populations) was resistant to an isolate of *Erynia neoaphidis*.[329] This resistant biotype as well as susceptible biotypes of *A. pisum* were both susceptible to other isolates of *E. neoaphidis*.

Variation in efficacy and viability of entomogenous fungi also might be influenced by the host plant. Ramoska and Todd[330] working with *Beauveria bassiana* saw differential mortality and fungal development on host cadavers where the living host (chinch bug) had fed on either corn or sorghum. These results indicate the presence of a plant-produced fungal inhibitor fortuitously protecting the insect host.

Entomogenous fungi infect insects by either penetrating the integument or digestive tract. The inability of a fungal pathogen to adapt to and utilize nutrients in vivo may lead to an aborted infection. Potential host defense mechanisms thus may be operating at any stage(s) of the infection cycle. Although the process that determines or regulates the specificity of fungal entomopathogens is not understood, the host cuticle (the initial site of the pathogen-host interaction) plays a key role in determining the host range of a fungal pathogen.[194,298,331-337] Zebold et al.[298] were able to correlate host specificity of *Coelomomyces psorophorae* to ability of zygotes to attach to and encyst upon the cuticle of mosquitoes. Similarly, Al-Aidroos and Roberts[333] attributed the hypovirulence of certain *Metarhizium anisopliae* mutants to their inability to attach to the perispiracular valve of *Culex pipiens* larvae. The wax layer of the epicuticle, reported to contain antifungal substances, is believed to prevent many fungi from being entomopathogenic.[194] Ether extracts from lepidopteran larval cuticle inhibited germination of *Aspergillus flavus* and *Beauveria bassiana*.[334-336] The medium-length saturated fatty acids (C:8-C:12), present in cuticle extracts, greatly inhibited conidial germination.[336,337] Longer chain fatty acids (C:14) induced germination and/or enhanced germ tube growth of *B. bassiana*, *Paecilomyces fumoso-rosea*, and *Erynia variabilis*.[215,337,338] The relative composition and levels of free fatty acids were suggested as playing a regulatory role in the specificity of *E. variabilis* to the adult stage of lesser houseflies, *Fannia canicularis*. The adult cuticle, containing five times the concentration of fatty acids as pupal cuticle, may contain levels of oleic and palmitoleic acid that initiate and induce growth of *E. variabilis* germ tubes.[338] Woods and Grula[339] demonstrated the presence of various amino acids and glucosamine on the cuticular surface of *Heliothis zea* larvae and indicated their potential role in the initiation and growth of *B. bassiana* germ tubes. In addition to epicuticular components certain microorganisms, associated with the epicuticular components certain microorganisms, associated with the epicuticular surface, may influence conidial germination. For example, Schabel[222] reported that the presence of bacteria and fungi on *Hylobius pales* inhibited germination of *M. anisopliae* conidia.

The ability of germ tubes to orient and attach to the epicuticle determines the relative virulence of certain entomopathogenic fungi. Pekrul and Grula[220] reported that highly pathogenic strains of *Beauveria bassiana* germinated very quickly and directly penetrated the cuticle and that low pathogenic strains took longer to germinate and grew extensively over the cuticle surface with only a limited degree of penetration. Milner[329] reported that conidia of *Erynia neoaphidis* germinated on both susceptible and resistant aphids but that penetration was inhibited in the resistant biotype. Similarly, both virulent and nonvirulent strains of *Conidiobolus obscurus* germinated on the cuticle of the pea aphid.[658] The absence of penetration, with nonvirulent strains, was attributed to their quasi-exclusive production of secondary conidia whereas conidia of virulent strains produced infectious germ tubes. Secondary spore formation of *E. variabilis* is induced by the presence of saturated fatty acids (C:14-

C:20).[275] Lipid extracts from the cuticle of the resistant pupal stage of *Fannia canicularis* induced the in vitro production of secondary spores from *E. variabilis* conidia.[338]

Fungal germ tubes must penetrate the cuticle to successfully cause a lethal infection in the host. The structure and composition of cuticle vary extensively between insects. Within a particular host, this barrier is heterogeneous varying from relatively thick, rigid, tergites to flexible intersegmental membranes. Molting, the sequential shedding of the old cuticle (fungal cells are discarded with the exuviae), is regarded as an important resistant mechanism.[223,340] Newly molted cuticle, possessing untanned cuticle with incomplete wax and cement layers, is considered especially vulnerable to fungal attack. Penetrating germ tubes can, via the cuticular phenoloxidase system, produce localized zones of melanization that abort the infection process.[341]

Two principal host-defense mechanisms from fungal infections (cellular and/or serological) are reported in insects.[321,342] Cellular response results in phagocytic encapsulation or nodulation of invading fungal cells.[342] In general, the encapsulation process results from hemocytes recognizing the foreign cells as "nonself". Initial contact of hemocytes with foreign particles can be accomplished by either a direct interaction between host cell receptors and surface components on the foreign particle or by coating the foreign body with an opsonizing substance which is then recognized by phagocytes. A rapid recruitment and adhesion of hemocytes usually follow initial contact, indicating that initial hemocyte attachment may cause elicitor cells to release substances that increase adhesion to other hemocytes.[343] The specificity of this recognition process, as it pertains to entomogenous fungi, was described by Dunphy and Nolan.[281] Granulocytes of *Choristoneura fumiferana* larvae adhered to sporiangiospores on nonentomopathogenic *Absida repens* and *Rhizopus nigricins* but did not attach to pathogenic *Entomophthora egressa* hyphal body preparations. Dunphy and Nolan[281] speculated that the lack of *E. egressa* could be due to absence of or masking of *N*-acetylglucosamine residues or the presence of inhibitory carbohydrates on the hyphal body cell wall.

Although, the production of a melanized layer is associated with cellular encapsulation, its immunological importance is unclear. In certain insects, melanization occurs very quickly after infection, whereas in other insects, the deposition of melanin is delayed and confers little or no protection. Precursors to melanization, released from the hemocytes, are believed to be activated by substances associated with the surface of the invading fungus. Studies by Unestam and Soderhall[344] and Soderhall and Unestam[345] demonstrated that the hemolymph phenoloxidase system in a crayfish, *Astacus astacus*, is activated by water-soluble β-1,3 glucans released from fungal cell walls. The pathway involves a multistep-enzyme cascade and a calcium-dependent serine protease that causes phenoloxidase, with four other proteins, to become "sticky".[346,347] At the cellular level, β-1,3 glucans will induce granulocytes to degranulate,[348] resulting in the release of cell-bound recognition factors (prophenoloxidase activating components) into the hemolymph which then trigger activation of adjacent hemocytes. Similarly, Ashida et al.[349] reported that the prophenoloxidase system in the silkworm, *Bombyx mori*, can be induced by β-1,3 glucans. The activated phenoloxidase system, in addition to producing a sticky substance, results in formation of fungicidal quinones and melanin from phenol substrates.[350]

In addition to the prophenoloxidase, several invertebrates produce proteolytic inhibitors[351-353] that are active against entomopathogenic fungi. Kucera[351,352] extracted inhibitors from *Galleria mellonella* hemolymph which were effective against both serine and sulfhydryl proteases of the fungal pathogen, *Metarhizium anisopliae*. Similarly, Hall and Soderhall[353] identified a serine protease inhibitor from cuticle preparations of *Astacus astacus* which was active against subtilisin and protease(s) produced by *Aphanomyces astaci*. They[353] suggested that the protease inhibitor was produced by hemocytes and deposited in the cuticle during molting. Such proteolytic inhibitors, depending upon their location, could be inhibiting both germ tube penetration of the cuticle or vegetative development of hyphae in the hemocoel.

Invertebrate specificity has applied implications since it is important that they not infect beneficial insects and mites. Entomogenous fungi considered as candidates for microbial control are tested against beneficial insects especially honeybees. No susceptibility was detected[354] when a commercial preparation of *Hirsutella thompsonii* was fed (dosages of 271 colony-forming units (CFU) per bee) to honeybees. *Nomuraea rileyi* had no detrimental effect on a number of predators and parasites.[355] King and Bell, however, reported that *N. rileyi* inhibited development of the parasite *Microplitis croceipes* within the host larvae if the larvae were infected with *N. rileyi* within 1 day of parasitization.

Vertebrate Host Specificity

All current experimental evidence indicates that the host range of entomogenous fungi excludes humans and other homeotherms. Safety tests against vertebrates, however, have not been conducted on all entomogenous fungi. Of the aquatic entomogenous fungi attacking mosquitoes, only *Lagenidium giganteum* and *Culicinomyces* sp., were tested against vertebrates. Fish, chickens, quail, and rats were not susceptible to *Lagenidium*.[47] Conidia suspensions of *Culicinomyces* sp. administered to rats, mice, guinea pigs, sheep, cattle, and two species of wild duck had no observable effect on the health of these test animals.[357]

Infection of man and other mammals by fungi of the genus *Entomophthora* has never been detected. *Conidiobolus*, however, has occasionally been reported as the causative organism of infection of the mouth, nasal mucosae, and respiratory tract of mammals.[86,87] Cases of nasal granuloma in man also have been found.[358] Lowe and Kennek[359] reported that mice were not susceptible to *C. coronatus* administered by injection, feeding, or introducing conidia into wounds. Fungal propagules could not be detected in mouse tissue[360] following the administration of an acute oral dose of a strain of *E. thaxteriana* (= *obscura*) resting spores to mice.

In a review, Ignoffo[361] cited a number of cases (both positive and negative) of toxicity, pathogenicity, or allergenicity of spore and mycelium preparations of *Beauveria bassiana*, *Metarhizium anisopliae*, and *Aspergillus* sp. against humans, vertebrates, and invertebrates. A case of fatal beauveriosis in a captive American alligator also was reported.[362] Popov et al.[363] exposed white mice to a water-spore suspension of *B. bassiana* (aerogenic, alimentary, and intraperitoneal) and detected spores in most organs but did not detect germinated spores or toxic manifestations. Although allergic sensitization associated with prolonged exposure to dry conidial preparations appears to be the most likely health hazard for *B. bassiana*, long-term studies with well-defined commercial preparations should be conducted to resolve any confusion in the literature.

A series of tests that involved gastric intubation, inhalation, eye sensitivity, acute dermal and acute oral administration of conidial preparations of *Nomuraea rileyi*,[364] and conidial/mycelial preparations of *Hirsutella thompsonii*[365] showed no harmful effects in gross and detailed diagnosis of rats, rabbits, and guinea pigs. These studies confirmed the previous lack of *per os* toxicity or pathogenicity of *H. thompsonii* to rats.[366] Human gastric juice also has been found to inactivate conidia of *N. rileyi* conidia in vitro.[367]

Hartmann et al.[368] established the safeness of *Cordyceps militaris* and *Paecilomyces fumoso-rosea* in acute oral tests with conidia against laboratory mice. Safety of large dosages of spore suspension of *Metarhizium anisopliae* and *P. fumoso-rosea* to Japanese quail also was confirmed.[369] Shadduck et al.[370] working with *M. anisopliae* detected no evidence of human or mammalian pathogenicity.

Virulence to Invertebrate Hosts

Virulence has been defined as the degree of pathogenicity against a specific host under controlled conditions.[321] Virulence like specificity is difficult to assess within the entomogenous fungi since it varies considerably under different physical (temperature and humidity) and cultural (nutrition) conditions. Innate virulence of a fungal pathogen only can be measured under conditions that allow its full expression. These conditions, which must be treated empirically via bioassay, are extremely difficult to assess with many fungi. Virulence is undoubtedly an important factor in mass production and application of entomogenous fungi as microbial control agents. If the innate virulence of a candidate fungi is unknown, then methods of production and formulation cannot optimize this virulence and field results can be unpredictable.

Differences in virulence among natural strains have been reported for *Beauveria bassiana*,[322,371] *Entomophthora* spp.,[328] *B. brongniartii*,[201] *Metarhizium anisopliae*,[371] *Paecilomyces* spp.,[371] and *Nomuraea rileyi*.[324] Selection of strains, with artificially modified virulence, has been achieved via mutagenesis, for species of *Beauveria*[372,373] and *M. anisopliae*,[333] and by hybridization or production of heterokaryous for *M. anisopliae*,[374,375] *B. bassiana*,[376] *B. tenella* (= *brongniartii*),[377] and *P. fumoso-rosea*.[378] Although these investigations did not result in more virulent strains, natural or artificially produced strains with improved virulence may be needed to stimulate more use of entomogenous fungi as microbial control agents.

Relationship Between Pathogenicity and Pathogen Dosage

Historically, the pathogenicity or virulence of entomogenous fungi has been based upon imprecise bioassay. Only recently have bioassay techniques been developed to accurately quantify the pathogenicity of fungi. Rather laborious techniques for evaluating species of *Entomophthora* against aphids and spruce budworm, have been developed by Wilding,[379] Papierok and Wilding,[328] Vandenberg and Soper,[380] and Milner and Soper.[381] Since *Entomophthora* species discharge spores, a spore shower method was developed. The LC_{50} values for aphids ranged from 1.0 to 26.9 spores/mm² with a slope of 1.07 while LC_{50} values for spruce budworm larvae ranged from 11 to 18 spores/mm² with a slope of 0.92 to 1.87.

Hall[382] developed a reliable bioassay technique for *Verticillium lecanii* against the aphid, *Macrosiphoniella sanborni*. Mortality was a function of the concentration of the conidial inoculum. Aphid mortality rose while killing time decreased with increasing conidia concentrations. The weighted mean LC_{50} value was 2.33×10^5 spores/mℓ of inoculum with a slope of 2.26. Similar bioassays also have been conducted using suspensions of blastospores of *V. lecanii* compared with conidia. Blastospores were twice as pathogenic as conidia in tests against *M. sanborni*.[383]

Quantitative assays for *Beauveria bassiana* have been developed against a number of hosts. Ferron and Robert[384] found that mortality of adults of *Acanthoscelides obtectus* was a function of the concentration of the inoculum. Infection was observed when hosts were exposed to concentrations ranging from 5×10^6 spores/mℓ to 1×10^9 spores/mℓ. In similar studies, Barson[385] also found that the mortality of adult *Scolytus scolytus* was a linear function of the concentration of the inoculum. The LD_{50} after 5 days of exposure at 23°C and 100% relative humidity (RH) was calculated as 1×10^6 spores/mℓ. Mortality declined with a decrease in temperature at 100% RH.

The only repetitive bioassay, designed to measure the stability of a commercial fungal formulation, was developed in the U.S.S.R. for boverin, a formulation of *Beauveria bassiana*.[386] The common housefly, *Musca domestica*, was the insect of choice and activity was presented as LC_{50} values. Interestingly, small doses of the boverin formulation were used to avoid obtaining a steep slope for the dose-mortality curves.

Leaf-treatment techniques for bioassay were developed for *Nomuraea rileyi*[324,387] and also have been used for assaying the insecticidal activity of boverin[388] and conidia of *Beauveria bassiana*.[389,390]

STABILITY AND PERSISTENCE IN THE ENVIRONMENT

Sunlight, temperature, humidity or free water, substrate, and chemicals all influence the stability, sensitivity, and persistence of different developmental stages of entomogenous fungi. These factors independently or collectively affect fungi that are naturally present as well as those introduced as microbial insecticides. These factors affect growth and survival of the fungi directly or indirectly via its host and usually the effect differs among fungal species.

Effect of Light

The effect of light on fungi may be morphogenetic, i.e., induces or inhibits the formation of a structure or nonmorphogenetic, i.e., influences the rate or the direction of movement or growth of a structure. Light induces conidial discharge and germination in *Entomophthora* species.[391-393] However, it retards mycelial growth of *E. sphaerosperma* and stimulates mycelial growth of *E. thaxteriana*.[394] Sporangial and gametangial formation can also be stimulated or inhibited depending on the species of *Entomophthora*. Wallace et al.[395] reported that light had a direct effect on germination of resting spores of *E. aphidis* while the opposite is true for *E. apiculata*.[394] Mycelial growth of *Beauveria bassiana* and *Paecilomyces farinosus* is stimulated by light but is retarded by light in *Metarhizium*.[394] The formation of reproductive structures by *Cordyceps militaris* is also inhibited by light.[396] Viability of conidia is seriously decreased by light for *Beauveria* spp., *Metarhizium anisopliae*, and *P. farinosus*.[394]

Different forms of radiation may be damaging, beneficial, or innocuous depending on the fungus and the nature of radiation. According to Ignoffo et al.[397] conidia of entomogenous fungi are more resistant to sunlight than protozoan spores but less resistant than bacterial spores. Kreig et al.[398] found far ultraviolet (UV) more detrimental than near UV (285 to 380 nm) to *Beauveria bassiana* conidia while Tuveson and McCoy[399] found that *Hirsutella thompsonii* is reactivated under far UV. The persistence of *B. bassiana* and *Nomuraea rileyi* on soybean foliage also was reduced under solar radiation.[400] The half-life of *N. rileyi* conidia on foliage is only 2 to 3 days. Conidia exposed to a simulated sunlight source on glass petri plates has a half-life of only 2 to 3 hr.[397,401] *Metarhizium anisopliae* conidia, exposed on water to direct sunlight, also lose viability after exposure to UV.[394] Exposure to artificial sunlight under field conditions, 24 hr incubation at 25°C, resulted in a conidial half-life of about 2 hr.[402]

Effect of Temperature

Temperature has a major effect on all cellular activities. Below 0°C, fungal cells generally survive but rarely grow and above 40°C most cells stop growing and soon die. Within this range, fungal activity may either increase or decrease depending upon the species and temperature.[403] Entomogenous fungi are mesophiles, that is, the optimum temperatures for development, pathogenicity, and survival of the pathogen generally falls between 20 to 30°C.[394,404]

As with light, the thermal death point of conidia and spores of entomogenous fungi varies with length of exposure and species. However, resting spores of *Entomophthora* species are considerably tolerant to high temperature. Most species survive temperatures as high as 80 to 100°C for 5 to 60 min,[292,394,405] while the optimum germination temperature for *Entomophthora* resting spores is about 16°C.[394,406] Azygospores of *E. obscura* (formed in vivo in aphids) survived and germinated after exposure in soil to winter temperatures of 7°C.[407] Zygospores of *E. fresenii* (= *Triplosporium*) also were able to survive and germinate at temperatures of 9 to 10°C.[269]

Generally, subzero temperatures are used to store fungi.[394,408] Suspensions of conidia of *Metarhizium anisopliae* were virtually unchanged in viability after freezing at −20°C.[394]

Air-dried conidia of *Hirsutella thompsonii* stored at $-20°C$ were more viable than the same conidia stored at 27°C.[409] Conidia of *Nomuraea rileyi* were stored on silica gel for 2 years at temperatures between 5 and $-20°C$ without any apparent loss in infectivity.[410] *Beauveria bassiana* conidia stored at 21°C, lost all viability after only a few months, whereas conidia stored at 8 and 4°C remained viable for 1 and 3 years, respectively.[411,412] Viability of *Entomophthora* spp. was maintained by storing mummified cadavers at temperatures of 0 to 10°C and a low humidity of 20%. Viability was maintained for 2 to 10 months under these conditions.[413] In the case of *E. gammae*, the pathogen was inactivated in less than 1 hr after freezing the cadavers.[292]

Effect of Moisture

High relative humidity and/or free water is generally required by entomogenous fungi for germination of infective propagules and formation of reproductive structures outside the host. Moisture also affects the stability and survival of fungus in other ways. Moisture conditions favoring epizootics of Entomophthorales generally required 8 to 10 hr of relative humidity greater than 90%, with prolonged periods of dew or fog occurring over a period of 2 to 3 days.[414-416] Similar conditions are required for *Beauveria bassiana*, *Metarhizium anisopliae*, and other deuteromycetous species.[400,411] High relative humidity in glasshouses enhanced the effectiveness of *Verticillium lecanii* as a microbial insecticide.[170] Free water for 4 hr, is important in the development of artificial and natural epizootics of *Hirsutella thompsonii* and dew appears to be important for spore dispersal.[409]

Generally, relative humidity declines with an increase in wind velocity, an important factor in spore dispersal. Wilding[414,416] and Milner and Bourne[417] found that still humid air, usually occurring in the early morning and afternoon, ensured maximum survival of *Entomophthora* spores. In contrast, dry gusty winds and dry foliage produced the highest levels of airborne conidia of *Nomuraea rileyi*.[418] Even low wind velocities will dislodge conidia of *N. rileyi* from lepidopteran cadavers.[418,419] Rainfall, during a period of spore dispersal, may actually clean the air of spores. Kish and Allen[418] found that airborne conidia of *N. rileyi* were significantly reduced following rain. Similar reports have been made for *B. bassiana*[400] and *V. lecanii*.[170] On the other hand, sprinkler-irrigation technology was considered as an artificial means for inducing and enhancing the development of epizootics of *E. aphidis*.[420] A saturated atmosphere or free water is important to the development of specialized structures such as azygospores in *E. obscura*.[421]

Relative humidity also will affect the longevity of conidia. Hall[170] found that conidia of *Verticillium lecanii* attached to the parent conidiophore survived 13 days (at 58% RH) and when detached survived less than 4 hr. Maximum survival, in storage, only occurred at high humidities. Survival of *Metarhizium anisopliae* conidia in storage is best at humidities of 10 and 90%.[422]

Effect of Substrates

The surface (air, soil, water, foliage, bark, organic matter, and the host) upon which the fungal spore or mycelium is deposited influences its ability to survive and subsequently induce disease.[394] Certain fungal species, i.e., *Coelomomyces* are specific to hosts associated with a specific habitat while more facultative forms, i.e., *Beauveria bassiana* are found everywhere and attack stages of many different hosts.

Effect of Soil

Research on the survival of the propagules of entomogenous fungi in soil, away from their host, has produced different results. According to Huber,[320] the soil has fungistatic agents which hinder the development of conidia of fungi (*Beauveria bassiana*). Reduced levels of conidial germination were reported for *B. bassiana*,[411] *Metarhizium anisopliae*,[411] and *Nomuraea rileyi*.[423] In some cases, this reduction appeared to be caused by soil microorganisms for heat sterilization or filtering of the soil removed the inhibitory activity.[411,424] Lingg and Donaldson[425] found *B. bassiana* persisting and growing in sterile soil but further research showed that patulin, a metabolite of the common soil fungus, *Penicillium urticae* inhibited conidial germination and growth.[426] Wartenburg and Freund[427] observed that *B. bassiana* conidia inhibited by unsterilized soil could be stimulated to germinate by exposure to the insect cuticle. Since the conidia remained viable for 2 years in the soil, they concluded that the fungistatic effect was beneficial, acting as a preservative.

In addition, *Beauveria bassiana* and *Paecilomyces* conidia, active against the Colorado potato beetle, appeared to remain active for 3 years.[428] Bell and Hamalle[429] found that *Metarhizium anisopliae* and *B. bassiana* conidia remained viable for 40 days in unsterilized soil. Latteur[430] considered unsterilized soil important in the conservation of *Entomophthora ignobilis*. Latge et al.[431] successfully stimulated azygospore formation by *E. obscura* by placing mycelia in the soil.

Most researchers believe that the development of entomogenous fungi in soil is dependent on adequate humidity, temperature, aeration, soil characteristics, and soil depth. Muller-Kogler[1] concluded that the highest pathogenic effect of fungi on insects occurred in the first 15 cm of the surface layer of soil. Ignoffo et al.[432] found that conidia percolated more readily in sand than in silt-loam soil where over 90% of the conidia remained in the upper 2 cm. In loam soil, free conidia were still infective after 372 days, but the half-life was only 40 to 65 days.[423] Infectivity, used to measure viability, was similar at 0.9 and 13 cm in sand. No infection was obtained below 4.6 cm in a silt-loam soil. Milner and Lutton[433] found that the survival of *Metarhizium anisopliae* conidia in the soil was reduced above and below 16°C but would survive many years at 16°C.

Effect of Water

Salinity, pollution, and pH of water are important to the germination and survival of zoospores and conidia of different aquatic, entomogenous fungi.[434] The presence of organic pollution and salinity levels of 1.5 ppt NaCl completely inhibited zoosporogenesis and infection of mosquitoes by *Lagenidium giganteum*.[434-436] Some isolates of *Beauveria bassiana* require water with nutrients for germination while others do not.[200] *Metarhizium anisopliae* conidia appeared to remain more active in clear than in muddy water ponds.[437]

Effect of Foliage

Although many entomogenous fungi are found on foliage, very little is known concerning their persistence on this substrate.[394] Burleigh[438] found a significantly higher number of *Heliothis* spp. larvae infected with *Nomuraea rileyi* in closed canopies of several cotton varieties. *Hirsutella thompsonii* persists in cadavers and outside the host as mycelial strands attached to leaves of the previous summer and fall flushes of citrus trees.[439] This characteristic form of survival on foliage also has been noted for *H. tydeicola*.[440]

Effect of Host

The Entomophthoraceae generally survive adverse environmental conditions by persisting in the host in the zygospore stage. Interestingly, the size of the host affected the type of development for *Entomophthora gammae*.[292] Primary spores were the predominant reproductive form in larvae <1.5 cm while resting spores were predominant in larvae > 2.5 cm. The stage and size of the infected larvae also are important in determining the conidial load of *Nomuraea rileyi*[441] and the dispersal of diseased larvae influences its persistence in soybeans.[442]

Effect of Chemicals

During the past 40 years, numerous commercial organic fungicides, insecticides, acaracides, herbicides, and more recently, insect growth regulators have been used widely in agriculture and other areas. These compounds interfere with fungal epizootics by reducing the available host population or inhibit spore germination and vegetative development of the fungi. In some cases, however, sublethal doses of insecticides may enhance the insecticidal activity of fungi.[38]

In a series of tests,[443-446] both in the laboratory and field, Zimmerman reported the effects of systemic fungicides on ten entomogenous fungi. In laboratory tests, benomyl and saprol severely suppressed spore germination of *Beauveria bassiana*, *B. bronginartii*, *Metarhizium anisopliae*, *Paecilomyces farinosus*, *Entomophthora virulenta*, and *E. thaxteriana* (= *obscura*). Benomyl, calixin, imugan, plantvax, and milstem all inhibited germination of *E. aphidis* while saprol had no effect. The lowest deleterious effects were caused by plantvax and milstem against deuteromycetes species. Cercobin M had virtually no effect on the *Entomophthora* spp. and was minimally effective on the others. Inhibition was generally more tolerant of the fungicides.

In field tests, the same systemic fungicides[443,445] had no effect on the spores of *Entomophthora aphidis* and *E. obscura* if applied before the fungus was applied.[445] There were no differences in mortality of aphids caused by Entomophthoraceae from treated and untreated plots when tridemorph, thiophanate-methyl, triadimeform, benomyl, and captafol were used.[446] Delorme and Fritz,[447] reported that maneb, sulfur, mancozeb, dinocap, ditalimfos, and ziram prevented development of *E. aphidis*, however, none of the materials eliminated the fungus. Systemic fungicides had negligible activity against *E. aphidis*.[448] Wilding[449] found that both systemic and nonsystemic fungicides had little effect on *Erynia neoaphidis* and Livingston et al.[450] reported that benomyl did not affect *Entomophthora gammae*. In soil studies, Oncuer and Latteur[451] found that the fungicides, benapacryl, captan, maneb, zineb, and sulfur were toxic to the spores of *E. obscura* and that toxicity persisted beyond 20 days. Spores, however, were not affected by copper oxychloride, dodine, benomyl, and carbendazimo. Fritz[452] found that systemic fungicides were more deleterious to the mycelial growth of *Basidiobolus ranarum*, *Conidiobolus osmodes*, and *E. virulenta* than were nonsystemic fungicides.

There was a reduced incidence of *Entomophthora* species attacking *Myzus persicae* four to five times and increased spotted alfalfa aphid populations while decreasing the incidence of *Entomophthora* in alfalfa field plots treated with mancozeb, captafol, and Bordeaux mixture. The insect growth regulator, Dimilin, had no effect on the growth of *E. aphidis*, *E. culicis*, *E. spaerosperma*, and *E. thaxteriana* (= *obscura*) and actually stimulated germination of spores and the formation of secondary spores of *E. aphidis* and *E. thaxteriana*.[453] Similarly, Dimilin had only a slight effect on growth of *Beauveria bassiana* and *Metarhizium anisopliae*[453] and no effect on growth and conidial germination of *Hirsutella thompsonii*.[454] Nolan and Dunphy[455] found that the fungal sex hormone, trisporic acid, and a synthetic insect juvenile hormone affected morphogenesis of *E. egressa*. Juvenile hormone had no effect on *Nomuraea rileyi* growth but promoted the mycelial growth of *B. bassiana*.[456] Ecdysone also stimulates the mycelial growth of *N. rileyi*.[456]

In Poland, the herbicide chlorfenvinphos restricted the growth of *Paecilomyces farinosus*, *P. fumoso-rosea*, and *Beauveria bassiana* in culture but had no effect when incorporated into the soil.[457] The carbamide herbicides, linuron and monolinuron, however, did affect either growth or pathogenicity of the same fungi.[458] The residual activity of the insecticide Tritox 30, used for control of the Colorado potato beetle in Poland, did not reduce the number of soil-inhabiting fungal species.[459] However, fungicides used on potatoes in the northeast U.S. did inhibit *B. bassiana*.[460]

Ignoffo et al.[461] used an in vitro paper-disk technique to determine the sensitivity of conidia to 44 chemical pesticides. Seven of eight fungicides inhibited growth of *Nomuraea rileyi*. Chlorothalonil and ferbam were the most active followed by fentin hydroxide, zinc-iron maneb, sulfur plus zineb, maneb, and benomyl. Pyroryclor did not inhibit *N. rileyi*.

Johnson et al.[462] found that benomyl alone and in combination with methyl parathion and carbaryl delayed the onset of an epizootic by *N. rileyi*. Benlate, DuTer, and Bravo were less disruptive to *N. rileyi* than Benlate and carbaryl in soybeans.[463] Inhibition was greatest when treatment was applied during the early stages of an epizootic. Ignoffo et al.[461] also found that 13 or 25 insecticides inhibited the growth on *N. rileyi* but none was as active as the fungicides. The three most deleterious insecticides were monocrotophos, phenthoate, and methyl parathion. The herbicides, dinoseb and benomyl at field dosages, were found to inhibit the ability of *N. rileyi* to infect *Trichoplusia ni* on soybean and in the soil.[461] Garcia and Ignoffo[464] reported that the antibiotics, methenamine, mandelate, and nystatin inhibited growth on *N. rileyi*. Oho and Satoh[465] found that fungicides (Dithane, Fumiron), lime sulfur, wettable copper, and the insecticides, endrin, parathion, and DDVP, all at recommended dosages prevented germination of conidia of *Aschersonia* sp. Hall[466] reported that captan, chlorothalonil, dichlorofluanid, quinomethionate, and thiram were incompatible with *Verticillium lecanii*. McCoy et al.[467] found that the conidial viability of *Hirsutella thompsonii* was adversely affected by certain fungicides and nutritional elements when tested in vitro. Field rates of copper fungicides, Difolatan, benomyl, and ferbam significantly affected mycelial growth and conidial germination of *H. thompsonii* whereas light and medium oils had no effect. In contrast, manganese oxide and zineb significantly affected conidial viability. Field plots receiving oil as a fungicide always maintained a high incidence of disease by *H. thompsonii*.[468,469] In further field studies, methidathion and chlorobenzilate, alone and in combination, reduced the incidence of disease to the citrus rust mite.[470] In vitro studies also confirmed the harmful effect of Bordeaux mixture on *V. lecanii* on the growth and conidial germination.[471]

Effect of Physical Factors on the Infection Process

The ability of an entomogenous fungus to inflict mortality in an insect population is dependent on germination of the infective propagule. Water, either liquid or vapor, has been recognized as being essential to this physiological process. It is important to recognize that liquid water or a high relative humidity must persist at the host integument or more specifically at the site of spore attachment or contact.

Aquatic fungi such as *Lagenidium*, *Coelomomyces*, and *Myiophagus* require free water for both zoospore dispersal and germination.[47] Free water or near-saturated air (95 to 100% RH) is essential for both the discharge of spore from its host and the germination of spores for most *Entomophthora* species and related genera.[292,414,472] Numerous researchers have associated high relative humidity with increased fungal infections and have speculated that increased mortality was probably due to increased germination.[473] The muscardine fungi, *Beauveria bassiana*, *Metarhizium anisopliae*, and *Paecilomyces farinosus* generally required 92% or greater relative humidity to obtain appreciable germination of conidia.[411,474,475] However, Ferron[476] showed that *Acanthoscelides obtectus* adults could be infected with *B. bassiana* by inoculation of the integument with conidia irrespective of relative humidity implying that phenomena at the boundary layer of the insect integument regulated conidial germination. High humidities were apparently not critical for conidial germination of *Aspergillus flavus*, *Fusarium solani*, and *Beauveria*.[477,478] *Verticillium lecanii* conidia required a high humidity to germinate and may only germinate in a water film.[170] Conidia of *Hirsutella thompsonii* prefer free water for germination but will germinate at RH of 90% or greater.[33,479] Conidial germ tubes survived 8 hr of exposure at 3 to 5% RH and at 60% RH but subsequently grew poorly when placed at 100% RH.[480] Conidia of *Nomuraea rileyi* require a high relative humidity and nutrient (maltose, neopeptone, insect extract) for maximum germination.[164] The lower limit of relative humidity for *N. rileyi* however, appeared to be 40 to 60%.[481]

Chemical constituents in the water may also affect spore germination. Conidial germination of *Culicinomyces*[482] decreased from 63 to 2% in fresh water compared to 200% in sea water.

Salinities as low as 0.5% and water pH less than 4 and greater than 9 decreased infection by zoospores of *Lagenidium*.[47] When conidia of *Metarhizium anisopliae* and *Beauveria bassiana* were placed in distilled water, germination did not occur or was greatly reduced.[411]

Temperature is an important factor influencing spore survival on the host, spore germination, and the infection process. The optimum temperature for germination of *Culicinomyces* was 27.5°C and infectivity occurred between 15 to 27.5°C.[47] For most species of *Entomophthora*, spores were discharged between 5 and 30°C.[261,405] Optimum temperature for germination of *Entomophthora* was between 16 and 27°C.[292,405,411,483] Optimum germination for other species were as follows: *Metarhizium anisopliae* and *Beauveria bassiana* between 25 to 30°C;[15,411] *Hirsutella thompsonii* 25 to 30°C;[409,479] *Verticillium lecanii* 20 to 25°C;[170] and *Nomuraea rileyi* 15 to 25°C.[164] Thermal death point for *M. anisopliae* and *B. bassiana* was near 50°C.[411] The upper limits for *V. lecanii* varied greatly among strains ranging from 31 to greater than 36°C.[170]

The rate of mycelial development and therefore mycoses is dependent on temperature.[484] Development of mycoses generally increases with an increase in temperature within the range of 20 to 30°C. As an example, *Entomophthora* killed *Tetranychus urticae* in about 4 days at 25°C but took 11 days at 15°C.[271]

Other factors such as pH and light also affect the various phases of fungal development. For many *Entomophthora* species, more spores are discharged in light,[261,485,486] however, spore germination appears less affected by light.[405,472] Light, however, will enhance secondary spore formation in some species of *Entomophthora*.[394] Gabriel[487] found that the pH of intestinal content of silkworm appeared to affect germination of conidia of *Beauveria bassiana* and *Metarhizium*.

PRODUCTION

Mass production, a basic requirement for the development of an entomogenous fungus as a microbial control agent, is dependent on man's ability to isolate and culture candidate fungi in a living organism or on a nonliving substrate.[488,489] The success in culturing these fungi differs considerably among taxa and even within genera. For example, the Mastigomycotina and Zygomycotina have more complex nutritional requirements for growth than the Deuteromycetes.

Biological specificity and virulence of the fungus appear to be the key genetically controlled mechanisms explaining culture variability. Simply obligate fungi with a narrow host range and high virulence are more difficult to culture than facultative species with both a parasitic and saprophytic stage in their life cycle. For example, many species within the Entomophthorales are impossible to culture on a nonliving substrate.

In nature, fungi satisfy basic requirements for growth by digesting various components of their host through enzymatic action. The same basic nutrients available from the living host should therefore be provided in optimum quantities in the culturing medium to achieve maximum growth and sporulation. Frequently, however, the nutritional components responsible for maximum mycelial growth of the fungus differ from those favoring sporogenesis. The requirements can be highly specific.[490,491] Often depletion of the carbon and/or nitrogen source or water will stimulate sporulation.[492] In addition, certain environmental factors govern sporulation of many entomopathogenic fungi. Unfortunately, less is known about the factors governing sporulation than those involved in growth of the fungus.

Another important consideration for the in vitro mass culture of a fungus control is the selection of strains.[493] Highly virulent strains adapted to a primary host but with a broad host spectrum are generally preferred for industrial development. Presently, technology is available to commercially mass produce entomogenous fungi on either living[494] or artificial substrates.[488,489,495]

Living Systems

In vivo production of fungal insecticides is generally considered costly and none is presently being produced by industry. Ignoffo and Hink[494] recognized eight species of entomogenous fungi that have been successfully cultured in vivo. The propagated fungi, in the form of live or dead diseased insects, are usually introduced or colonized in the immediate habitat of the host. This was successfully implemented with different species of *Entomophthora* against the spotted alfalfa aphid in California[496] and other obligate fungi such as *Coelomomyces* against mosquitoes.[497]

Because of their potential as microbial control agents of mosquitoes, research on the mass culture of *Coelomomyces* has received considerable attention.[47] Although in vitro culture has not been successful, Shapiro and Roberts[498] reported growth of the sporophyte of *C. psorophorae* in a fortified mycoplasm medium and Nolan[499] stimulated germination of resistant sporangia of this species with plant hormones. Castillo and Roberts[500] successfully grew mycelium of *C. punctatus* in a defined tissue culture media and sporangia were observed in several media. Numerous species of *Coelomomyces* were produced in vivo using both mosquitoes and copepods as hosts.[47] However, the number of fungal stages produced was limited and quantities usually were only adequate to maintain the culture. The copepod *Cyclops vernalis* was used as the intermediate host and larvae of *Anopheles quadrimaculatus* were used as the definitive host in laboratory in vivo culture of *C. dodgei*.[501] The culture was perpetuated by infecting copepods and mosquitoes using separate procedures. Federici[502] significantly improved the in vivo culture of *C. dodgei* by infecting large populations of synchronously developing nauplii (immature copepods).

Surface Culture

All facultative species of entomogenous fungi initially are grown in surface culture where they generally produce conidia or spores on aerial mycelium. Different species may grow or sporulate on one or more defined agar base media (i.e., potato-dextrose, Sabouraud, Czapeck-Dox, etc.) or on natural substrates (i.e., wheat, bran, rice, egg yolk, potato pulp, etc.). Although culturing fungi on solid substrates in flasks or flat pans is a disadvantage in mass production (a relatively small spore mass is generally produced per surface unit), several candidate microbial control agents only can be produced in this manner (Table 4).

Culicinomyces clavisporus was produced on a media consisting of 0.3% beef extract, 0.5% peptone, and 1.5% agar in 1-ℓ Roux flasks.[503] Conidia, harvested after 7 days by washing with distilled water, yielded an average of 10^5 conidia/mℓ. Increased salinity of the medium caused a decrease in conidial germination and hyphal growth.[47] *Lagenidium giganteum*, a facultative parasite of mosquito larvae, has been produced on several nondefined and defined solid media. Solid media containing water extract of crushed whole hemp seed or soybean produced about 5000 zoospores/mℓ.[504] Nutritional studies show that *L. giganteum* required an exogenous source of sterols to induce reproductive development.[490,505] Oosporogenesis involves a series of complex developmental events which are regulated in part by cyclic nucleotides[506] and calcium metabolism.[507]

Latge[508] reviewed culture and production methods for the Entomophthorales. According to Wilding,[405] the more obligate species of the genus *Entomophthora* grow readily on coagulated egg yolk or Sabouraud-dextrose or -maltose agar fortified with egg yolk. Some species of the *Entomophthorales* also can be produced on undefined media (i.e., liver, fish, veal, potato, etc.).[509] The protoplasts of *E. egressa*[282,510] and *E. grylli*[511] were grown in pure culture on Grace's tissue culture media with supplements. Latge and de Bievre[512] confirmed the earlier work by Gustafsson[513,514] by showing that growth and sporulation of entomophthoraceous fungi on egg yolk were not stimulated by specific factors but were stimulated by a complex high nitrogen and carbon source. Other studies[515-518] showed that concentrations of nutrients and the carbon/nitrogen ratio were important to the growth and sporulation of entomophthoraceous species on solid medium.

Table 4
PRODUCTION AND STORAGE INFORMATION ON SELECTED ENTOMOGENOUS FUNGI AS CANDIDATE MICROBIAL CONTROL AGENTS

Pathogen species	Production method	Media	Yield	Stability	Formulation	Ref.
Coelomyces spp.	Living copepods and mosquito larvae	Host rearing medium	Variable	—	Sporangia or infected copepods	47, 501, 502
Lagenidium giganteum	Surface or broth culture	Different solid or liquid media that include a sterol	5000 zoospores/mℓ	16 days at 4°C	Zoospore prep.	47, 490, 505
Entomophthora spp.	Submerged culture	Egg yolk, dextrose, protein hydrolysates, yeast extracts, fats, triglycerides	10 resting spores/mℓ	1—2 years	Resting spores	291, 508, 546—552
Beauveria spp.	Surface and semisolid culture	Potato glucose agar, sterile wheat rice	1×10^{10} conidia/m	18 months at 5°C	Conidia in bentonite	520—522
	Submerged culture	Yeast-peptone, malt sprout extract	2×10 blastospores/10 conidia/g	8 months at 5°C	Blastospores and conidia	533, 542
Culicinomyces clavisporus	Surface culture	0.3% beef extract, 0.5% peptone, 1.5% agar	10 conidia/mℓ	7 days at 4°C	Conidial prep.	47, 482
Hirsutella thompsonii	Semisolid culture	2.0% molasses	1×10^{10} conidia/g	12 months at 4°C	Conidial dust or wettable powder	543, 556
Metarhizium anisopliae	Submerged culture	2.0% soyflour		6 months at 4°C	Mycelial slurry	543, 545
	Surface culture	Boiled rice	10 conidia/mℓ	12 months at 7°C	Conidial prep.	514, 525
	Submerged culture	3% corn steep liquor, 4% glucose, 4% yeast extract	Unknown	Unknown	Unknown	535
Nomuraea rileyi	Surface culture	Sabouraud's maltose agar + 1% yeast extract	6.3×10 conidia/cm	2 years at 5 to −20°C	Conidial prep.	164, 527, 528
Paecilomyces spp.	Surface culture	Lactose, autolyzed yeast, maize, beef extract, potato	Unknown	Unknown	Conidial prep.	520, 522
Verticillium lecanii	Surface culture	Czapeck-Dox malt extract, Sabouraud dextrose	41×10 conidia/g	Unknown	Conidial prep.	170
Aschersonia spp.	Surface culture	Potato sucrose agar, beer wort	1×10^{10} conidia/mℓ	Unknown	Conidial prep.	530, 531

Large quantities of conidia of *Beauveria bassiana* and *B. brongniartii* were produced in surface cultures[519] containing media such as potato-glucose agar, sterile wheat, and rice.[512,520] Detailed production methods, using different cultivation vessels, have been described by Goral and Lappa,[521] Bajan et al.,[522] Farques et al.,[523] and Bertatlief.[524] Yields generally ranged from 10^9 to 10^{12} conidia/m².

Surface culture has been used to mass produce conidia of *Metarhizium anisopliae* for experimental and commercial use in Brazil.[525,526] This fungus was cultivated on a boiled-rice substrate for 15 to 20 days in polypropylene bags, dehydrated at 25°C and 35% RH for 72 hr, and then milled to obtain a conidial powder.

Although *Nomuraea rileyi* has been successfully grown vegetatively in broth culture, mass production methodology has been dependent on surface culture techniques.[164,527] Conidia were produced on Sabourauds-maltose agar plus yeast extract, or on potatoes, corn meal, bajra meal, or soybean.[528] Yields, after 21 days, have been as high as 6.3×10^8 conidia/cm² of medium.

Verticillium lecanii was successfully mass cultured on numerous mycological media such as Czapek-Dox, malt agar, or Sabourauds and potato-dextrose agar.[170] Yields averaged 41×10^7 conidia/g. Interestingly, conidia produced on solid media were more virulent than blastospores produced in fermentation cultures.[170,529]

Paecilomyces farinosus and *P. fumoso-rosea* have been produced in solid culture in Erlenmeyer flasks and petri plates on both defined and undefined media.[520,522] Maximum virulence and production of the two species were obtained with lactose or sucrose as a carbon source and peptone, autolyzed yeast, or beef extract as a nitrogen source.

Numerous researchers have mass produced the whitefly fungus, *Aschersonia*, on solid media containing wort beer[530,531] or potato-sucrose agar. A commercial preparation of *A. placenta* produced in the U.S.S.R. yielded 10^6 to 10^7 conidia/mℓ.[530]

Submerged Culture

Submerged or fermentation culture, whereby a microbe is produced in a liquid medium under defined conditions, is the standard technology used by industry for production of many bacteria and fungi.[488,489,495] In the case of entomogenous fungi, however, few species produce infective spores or conidia in liquid culture and thus, a diphasic fermentation is required.[489] Exceptions are selected strains of *Aspergillus ochraceus*,[532] *Beauveria bassiana*,[533] and *Hirsutella thompsonii*[534] that will produce conidia in submerged culture.

The dimorphic deuteromyceteous fungi such as *Metarhizium anisopliae*,[535] *Beauveria bassiana*,[536] *B. brongniartii*,[537] *Paecilomyces farinosus*,[520] *Verticillium lecanii*,[170] and *Nomuraea rileyi*[527] generally produce yeast-like blastospores in submerged culture. Extensive technology was developed in the U.S.S.R. and Czechoslovakia for production of blastospores of *B. bassiana* in both batch and continuous submerged culture.[536,538] Similarly, *B. brongniartii* yielding 8.0×10^9 to 1.3×10^{10} viable blastospores/g was produced in 250-ℓ fermentors in France.[537] Culture media with defined carbon and nitrogen source have been identified for maximum mycelia growth and sporulation of *B. bassiana* in submerged culture.[539-541] In the above cases, however, blastospores were difficult to conserve after production and emphasis on their use for field application was abandoned.[15]

Conidia of selected strains of *Beauveria bassiana* were successfully grown in a submerged culture containing yeast-peptone and a malt-sprout extract.[533] True conidia are formed from phialides produced from germinating blastospores toward the end of culture when glucose and nitrogen have been exhausted. In a detailed account of the production technique, Belova[542] of the U.S.S.R. reported an average yield of 10 to 20 kg of biomass per mm³ of broth or 10 to 15 billion viable conidia/g after 72 hr of incubation. Although this research may be important to the economical production of infective units of entomogenous fungi, the feasibility of this methodology still is untested.

Strains of *Hirsutella thompsonii*, that produce only mycelia in submerged culture, were mass-produced in a defined liquid medium using sterilizeable solution bottles and a specific aeration system.[543] Media containing dextrose, yeast-extract peptone mixture, nutritional salts, or molasses and soybean gave excellent mycelial growth at a pH of 7.5.[544,545] Utilization of mycelium as the microbial agent eliminated the need for the sporulation phase of production. Storage of the relatively unstable mycelial mat, however, produced problems that limited this novel approach to use of mycelium as a microbial insecticide.[36,409]

A specific strain, *Hirsutella thompsonii* var. *synnematosa* from the Ivory Coast, will produce true conidia in liquid culture containing a concentration of 10 g/ℓ of corn steep liquor and 0.2% Tween 80.[534] The fungus began producing infective conidia after 3 days incubation reaching a peak ranging from 6.8×10^5 to 9.7×10^7 conidia/mℓ between 6 and 11 days. Germination of submerged conidia, however, was low ranging from 5 to 13%.

The characteristic of the infective spores of *Entomophthora* spp., that is, short life in storage and its inability to form in liquid medium, excluded their being produced via submerged fermentation. However, resting spores are suitable and considerable research was directed to their growth, germination, and mass production in liquid culture. Various animal and vegetable fats, triglycerides, dextrose corn syrup, protein hydrolysates, yeast extract, and egg yolk were excellent carbon and nitrogen sources for production of several *Entomophthora* sp.[291,546-550] Chemically defined media, of optimum concentration and physical conditions, also were developed for certain species of *Entomophthora*.[551]

Soper et al.[552] and Yegina[549] described techniques for the mass production of resting spores of *Entomophthora thaxteriana*. The former produced approximately 1 kg of resting spores from 5 dozen eggs. More recently, azygospores of the aphid pathogen, *E. obscura* (= *Conidiobolus*), were produced using industrial media containing corn-steep liquor and Ambrex or unrefined corn oil.[553] More than 4.0×10^6 azygospores/mℓ were produced at the end of the growth phase using this media.[553,554]

The production of sexual spores of lower fungi represents probably the most difficult example of mass production because of the need of contact between gametes. In this case, the nutritional and hormonal requirements of the fungi are extremely important. Recently, *Lagenidium giganteum* oospores were successfully produced in liquid culture.[555] In addition to a basic requirement of yeast extract, consistent yield of viable oospores in liquid culture required exogenous sterols and unsaturated fatty acids, calcium, and magnesium. Current yields of oospores are 5.0×10^4 oospores/mℓ.

Semisolid Culture

Semisolid fermentation or diphasic fermentation (submerged fermentation for mycelial growth followed by incubation in shallow pans for spore or conidial production) has been the most commonly used large-scale mass production techniques for such facultative fungal species as *Beauveria* spp.,[533] *Metarhizium anisopliae*,[521] *Hirsutella thompsonii*,[409] *Nomuraea rileyi*, and *Verticillium lecanii*.[170] The diphasic method appears to be most widely used for boverin[533] production in the U.S.S.R., although other production methods also have been used for *B. bassiana*. Blastospores produced in submerged culture were poured onto flat surfaces and bran or cereal was added to promote mycelial and conidial formation.

Diphasic fermentation also was developed for conidial production of *Hirsutella thompsonii* and *Nomuraea rileyi* by Abbott Laboratories[556] and *Verticillium lecanii* by Tate and Lyle Ltd. (England).[170] Production technology has been standardized, optimum media for growth developed, and the growth cycle identified. Yields of 10^5 to 10^6 conidia per gram were consistently produced via diphasic fermentation.

In view of various difficulties associated with the production and storage of conidia of both the Entomophthorales and Deuteromycotina used in pest control, various production methods utilizing fungal mycelia as a field inoculum (i.e., *Hirsutella thompsonii*) have been developed. Recently, the marcescent process where mycelia produced in submerged culture is dried and stored at refrigerator temperatures for long periods, was developed for the Entomophthorales fungi.[557] Mycelium is produced in aerated liquid medium, harvested by filtration, and dried with a sugar desiccation protectant. The dried mycelium is milled and stored at low temperatures.

Formulation and Stability of Fungal Preparations

Unlike chemical pesticides, mycelia, conidia, resting spores, blastospores, zoospores, or sporangia, produced by various species of entomogenous fungi, are living propagules. Therefore, the type of formulation and selection of additives for a given formulation are critical to their stability. The objective of the formulator is to maximize stability of the infective unit so as to enhance the infection of the host following application in the field. Additives that have hydroscopic and adhesive properties are important to a formulation since the mode of action for fungi generally involves integumental contact of the host by the infective unit under conditions of high moisture. As with other pesticides, wetting agents, that reduce the surface tension of the spray droplet, are important in maximizing distribution of the infective unit.

Conidia have been generally formulated as a wettable powder or dust for those fungi that have been successfully mass produced (Table 4). Conidia of wettable powders are generally diluted with inert filters, wetting agents, and spreaders. Water is often used as the carrier. Where conidia germinate quickly in water, a dust is the preferred formulation. Talc, flour, and milk powder have served as suitable diluents for dusts. Liquid formulation, generally not acceptable for conidial or blastospore preparations, was successfully used when *Hirsutella thompsonii* was applied as a fragmented, mycelial slurry.[165,543] In this case, molasses and Dacagin were added as nutrients to stimulate conidiation of the fungus after application. Granular formulations of *Beauveria bassiana* containing corn meal also have been successfully used as a nutrient supplement.[558] Fargues et al.[559] demonstrated that a clay coating extended the soil viability of *B. bassiana* blastospores. *B. brongniartii* conidia coated with bentonite and then spray dried proved to be a better formulation than blastospores lyophilized with powdered milk supplemented with glycerin.[516] Daoust et al.[560] found that a liquid-oil formulation was more detrimental to conidia of *Metarhizium anisopliae* than were dry, granular, or dust formulations. Thixcin-R and bentonite additives in dry formulated *M. anisopliae* conidia maintained virulence for a longer period than unformulated conidia.[561]

The stability of standardized preparations of fungal strains and formulated microbial insecticides is critical to their success as microbial control agents. *Entomophthora* sp. are readily stored in liquid nitrogen[405] and most deuteromycetes can be maintained at low or ultra-low temperatures. High and low relative humidity appears vital for maximum stability. Most candidate fungi do not remain stable beyond 12 months at 4°C (Table 4).

Change in Virulence by Subculture and Passage Through Insects

Virulence and viability of spores and conidia of many entomogenous fungi, particularly the facultative hyphomycetes species such as *Metarhizium anisopliae*, decline after continual maintenance on artificial media under normal temperatures for growth and sporulation.[518,562] Certain strains of *Nomuraea rileyi* begin to lose virulence and viability of conidia after only 2 to 3 days on an artificial media while others are considerably stable when serially passed on media or through the living host.[164] Strains of *Verticillium lecanii* also remain relatively stable on artificial media (even after as many as 98 subculturings) and passage through the aphid host did not increase virulence.[563] Samsinakova and Kalalova[564] also found that repeated subculture of *Beauveria bassiana* mutants had no effect on virulence. Hall et al.[565] found aphid host did not increase virulence.[563] Samsinakova and Kalalova[564] also found that repeated subculture of *Beauveria bassiana* mutants had no effect on virulence. Hall et al.[565] found that when the virulence of *B. bassiana* was lost it could be restored after three successive passages through a susceptible host. Hartman and Wasti[566] and Krejzova[567] claimed to have increased the infectivity (in vitro) of entomophthoraceous fungi by serial transfer through the living host, although comparisons were not made with cultures of a fresh isolate. Recently, Fargues and Robert,[568] working with two pathotypes of *M. anisopliae* found that virulence, which was lost, was almost completely restored after one passage through the insect. Virulence of the host-adapted fungus increased 10- to 100-fold. As soon as the parasitic selection pressure ended, however, virulence decreased. Fargues and Robert[568] also found that successive subculturing on artificial medium did not result in a loss of virulence. Daoust and Roberts[422] also found that virulence could be restored via passage through a living host.

These studies strongly suggest that the virulence of entomogenous fungi is influenced by nutrition and that suboptimum conditions of artificial culture can impair normal physiological development of the fungus. Goral[569] recently showed that nutrition and cultivation conditions for *Beauveria bassiana*, *Metarhizium anisopliae*, and *Paecilomyces farinosus* greatly affect their virulence. Maximum virulence of conidia is generally obtained in a complete nutrient media rich in organic substances at optimum pH values.

BIOCONTROL OF ARTHROPOD PESTS

Although extensive ecological information is lacking, it is apparent that endemic mycosis significantly contributes to the seasonal natural mortality of many populations of arthropod pests. In addition, spectacular epizootic declines in insect and mite populations, caused by a number of entomophthoraceous[405,570] and deuteromycetous fungal species,[409,571] also are relatively common. The degree of natural mortality via naturally occurring fungi is usually dependent upon agrotechnological practices. It is within this context that integrated pest management (IPM) has become important and why fungi are considered significant biotic agents for protection of crops. The common observations of fungal epizootics of arthropods cannot be overlooked as a motivating force in research and development of fungi as microbial control agents.

Three strategies have been recognized for the use of entomogenous fungi as biological control agents.[34,572] Of these, the more classical method (introducing exotic fungal strains of greater virulence or a new species from one geographical location to another) has received the least attention. Fungi also can be produced in quantity, formulated, and applied to induce an artificial or premature epizootic or used like a chemical pesticide.[9,573] Finally, control of pests by endemic fungal species can be enhanced (in their natural environment) by modifying the physical environment or management practices that favor survival of the fungus and/or infection of the host. Raising the economic injury thresholds for a given pest may stimulate fungal mycosis by supplying more available hosts.[467]

As previously mentioned, entomogenous fungi were the first microorganisms utilized by man as a microbial insecticide. During the past century however, only a few species have been mass produced and applied to control arthropod pests. In most cases, these fungi have been produced by governmental agencies rather than by private industry. Presently, only Brazil, the U.S.S.R., and China are regularly using fungi as microbial insecticides. The fungus *Hirsutella thompsonii* (under the trade name Mycar) and *Verticillium lecanii* (named Vertalec) have been registered as microbial control agents in the U.S. and England, respectively (Table 5).

Since entomogenous fungi are ubiquitous, generally only regulate invertebrate populations at high densities,[15,574] with their mode of action dependent upon environmental conditions, it is not surprising that their potential use as mycoinsecticides has been controversial.[34,572] In addition, they are known allergens and in a few instances pathogens of vertebrates.[574] Yet, many researchers are convinced that fungi can be effectively used as biological insecticides.[33,572,573]

Table 5
TRADE NAMES OF COMMERCIAL OR EXPERIMENTAL PREPARATIONS OF FUNGI FORMULATED AS MICROBIAL INSECTICIDES

Fungal species	Trade name	Producer
Aschersonia aleyrodis	Aseronija	All Union Inst. (U.S.S.R.)
Beauveria bassiana	Biotrol FBB	Nutrilite Products (U.S.)
	Boverin	Glavmikrobioprom (U.S.S.R.)
	ABG-6178	Abbott Labs. (U.S.)
Hirsutella thompsonii	Mycar	Abbott Labs. (U.S.)
Metarhizium anisopliae	Biotrol FMA	Nutrilite Products (U.S.)
	Metaquino	CODECAP (Brazil)
Verticillium lecanii	Vertalec	Tate and Lyle (England)
	Mycotol	Tate and Lyle (England)

Table 6
CURRENT ENTOMOGENOUS FUNGI BEING CONSIDERED AS POTENTIAL MICROBIAL INSECTICIDES

Fungus	Infective stage	Insect host	Habitat
Culicinomyces	Conidia	Mosquitoes	Aquatic
Lagenidium	Motile zoospores	Mosquitoes	Aquatic
Entomophthora	Conidia or resting spores	Aphids	Foliage
Aschersonia	Conidia	Whitefly	Foliage
Beauveria	Conidia	Beetles, caterpillars	Foliage
Hirsutella	Conidia	Mites	Foliage
Metarhizium	Conidia	Froghoppers	Foliage
		Beetles	Soil
Nomuraea	Conidia	Caterpillars	Foliage
Verticillium	Conidia	Aphids	Glasshouse
		Whitefly	Foliage
		Scales	

Certain intrinsic characteristics are needed before an entomogenous fungus can be considered as a potential microbial insecticide. These are high virulence, rapid mode of action, a broad host range, stability in culture and storage, amenability to submerged fermentation, and amenability to quantitative bioassay.[574]

The overall criterion used by industry to evaluate fungi as potential candidates for development into microbial insecticides is similar to that for other microbes. The major criterion is commercial feasibility, i.e., profit. Four technical primary prerequisites are (1) effectiveness against economically important pests; (2) safety to humans, other vertebrates, beneficial invertebrates, and plants; (3) production feasibility; and (4) product formulation stability.[575] Private industry must also consider other factors such as marketing potential, reliability, predictability, production cost, and grower acceptance of product.

There are nine genera of entomogenous fungi that have been used, in different parts of the world, as microbial insecticides (Table 6). Two genera contain species that have potential for control of mosquitoes while the remaining seven are directed against pests associated with agriculture. Only one species has been developed for control of phytophagous mites.

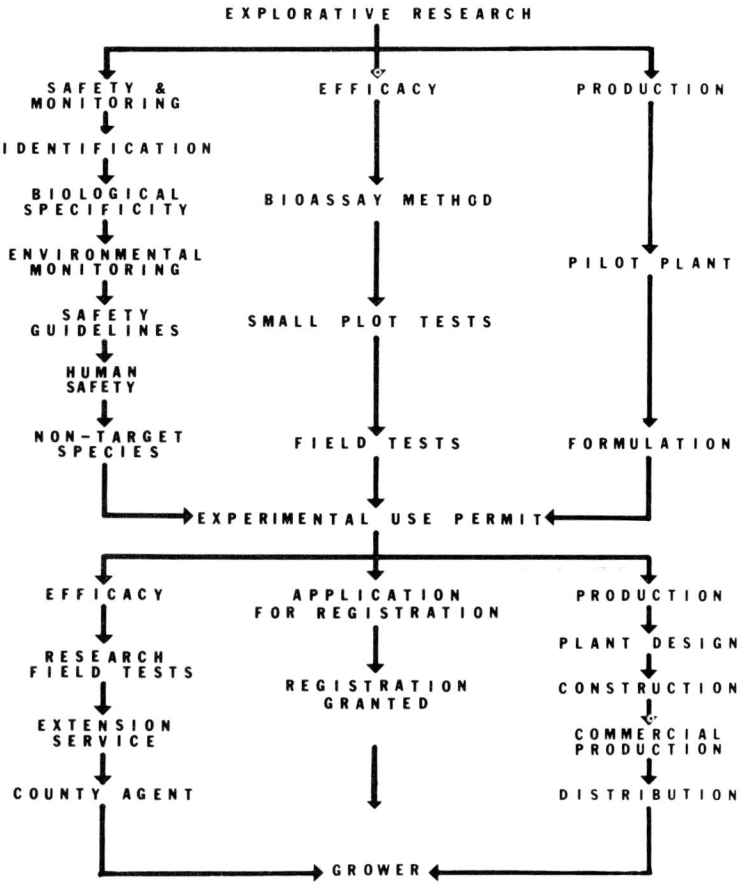

FIGURE 21. A flow-diagram indicating same requirement for the development and registration of a candidate fungal microbial insecticide in the U.S.

A flow-diagram, depicting requirements for the development of a fungal, microbial insecticide is shown in Figure 21.

Culicinomyces

Two strains of *Culicinomyces clavospirus* are effective against a wide range of mosquito species of the suborder Nematocera order Diptera.[47] Isolates of this fungus are mainly effective at low salinities and moderate temperatures thereby excluding their use in the tropics or in brackish waters.[482,503] Since conidia are ingested and infection occurs via the lining of the foregut or midgut, external microclimatic conditions for conidial germination are not critical.[174] Presently, *C. clavospirus* is considered a candidate as a microbial insecticide because it is readily produced in both surface and submerged culture and has the potential to recycle within the host population.[576] It has been used in both Australia and the U.S. In small-plot outdoor trials, *C. clavospirus* was introduced into rock pools where low numbers of *Aedes rupestris* were breeding. All larvae died within 5 days of treatment.[577] In artificial ponds, it gave 100% mortality of *Culex quinquefasciatus* larvae and 86 to 100% mortality of *Anopheles amictus*.[577] It gave 90 to 95% control of *C. australicus* and *C. annulirostris* in a natural pond.[578,579] Conidia and hyphae of *C. clavospirus* applied at a rate of 1×10^{10} conidia/m^2 to the surface of salt marsh ponds (U.S.), produced 100% mortality of field-collected *A. taeniorhynchus* larvae (in sentinel cages) within 24 hr after application.[580]

Lagenidium

Lagenidium giganteum is a facultative parasite of mosquito larvae particularly culicines.[47] Vertebrates (fish, birds, and rats) are not susceptible. The infectious zoospores are sensitive to pH values below 5.5 and above 8.0 but can tolerate salinities lower than pH 1.5.[54] The fungus can be maintained in vivo in the laboratory[54] or grown in vitro on a variety of undefined and defined solid or liquid media.[504,505] The major problem with commercialization of this fungus is production of zoospores in artificial media. Successful production of oospores in a defined liquid media could overcome this obstacle.

Lagenidium giganteum, introduced as diseased cadavers into ponds in North Carolina, induced 43% infection of *Culex restuans* within 3 days.[54] Introduction of *L. giganteum* as a spray onto seepage ditches adjacent to rice fields in California (dosage of about 3,000,000 zoospores/m^2) eliminated *C. tarsalis* from ditches where environmental conditions were not favorable and produced low rates of infection in ditches where environmental conditions were not favorable.[581] *L. giganteum* also survived seasonal droughts and land-management practices. Recent field studies[582] showed that *L. giganteum* could be established in freshwater ponds in North Carolina resulting in 100% mortality of *Culex* spp.; however, Merriam and Axtell[580] showed that *L. giganteum* would not infect *Aedes taeniorhynchus* larvae in brackish water.

In California rice fields, fermentor-grown cultures of sexual and asexual stages of *L. giganteum* gave immediate control of sentinel *Culex tarsalis* and *Anopheles freeboni* larvae when introduced by ground and aerial application.[583] Asexual stages were effective throughout the 4-month mosquito breeding season while oospores which were desiccated in the field following application provided consistently high larval infection levels after reflooding of the fields.

Entomophthoraceae

Numerous species of *Entomophthorales* cause spectacular epizootics in aphids and grasshoppers.[65,570,584-586] However, the use of the Entomophthoraceae as microbial control agents has been hindered by the poor survival of conidia. Yet, *Conidiobolus obscurus* and *Erynia neoaphidis* have received considerable attention as microbial control agents of aphid because of their resistance to environmental conditions.

In a successful colonization attempt, *Erynia radicans* (*E. sphaerosperma*), collected from Israel, was introduced into five Australian populations of the spotted alfalfa aphid.[587] An extensive epizootic resulted with up to 84% infection being recorded within 3 m of the point of introduction. *Entomophthora* was utilized in integrated control programs in the U.S.S.R. to control the pea aphid.[542] Extensive ecological studies, conducted from 1962 to 1969, showed that rainfall was a key factor in the development of an *Entomophthora* mycosis in the pea aphid. By dividing the country into regions (based on hydrothermal coefficients), epizootics could be accurately predicted thereby resulting in reduced or no treatment for control of aphids.

The use of sprinklers to increase the relative humidity in glasshouses, resulted in the initiation of *E. fresenii* epizootics in *Aphis fabae*.[588] The same fungus introduced into field populations of *A. fabae* in England, however, failed to spread.[589] In Latavia, aqueous suspensions of conidia, hyphal bodies, and resting spores of *E. thaxteriana* and *E. virulenta*, sprayed in glasshouses for aphids and mites,[590] provided rapid control possibly due to the presence of a toxin. A suspension of *E. aphidis* (6×10^6 conidia/mℓ), obtained from bran and water, gave 76 to 100% mortality of several aphid species.[405]

The use of resting spores for the field application of *Entomophthora* generally has been unsuccessful. These programs involved the use of *E. obscura* against cereal aphids in France[591] and *E. virulenta* against green peach aphid in the U.S.[592] A major limitation to the performance of these species in the field appears to be their inability to spread.

Beauveria

During the last 15 years, special attention has been given in the U.S.S.R. and eastern Europe to the use of *Beauveria bassiana* as a component of an integrated pest management strategy against the Colorado potato beetle, *Leptinotarsa decemlineata* and to a lesser extent, the codling moth *Cydia pomonella*. An excellent summary of this research has been recently published by Ferron.[593]

Large-scale studies against *Leptinotarsa decemlineata* in the U.S.S.R. were begun by Pospelov[594] and followed by mass production and field application (along or combined with insecticides) by numerous workers.[15,38,595,596] Generally, two treatments made at 15-day intervals with 1.5 kg/ha concentrated boverin (30 × 10^9 conidia/g) combined with a reduced dosage of chlorophos were used against larvae stage.[593] A similar treatment scheme was developed in Poland, combining *Beauveria bassiana*, *Paecilomyces farinosus*, and a parasitic nematode. Results, however, were no better than with *B. bassiana* alone.[597] *Beauveria bassiana* plus *Bacillus thuringiensis*[598] and *Beauveria bassiana* plus despirol[599] also have been used in the U.S.S.R. with varied results. In France, two to four applications of *B. bassiana* at 10^4 blastospores/hectare significantly reduced larval populations of *L. decemlineata*.[600,601] Conidia at 2.0 to 4.5 × 10^9/g compared to 2.0 to 2.5 × 10^9 blastospores/g provided superior control of *L. decemlineata* larvae in Romania after 12 days.[602] In the U.S. *B. bassiana* applied to the soil, reduced pupal populations of Colorado potato beetle.[603] Generally, high doses of boverin will work as a prophylactic measure against younger larvae of *L. decemlineata* but an insecticide must be included to control older larvae.

Strategies for the use of boverin in the Ukraine, usually with a reduced dosage of chemical insecticides (malathion), were successful by reducing the following spring population after treatment of the summer generation of codling moth (*Cydia pomonella*).[593,604] Introduced strains of *Beauveria bassiana* have appeared most effective.[605] Where fungicide applications do not interfere, three applications of boverin at 1 kg (24 × 10^9 spores/g) per hectare were effective at controlling *C. pomonella* in the U.S.S.R.[14] In France, conidia of *B. bassiana* applied at 6 × 10^9 conidia per tree, against late instar larvae of *C. pomonella*, gave a 50% reduction in population and also reduced the following spring population.[593,606]

In the Peoples Republic of China, *Beauveria bassiana* has been used for many years against the European corn borer.[607] Likewise, a specific mutant has been effective in experimental trials against *Ostrinia nubilalis* in France.[608] In the U.S., field trials with conidial dust have been conducted against lygus bugs,[609] stored products pests,[609] and caterpillars of cob crops.[388] Soil-inhabiting weevils have also been treated with *B. bassiana*.[411,610-613]

The other closely related *Beauveria* species, *B. brongiartii* (= *tenella*) has been tested extensively against soil-inhabiting scarabeid beetles, primarily *Melolontha melolontha* in Western Europe.[593,614,615] Soil treatment with 20 × 10^9 conidia/m² caused an epizootic and the muscardine appeared to persist in the soil for long periods of time.[593,616] Blastospore preparations were inferior to conidia and sublethal dosages of insecticides combined with conidia did not enhance control.[617]

Metarhizium

The green muscardine, *Metarhizium anisopliae* var. *anisopliae* and var. *major* like the white muscardine fungi have been studied for 100 years. However, promising field results with this fungus have occurred only within the last 15 years.[15,593]

The rhinoceros beetle, *Oryctes rhinoceros*, a serious pest in coconut growing regions of the South Pacific, has been successfully controlled in Fiji, Tonga, and Western Samoa by *Metarhizium anisopliae* var. *major*.[618-620] This species was propagated on grain and introduced by hand to breeding sites of the larvae. About 50 g of conidial preparation/m² of soil gave 80% mortality after 15 to 24 months and improved coconut yields by 25% within 1 year.[14,620]

The use of Metaquino in Brazil, a conidial formulation of *Metarhizium anisopliae*, was successfully used to control spittlebug on sugarcane and in pastures for nearly 8 years.[593,621,622] As much as 50,000 ha in a single province were aerially treated at 50 ℓ/ha with 6×10^{11} to 1.2×10^{12} conidia/ℓ.[593,622] Although *M. anisopliae* occurs in natural epizootics, introducing the fungus early in the season appears to keep spittlebugs below damaging levels.[593] *M. anisopliae* has also been effective in field trials against mosquitoes.[561] In Australia, it has reduced pasture cockchafer populations when applied as a spore bait.[623] In China, *M. anisopliae* was pathogenic to *Colasposoma* when applied to the soil at a rate of 75 kg/ha.[624] From 50 to 60% mortality of pecan weevil also has been obtained in the U.S. with applications of *M. anisopliae*.[625]

At this time, there is considerable interest in developing *Metarhizium anisopliae* into a commercial product and the program is expanding to include control of other insects.[626]

Nomuraea

Epizootics of *Nomuraea rileyi* in noctuid caterpillars occur regularly in agroecosystems in the U.S. and other countries.[164,571,627] Extensive research has been conducted on the survival of the fungus in soybean fields[164] and a model has been proposed for predicting the level of infection in velvetbean caterpillar populations in Florida.[418]

Early close planting of soybeans and cotton, that influences microclimatic field conditions, also enhanced the development of *N. rileyi* in caterpillar populations.[438,628] Since natural epizootics caused by *N. rileyi* generally occur after catepillars have caused damage and too late to prevent losses in soybean yield, the prophylactic use of *N. rileyi* has been tested using high conidial doses with encouraging results.[164] A dust formulation of *N. rileyi* with 12.4×10^{10} conidia/g applied early in the season at 1.1×10^{13}/acre to soybeans caused a premature epizootic.[571,629] These studies suggest that a prophylactic treatment 2 to 4 weeks before economic levels of injury are anticipated, may provide control of defoliating caterpillars. Data corroborating these suggestions have been published using infected larvae of *Heliothis virescens* in soybeans in North Carolina.[630]

Verticillium

Verticillium (= *Cephalosporium*) *lecanii*, reported from scale insects for some time, has become important as a microbial insecticide for use in glasshouses.[170] In the U.S.S.R., *V. lecanii* has been successfully cultivated on a synthetic medium containing grain or bread cubes.[631] A spore preparation containing 41×10^7 active spores/g either undiluted or as a 10% concentration (diluted with talc or water) was applied to citrus leaves infested with *Coccus hesperidum* L.[631] In the greenhouse, 85 to 100% mortality was achieved within 3 weeks. Although *V. lecanii* killed all stages of the brown soft scale, it was unable to complete its development because of lack of nutrients. Hence the fungus, unable to survive and spread within the scale population, must be repeatedly introduced.

Hall and colleagues,[170,632,633] in a well-conceived research program, achieved excellent control of *Myzus persicae*, a major pest of chrysanthemums in glasshouses with *Verticillium lecanii*. Conidia (10^5 to 10^7/mℓ) and blastospores (10^7 to 10^8/mℓ), suspended in phosphate buffer containing 0.02% Triton X-100 as a wetting agent, sprayed to run-off gave consistent results when sprayed in the evening and when plants were covered with polyethylene blackout sheets to boost humidity.[170,633] In similar trials on chrysanthemums, *Macrosiphoniella sanborni* and *Brachycaudus helychrysi* were also controlled at low to high population densities during all seasons.[634] Although differences in species' susceptibility and effect of spore concentration were detected in laboratory assay, no differences appeared in glasshouse trials.[635]

In trials on cucumbers, *Verticillium lecanii* was also effective against *Aphis gossypii* (within 14 days of treatment),[636] and the whitefly, *Trialeurodes vaporariorum* in England,[34,637]

Sweden[638] and the U.S.S.R.[639] Multiple sprays were required for control of the whitefly probably because stages of this host are nonmotile.

Since *Verticillium lecanii* appears to have no effect on the whitefly parasites, *Phytoseiulus persimilis* or *Encarsia formosa* or predators of *Tetranychus urticae*, it fits nicely into an integrated control scheme.[170] In addition, dioxathan, commonly used for leaf miner control, is innocuous to *V. lecanii*. The incorporation of *V. lecanii* into an IPM system and its efficacy in humid glasshouses certainly were important factors leading to its commercialization as Vertalac.

Aschersonia

Aschersonia aleyrodis has been recognized for nearly a century as an effective biological control agent of citrus whitefly particularly in Florida, where it was colonized successfully in the early 1900s.[31] From 1958 to 1973, 11 strains of several species of *Aschersonia* were introduced from India, China, Japan, Vietnam, the U.S., and Cuba for use in biological control of the citrus whitefly (*Dialeurodes citri*) in the U.S.S.R. The citrus whitefly was accidentally imported in the U.S.S.R. in 1957 and had no natural enemies. Citrus plantings in Azerbaijan were sprayed with *Aschersonia* to control whitefly during the period 1961 to 1964.[190] Approximately 80% larval mortality was obtained (under favorable environmental conditions) and the introduced fungus adapted well and spread to new citrus plantations. A strain of *A. placenta* of Vietnamese origin proved to be particularly virulent, producing 90% larval mortality.[15] This isolate was applied, as an aqueous suspension containing 10^6 to 10^8 spores/mℓ, at the rate of 900 to 1000 ℓ/ha.

Tests with *Aschersonia* to control the greenhouse whitefly (*Trialeurodes vaporariorum*) began in 1973 in the U.S.S.R. Moldavian, Cuban, and Japanese isolates of *Aschersonia* were the most effective against larvae, but effectiveness was highly dependent upon humidity. Best results (83% mortality by the Cuban isolate) were obtained in the greenhouse on cucumbers when humidity reached 85 to 95%.[640,641] Aqueous suspensions containing 5×10^6 to 5×10^8 spores/mℓ were the most effective rates, and applications had to be repeated after 10 to 14 days. In Bulgaria, good control of *T. vaporariorum* on cucumbers was obtained using *A. aleyrodis* by spraying the larvae on the underside of the leaf. Conidial preparations gave 71.3% mortality compared to 14.3% mortality when dried and ground mycelium was used.[642] Field tests for the control of *Dialeurodes citri* on citrus with *Aschersonia* sp. in Japan were inhibited by fungicides.[643] In Holland, *A. aleyrodis* gave acceptable control of *T. vaporariorum* larvae on greenhouse cucumbers in combination with the parasite *Encarsia*.[644,645]

Hirsutella

Hirsutella thompsonii causes seasonal epizootics in citrus rust mite populations lasting 2 to 3 weeks during the summer when mite densities reach injurious levels in Florida.[409,467] Nonselective pesticides can reduce the incidence of *H. thompsonii* through direct and indirect action resulting in a greater dependence on chemical controls for the mite. By using a selective fungicide (oil) for phytopathogenic fungi and raising the spray threshold for the citrus rust mite, the incidence of *H. thompsonii* in mite populations was significantly increased thereby decreasing the number of spray applications from 4 to 2/year.[409] During the last 12 years, *H. thompsonii* was mass-produced, formulated as a mycoacaricide, and applied by government agencies and industry for control of the citrus rust mite on citrus and other mite pests of turf, coconut, and greenhouse crops. A detailed summary of the results of much of this field research has been published.[409]

The fungus, in early tests, was applied as a fragmented-mycelial slurry to citrus trees (30 ℓ/tree; 1.0 to 10.0% w/v) with a high pressure sprayer with excellent results in Florida.[646,647] In China, laboratory-produced mycelium was applied at dosages of 0.5 to 1.0 g/ℓ to citrus with 90% mortality in 3 days.[648,649] In Surinam, sprays of 0.025, 0.05, and 0.1% spore-mycelial suspension on fruit prevented an increase in mite populations during the dry season.[650] Also, a spore-mycelia rainwater formulation at 0.05 to 0.1% controlled mites on fruit for at least 3 weeks in dry weather.

In greenhouses, Gerson et al.[479] sprayed a conidial-bran wettable powder (2.5×10^6 conidia/mℓ) on ground-nut plants infested with the carmine spider mite; however, extreme temperatures (35°C) and low humidity prevented infection. The inoculum, however, survived and subsequent mite exposure resulted in infection. Gardner et al.[651] obtained up to 99% mortality to *Tetranychus urticae* in laboratory bioassays with *H. thompsonii*; however, rates of 1.2 to 9.6 g/ℓ applied to plants in the greenhouse were ineffective.

In 1976, Abbott Laboratories produced the first commercial formulation of *H. thompsonii* and labeled it Mycar. Initial small-scale field trials with a wettable powder at 2.5, 5.0, and 10.0 g/500 mℓ (1.9×10^5 spores/g) and about 30 ℓ/tree provided good control of mites (in Florida).[33,409] Additional field studies with both a wettable powder and dust formulation at 2 to 4 lb and 25.0 lb/acre, respectively, provided good control of mites.[36] Mycar also was effective in field trials against the citrus rust mite in the U.S.S.R.,[652] Brazil,[653] and Colombia.[654]

In the U.S. *Hirsutella thompsonii* was registered for control of eriophid mites on citrus and turf. After the sale of several hundred kilograms of technical material, commercial production was terminated in 1985. Numerous factors such as the sensitivity of the primary conidia (infective unit) to available water and fungal survival in the field influenced its reliability as a commercially acceptable control practice. Subsequently, limitations in storage and transportation to maintain fungal stability resulted in a discontinuation of commercial sale in the U.S.

CONCLUSIONS AND FUTURE DEVELOPMENTS

It is evident that many entomogenous fungi naturally control a wide range of arthropod pests attacking many different commodities in parts of the world. Many of these arthropod pests are major economic pests important to agriculture and human health. Entomogenous fungi have been used as microbial control agents to combat some of these pests. The practical use of fungi as microbial insecticides seems to have developed more rapidly in countries where organic chemicals are not widely used and in areas where chemical control is impractical or has been restricted due to environmental concern. Skepticism, however, about the use of fungi as microbial insecticides still exists, probably due to inconsistency in effectiveness and slow kill of the pests. It is also obvious, as shown in this review, that large gaps in our knowledge still exist. This has been caused mostly by the lack of researchers working on entomogenous fungi and the complexity of host-pathogen interactions. Mycologists and entomologists with fundamental training in ecological principles must work more closely together to solve basic and applied problems common to these host-fungi interactions. The need for basic research in fungal genetics, as a precursor to more virulent pathotypes through genetic engineering is apparent. The use of fungal insecticides, however, is not a miraculous cure for pest control of all pests, even though they are a safer, more environmentally compatible option.

Research developments that enhance the use of fungi in pest control and their development as future microbial control agents should come in the following areas:

1. The selection of strains of fungi having increased virulence to target pest insects and improved epidemic characters is paramount for their implementation as cost-effective microbial control agents. Both classical and/or recombinant genetic techniques have been utilized for selecting improved strains of other commercially important fungi. Adaption of such technology to entomogenous fungi will depend upon research directed

at: (1) identification of gene products which are responsible for the pathogenic and/or epidemic potential of these fungi; (2) the establishment of genetic libraries of both chromosomal and nonchromosomal genetic elements associated with these organisms; (3) the development of a suitable transformation system; and (4) the construction of suitable shuttle vectors with appropriate "marker genes" required for selection of recombined prototype strains.
2. A more detailed understanding of fungal epizootics for key fungi via system modeling of major agroecosystems will make it possible to maximize and predict the effectiveness of applications of fungal insecticides.
3. Advances in fungal nutrition, mass-production, and formulation techniques to improve fungal stability in storage and in the field.

ACKNOWLEDGMENTS

The authors express their sincere gratitude to Peggy A. Hicks, Juanita H. Ray, Mary L. Kane, Oveida A. Nealy, Cynthia B. Evans, and Ernest R. Harben for their assistance in the preparation of this work.

Grateful appreciation is also extended to the following persons who supplied photographs, unpublished information, or critical review of the work: J. Paul Latge, C. M. Ignoffo, R. A. Humber, R. J. Milner, R. Fritz, R. A. Nolan, N. Wilding, B. Papierok, P. Ferron, G. Zimmerman, S. Oshima, J. Aoki, B. A. Federici, E. A. Grula (deceased), R. A. Hall, D. W. Roberts, H. Schabel, G. E. Cantwell, D. Tyrrell, L. P. Kish, J. V. Bell, T. McInnis, Jr., G. R. Carner, and M. Gilliam.

REFERENCES

1. **Muller-Kogler, E.,** *Pilzkrankheiten bei Insekten,* Anwendung zur Biologischen Schadlingsbekampfung und Grundlagen der Insektenmykologie, P. Parey, Berlin, 1965, 444.
2. **Ainsworth, G. C., Sparrow, F. K., and Sussman, A. S., Eds.,** *The Fungi: An Advanced Treatise,* Vol. 4a, Academic Press, New York, 1973, 592.
3. **Ainsworth, G. C., Sparrow, F. K., and Sussman, A. S., Eds.,** *The Fungi: An Advanced Treatise,* Vol. 4b, Academic Press, New York, 1973, 504.
4. **Muller, E. and Loeffler, W.,** *Mykologie, Grundriss fur Naturwissenschaftler und Mediziner,* Vol. 3, Aufl. G. Thieme, Stuttgart, 1977, 340.
5. **Alexopoulos, C. J. and Mims, C. W.,** *Introductory Mycology,* 3rd ed., John Wiley & Sons, New York, 1979, 704.
6. **Batra, L. R., Ed.,** *Insect-Fungus Symbiosis, Nutrition, Mutualism, and Commensalism,* Allanheld, Osmun & Co., New York, 1979, 276.
7. **Evlakhova, A. A.,** *The Entomogenous Fungi, Systematics, Biology and Practical Importance,* Leningrad, Nauka, Khokhria Kov, M. K., Ed., 1974, 260.
8. **Koval, E. Z.,** Guidebook to Entomophilic Fungi of the USSR, *Kiev. Nauka, Dumka,* 1974, 260.
9. **Burges, H. D., Ed.,** *Microbial Control of Pests and Plant Diseases 1970—1980,* Academic Press, New York, 1981, 949.
10. **Poinar, G. O., Jr. and Thomas, G. M.,** *Diagnostic Manual for the Identification of Insect Pathogens,* Plenum Press, New York, 1978, 218.
11. **Weiser, J.,** *An Atlas of Insect Diseases,* 2nd ed., Dr. W. Junk, B. V., Publ., The Hague, 1977, 240.
12. **Steinhaus, E. A., Ed.,** *Insect Pathology, An Advanced Treatise,* Vol. 2, Academic Press, New York, 1963, chap. 5—8.
13. **Madelin, M. F.,** Fungal parasites of insects, *Annu. Rev. Entomol.,* 11, 423, 1966.
14. **Ferron, P.,** Les Champignons Entomopathogens: Evolution des Recherches au cours des Dix Dernieres Annees. WPRS Bulletin 3, Swiss Federal Institute of Technology, Zurich, 1975, 54.
15. **Ferron, P.,** Biological control of insect pests by entomogenous fungi, *Annu. Rev. Entomol.,* 23, 409, 1978.

16. **Burges, H. D. and Hussey, N. W.**, Eds., *Microbial Control of Insects and Mites*, Academic Press, New York, 1971, 861.
17. **Bell, J. V.**, *Mycoses in Insect Diseases*, Vol. 1, Cantwell, G. E., Ed., Marcel Dekker, New York, 1974, 185.
18. **Roberts, D. W. and Humber, R. A.**, Entomogenous fungi, in *Biology of Conidial Fungi*, Vol. 2, Cole, G. T. and Kendrick, B., Eds., Academic Press, 1981, 201.
19. **Hall, R. A. and Papierok, B.**, Fungi as biological control agents of arthropods of agricultural and medical importance, *Parasitology*, 84(4), 205, 1982.
20. **Ainsworth, G. C.**, *Ainsworth & Bisby's Dictionary of the Fungi*, 6th ed., Commonwealth Mycological Institute, Kew, Surry, England, 1971, 663.
21. **Kobayasi, Y.**, The genus *Cordyceps* and its allies, *Sci. Rep. Tokyo Burnika Daigaku Sect. B*, 5, 53, 1941.
22. **Lloyd, C. G.**, *Cordyceps sinensis*, from N. Gist Gee, China, *Mycol. Notes*, 54, 766, 1918.
23. **Steinhaus, A.**, Microbial control — the emergence of an idea, *Hilgardia*, 26, 107, 1956.
24. **Cooke, M. C.**, Vegetable wasps and plant worms, *Soc. Promoting Christian Knowledge*, 364, 1892.
25. **DeGeer, C.**, *Memoires Pour Servir a' l'histoire des Insectes*, Vol. 6, Pierre Hesselberg, Stockholm, 1776, 75.
26. **Kirby, W. and Spence, W.**, Diseases of insects, in *An Introduction to Entomology or Elements of the Natural History of Insects*, Longman, London, 1826, 634.
27. **Pasteur, L.**, Observations sur la coescistence der phyllocera et du mycelium constate a cully, *C. R. Acad. Sci. Paris*, 79, 1233, 1874.
28. **Metchnikoff, E.**, Untersuchungen uber die mesodermalen phagocyten einiger wirbeltiere, *Zentralbl. Biol. Aerosol Forsch.*, 3, 560, 1883.
29. **Krassilstschik, I. M.**, La production industrielle des parasites vegetaux pour la destruction des insectes nuisibles, *Bull. Sci. Fr. Belg.*, 19, 461, 1888.
30. **Billings, F. H. and Glenn, P. A.**, Results of the artificial use of the white fungus disease in Kansas, *U.S. Dept. Agric. Bur. Entomol. Bull.*, 107, 58, 1911.
30a. **Snow, F. H.**, Contagious diseases of the chinch bug, *Annu. Rep. Div. Kansas University Exp. Stn.*, 5, 7, 1896.
31. **Berger, E. W.**, Natural enemies of scale insects and whiteflies in Florida, *Q. Bull. St. Plant Board*, 5, 141, 1921.
32. **Fawcett, H. S.**, Fungus and bacterial diseases of insects as factors in biological control, *Bot. Rev.*, 10, 327, 1944.
33. **McCoy, C. W.**, Entomopathogens in arthropod pest control programs for citrus, in *Microbial Control of Insect Pests: Future Strategies in Pest Management Systems*, Allen, G. E., Ignoffo, C. M., and Jacques, R. D., Eds., University of Florida, Gainesville, 1978, 211.
34. **Yendol, W. G. and Roberts, D. W.**, Is microbial control with entomogenous fungi possible?, Proc. 4th Int. Colloq. on Insect Pathol., College Park, Md., 28, 1970.
35. **Dulmage, H.**, personal communication, 1986.
36. **McCoy, C. W. and Couch, T. L.**, Microbial control of the citrus rust mite with the mycoacaricide Mycar, *Fla. Entomol.*, 65, 116, 1982.
37. **Hall, R. A.**, Control of whitefly, *Trialeurodes vaporariorum* and cotton aphid, *Aphis gossypii* in glasshouses by two isolates of the fungus, *Verticillium lecanii*, *Ann. Appl. Biol.*, 101, 1, 1982.
38. **Yevlakhova, A.**, Basic trends in the use of entomopathogenic fungi in the USSR, in Proc. First Joint US/USSR Conf. on Production, Selection and Standardization of Entomopathogenic Fungi, Ignoffo, C. M., Ed., 1978, 20.
39. **Matanomi, B. A. and Libby, J. L.**, The life stages of *Entomophthora virulenta* and *Conidiobolus coronatus*, *J. Invertebr. Pathol.*, 26, 125, 1975.
40. **Cole, G. T. and Samson, R. A.**, *Patterns of Development in Conidial Fungi*, Pitman Publ., London, 1979, 190.
41. **Hammill, T. M.**, Electron microscopy of phialoconidiogenesis in *Metarhizium anisopliae*, *Am. J. Bot.*, 59, 317, 1972.
42. **Hammill, T. M.**, Additional electron microscopy of phialoconidiogenesis in *Metarhizium anisopliae*: microtubules in phialidic necks, *Mycologia*, 69, 1058, 1977.
43. **Kulik, M. M. and Vincent, P. G.**, Pyrolysis — gas liquid chromatography of fungi: observations on variability among nine *Penicillium* species of the section Asymmetrica, subsection Fasciculata, *Mycopathol. Mycol. Appl.*, 51, 1, 1973.
44. **Tatarenko, E. S.**, Use of biochemical characteristics in taxonomy of molds, *Mikol. Fitopatol.*, 7, 447, 1973.
45. **Messias, C. L., Roberts, D. W., and Grefig, A. T.**, Pyrolysis gas chromatography of the fungus *Metarhizium anisopliae*: an aid to strain identification, *J. Invertebr. Pathol.*, 42, 393, 1983.
46. **Anon.**, Microbial insect control agents, in *Microbial Processes: Promising Technologies for Developing Countries*, National Academy of Science, Washington, D.C., 1979, chap. 5.

47. **Federici, B. A.**, Mosquito control by the fungi *Culicinomyces, Lagenidium* and *Coelomomyces*, in *Microbial Control of Pests and Plant Diseases 1970—1980*, Burges, H. D., Ed., Academic Press, New York, 1981, 555.
48. **Roberts, D. W.**, *Fungal Infections of Mosquitoes in Mosquito Control*, Bourassa, J. P., Ed., University of Quebec Press, Montreal, Canada, 1974, 143.
49. **Whisler, H. C., Zehold, S. L., and Shemanchuk, J. A.**, Alternate host for mosquito parasite *Coelomomyces, Nature (London)*, 251, 715, 1975.
50. **Federici, B. A.**, Experimental hybridization of *Coelomomyces dodgei* and *Coelomomyces punctatus, Proc. Natl. Acad. Sci. U.S.A.*, 76, 4425, 1979.
51. **Federici, B. A.**, Differential pigmentation in the sexual phase of *Coelomomyces, Nature (London)*, 267, 514, 1977.
52. **Federici, B. A. and Thompson, S. N.**, β-Carotene in the gametophytic phase of *Coelomomyces dodgei, Exp. Mycol.*, 3, 281, 1979.
53. **Willoughby, L. G.**, Pure culture studies on the aquatic phycomycete, *Lagenidium giganteum, Trans. Br. Mycol. Soc.*, 52, 393, 1969.
54. **Umphlett, C. J. and Huang, C. S.**, Experimental infection of mosquito larvae by a species of the aquatic fungus *Lagenidium, J. Invertebr. Pathol.*, 20, 326, 1972.
55. **Weiser, J. and Zizka, Z.**, The ultrastructure of the chytrid *Coelomycidium simulii* Deb. I. Ultrastructure of the thalli, *Ceska Mykol.*, 28, 159, 1974.
56. **Federici, B. A., Lacey, L. A., and Mulla, M. S.**, *Coelomycidium simulii*: a fungal pathogen in larvae of *Simulium vittatum* from the Colorado River, Proc. 45th Conf. Calif. Mosq. Vect. Cont. Assoc., 13, 1977.
57. **Martin, W. W.**, A new species of *Catenaria* parasitic in midge eggs, *Mycologia*, 67, 264, 1975.
58. **Martin, W. W.**, *Aphanomycopsis sexualis*, a new parasite of midge eggs, *Mycologia*, 67, 923, 1975.
59. **Martin, W. W.**, The development and possible relationships of a new *Atkinsiella* parasitic in insect eggs, *Am. J. Bot.*, 64, 760, 1977.
60. **Martin, W. W.**, Two additional species of *Catenaria* (Chytridiomycetes, Blastocladiales) parasitic in midge eggs, *Mycologia*, 70, 461, 1978.
61. **Martin, W. W.**, *Couchia circumplexa*, a water mold parasitic on midge eggs, *Mycologia*, 73, 1143, 1981.
62. **Karling, J. S.**, Chytridiosis of scale insects, *Am. J. Bot.*, 35, 246, 1948.
63. **Evans, H. C. and Samson, R. A.**, *Sporodiniella umbellata*, an entomogenous fungus of the mucorales from cocoa farms in Ecuador, *Can. J. Bot.*, 55, 2981, 1977.
64. **MacLeod, D. M. and Muller-Kogler, E.**, Entomogenous fungi: *Entomophthora* species with pear-shaped to almost spherical conidia (Entomophthorales: Entomophthoraceae), *Mycologia*, 65, 823, 1973.
65. **MacLeod, D. M.**, Entomophthorales infections, in *Insect Pathology*, Vol. 2, Steinhaus, E. A., Ed., Academic Press, New York, 1963, 180.
66. **MacLeod, D. M.**, Insect pathogens: species originally described from their resting spores mostly as *Tarichium* species (Entomophthorales: Entomophthoraceae), *Mycologia*, 62, 33, 1970.
67. **MacLeod, D. M., Muller-Kogler, E., and Wilding, N.**, *Entomophthora* species with *Entomophthora muscae*-like conidia, *Mycologia*, 68, 31, 1976.
68. **Gustafsson, M.**, On species of the genus *Entomophthora* Fres. in Sweden, *Lantbrukshoegsk. Ann.*, 31, 103, 1965.
69. **Zimmerman, G.**, Zur biologie, untersuchungs-methodik und bestimmung von Entomophthoraceen (Phycomycetes:Entomophthorales) an Blattlausen, *Z. Angew. Entomol.*, 85, 241, 1978.
70. **Batko, A.**, On the new genera: *Zoophthora* gen. nov. *Triplosporium* (Thaxter) gen. nov. and *Entomophaga* gen. nov. (Phycomycetes: Entomophthoraceae), *Bull. Acad. Pol. Sci. Biol.*, 12, 323, 1964.
71. **Batko, A.**, Remarks on the genus *Lamia nowakowski* 1883 vs. *Culicicola* Nieuwland 1916 (Phycomycetes: Entomophthoraceae), *Bull. Acad. Pol. Sci. Biol.*, 12, 399, 1964.
72. **Waterhouse, G. N. and Brady, B. C.**, Key to the species of *Entomophthora* sensu lato, *Bull. Br. Mycol. Soc.*, 16, 113, 1982.
73. **Remaudiere, G. and Keller, S.**, Revision systematique des genres d *Entomophthoraceae* a potentialite entomopathogene, *Mycotaxon*, 11, 323, 1980.
74. **Tyrrell, D. and Weatherston, J.**, The fatty acid composition of some Entomophthoraceae. IV. The occurrence of branched-chain fatty acids in *Conidiobolus* species, *Can. J. Microbiol.*, 22, 1058, 1976.
75. **King, D. S.**, Strain variation in *Conidiobolus*, in *Proc. 1st Int. Colloq. Invertebr. Pathol.*, Queen's University Press, Kingston, Canada, 1976, 277.
76. **King, D. S. and Humber, R. A.**, Identification of the Entomophthorales, in *Microbial Control of Pests and Plant Diseases 1970—1980*, Burges, H. D., Ed., Academic Press, New York, 1981, 107.
77. **King, D. S.**, Systematics of fungi causing entomophthoramycosis, *Mycologia*, 71, 731, 1979.
78. **Tyrrell, D. and MacLeod, D. M.**, A taxonomic proposal regarding *Delacroixia coronata*, *J. Invertebr. Pathol.*, 20, 11, 1972.

79. **Remaudiere, G. and Hennebert, G. L.,** Revision systematique de *Entomophthora aphidis* Hoffm. in Fres. Description de deux nouveaux pathogenes d' aphides, *Mycotaxon,* 11, 269, 1980.
80. **Latge, J. P., King, D. S., and Papierok, B.,** Synonymie de *Entomophthora virulenta* Hall et Dunn et de *Conidiobolus thromboides* Prechsler., *Mycotaxon,* 11, 255, 1980.
81. **Humber, R. A.,** An alternative view of certain taxonomic criteria used in the Entomophthorales (Zygomycetes), *Mycotaxon,* 13, 191, 1981.
82. **Humber, R. A.,** *Erynia* (Zygomycetes: Entomophthorales): validations and new species, *Mycotaxon,* 13, 471, 1981.
83. **Ben Ze'ev, I. and Kenneth, R. G.,** Features criteria of taxonomic value in the Entomophthorales. I. A revision of the Batkoan classification, *Mycotaxon,* 14, 393, 1982.
84. **Ben Ze'ev, I. and Kenneth, R. G.,** Features criteria of taxonomic value in the Entomophthorales. II. A revision of the genus *Erynia* Nowakowski 1881 (= *Zoophthora* Batko 1964), *Mycotaxon,* 14, 456, 1982.
85. **King, D. S.,** Systematic of *Conidiobolus* using numerical taxonomy. II. Taxonomic considerations, *Can. J. Bot.,* 54, 1285, 1976.
86. **Rippon, J. W.,** *Medical Mycology,* W. B. Sanders, Philadelphia, 1974, 587.
87. **Emmons, C. W., Binford, C. H., Utz, J. P., and Kwong-Chung, K. J.,** *Medical Mycology,* Lea & Febiger, Philadelphia, 1977, 592.
88. **Emmons, C. W. and Bridges, C. H.,** *Entomophthora coronata,* etiologic agent of a phycomycosis of horses, *Mycologia,* 53, 307, 1961.
89. **Humber, R. A.,** The systematics of the genus *Strongwellsea* (Zygomycetes: Entomophthorales), *Mycologia,* 68, 1042, 1976.
90. **Soper, R. S.,** The genus *Massospora* entomopathogenic for cicadas. I. Taxonomy of the genus, *Mycotaxon,* 1, 13, 1974.
91. **Soper, R. S., Delyzer, A. J., and Smith, L. F. R.,** The genus *Massospora* entomopathogenic for cicadas. II. Biology of *Massospora levispora* and its host *Okanagana rimosa,* with notes on periodical cicadas, *Ann. Entomol. Soc. Am.,* 69, 89, 1976.
92. **Soper, R. S., Smith, L. F. R., and Delyzer, A. J.,** Epizootiology of *Massospora levispora* in an isolated population of *Okanagana rimosa, Ann. Entomol. Soc. Am.,* 69, 275, 1976.
93. **Couch, J. N., Andreeva, R. V., Laird, M., and Nolan, R. A.,** *Tabanomyces milkoi* (Dudka and Koval) emended, genus novum, a fungal pathogen of horseflies, *Proc. Natl. Acad. Sci. U.S.A.,* 76, 2299, 1979.
94. **Remaudiere, G., Keller, S., Papierok, B., and Latge, J. P.,** Considerations systematiques et biologiques sur quelques especes D'*Entomophthora* du groupe *Sphaerosperma* pathogene d'insectes (Phycomycetes: Entomophthoraceae), *Entomophaga,* 21, 163, 1976.
95. **Latge, J. P. and de Bievre, C.,** Lipid composition of *Entomophthora obscura* Hall and Dunn, *J. Gen. Microbiol.,* 121, 151, 1980.
96. **May, B., Roberts, D. W., and Soper, R. S.,** Intraspecific genetic variability in laboratory strains of *Entomophthora* as determined by enzyme electrophoresis, *Exp. Mycol.,* 3, 289, 1979.
97. **Lichtwardt, R. W.,** Trichomycetes, in *The Fungi: An Advanced Treatise,* Vol. 4b, Ainsworth, G. C., Sparrow, F. K., and Sussman, A. S., Eds., Academic Press, New York, 1973, 237.
98. **Moss, S. T.,** Commensalism of the Trichomycetes, in *Insect-Fungus Symbiosis, Nutrition, Mutualism and Commensalism,* Batra, C. R., Ed., Allenheld, Osmun & Co., New York, 1979, 201.
99. **Benjamin, R. K.,** Laboulbeniomycetes, in *The Fungi: An Advanced Treatise,* Vol. 4a, Ainsworth, G. C., Sparrow, F. K., and Sussman, A. S., Eds., Academic Press, New York, 1973, 223.
100. **Spiltoir, C. F. and Olive, L. S.,** A reclassification of the genus *Pericystis* Betts, *Mycologia,* 47, 238, 1955.
101. **Bailey, L.,** The effect of temperature on the pathogenicity of the fungus *Ascosphaera apis* for larvae of the honeybee *Apis mellifera,* Proc. Int. Colloq. Insect Pathol. Microbial Control, Wageningen, The Netherlands, 1966, 162.
102. **Skou, J. P.,** Ascosphaerales, *Friesia,* 10, 1, 1972.
103. **Skou, J. P.,** Two new species of *Ascosphaera* and notes on the conidial state of *Bettsia alvei, Friesia,* 11, 62, 1975.
104. **Brady, B. L. K.,** *Ascosphaera apis,* in *CMI Descriptions of Pathogenic Fungi and Bacteria,* No. 601, Ferry Lane, Kew, Surrey, England, 1979, 1.
105. **Gilliam, M., Taber, S., and Rose, J. B.,** Chalkbrood disease of honey bees, *Apis mellifera* L.: a progress report, *Apidologie,* 9, 75, 1978.
106. **Skou, J. P.,** *Ascosphaera asterophora* species hora, *Mycotaxon,* 14, 149, 1982.
107. **Skou, J. P. and Hacket, K.,** A new homoethallic species of *Ascosphaera, Friesia,* 11, 265, 1979.
108. **Skou, J. P.,** *Microascus exsertus,* sp. nov., associated with a leaf-cutting bee, with considerations on relationships of species in the genus *Nicroascus* Zukal, *Antonie van Leeuwenhoek,* 39, 529, 1973.
109. **Mains, E. B.,** North American entomogenous species of *Cordyceps, Mycologia,* 50, 169, 1958.
110. **Kobayasi, Y. and Shimizu, D.,** *Cordyceps* species from Japan, *Bull. Natl. Sci. Mus. Ser. V. (Bot.),* 4(2), 43, 1978.

111. **Moureau, J.**, *Cordyceps* in Congo Belge, *Mem. Inst. R. Colonial Belge*, 7, 1, 1949.
112. **Chen, Z.**, Notes on new Formosan forest fungi. VI. Genus *Cordyceps* and their distribution in Taiwan, *Taiwania*, 23, 153, 1978.
113. **Samson, R. A., Evans, H. C., and Hoekstra, E. S.**, Notes on entomogenous fungi from Ghana. VI. The genus *Cordyceps, Proc. K. Ned. Akad. Wet.*, 85, 589, 1982.
114. **Petch, T.**, Studies in entomogenous fungi. II. The genera *Hypocrella* and *Aschersonia, Ann. R. Bot. Gard. Perade N.Y.*, 7, 167, 1921.
115. **Mains, E. B.**, Species of *Hypocrella, Mycopathol. Mycol. Appl.*, 11, 311, 1959.
116. **Rossman, A. Y.**, The phragmosporous species of *Nectria* and related genera, *Mycol. Pap.*, 164, 150, 1983.
117. **Booth, C.**, *The Genus Fusarium*, Commonwealth Mycological Inst., Kew, Surrey, England, 97, 1971.
118. **Stifler, C. B.**, A new genus of Hypocreales, *Mycologia*, 33, 82, 1941.
119. **Blackwell, M. and Gilbertson, R. L.**, *Cordycepioideus octosporus*, a termite suspected pathogen from Jalisco, Mexico, *Mycologia*, 73, 358, 1981.
120. **Rossman, A. Y.**, *Podonectria*, a genus in the Pleosporales on scale insects, *Mycotaxon*, 7, 163, 1978.
121. **Couch, J. N.**, *The Genus Septobasidium*, University of North Carolina Press, Chapel Hill, 1938.
122. **Coles, R. B. and Talbot, R. H. B.**, *Septobasidium clelandii* and its conidial state, *Harpographium cornelioides*, parasitizing female coccids, *Kew Bull.*, 31, 481, 1977.
123. **Evans, H. C. and Samson, R. A.**, Entomogenous fungi from the Galapagos Islands, *Can. J. Bot.*, 60, 2325, 1982.
124. **Dykstra, M. J.**, Some ultrastructural features in the genus *Septobasidium, Can. J. Bot.*, 52, 971, 1974.
125. **Malloch, D., Kane, S., and Lahaie, P. G.**, *Filobasidiella arachnophila* sp. nov., *Can. J. Bot.*, 56, 823, 1978.
126. **Samson, R. A.**, Identification: entomopathogenic deuteromycetes, in *Microbial Control of Insects and Mites*, Vol. 2, Burges, H. D., Ed., Academic Press, New York, 1981, 93.
127. **Sands, W. A.**, The association of termites and fungi, in *Biology of Termites*, Vol. I, Krishna, K. and Weesner, F. M., Eds., Academic Press, New York, 1969, 495.
128. **Blackwell, M. and Kimbrough, J. W.**, *Hormiscioideus filamentous* gen. et spec. nor., a termite-infesting fungus from Brazil, *Mycologia*, 70, 1274, 1978.
129. **Mains, E. B.**, Entomogenous species of *Akanthomyces, Hymenostilbe* and *Insecticola* in North America, *Mycologia*, 42, 566, 1950.
130. **Samson, R. A. and Evans, H. C.**, Notes on entomogenous fungi from Ghana. II. The genus *Akanthomyces, Acta Bot. Neerl.*, 23, 28, 1974.
131. **Samson, R. A.**, *Paecilomyces* and some allied Hyphomycetes, *Stud. Mycology*, 6, 1, 1974.
132. **Samson, R. A. and Evans, H. C.**, Notes on entomogenous fungi from Ghana. IV. The genera *Paecilomyces* and *Nomuraea, Proc. K. Ned. Akad. Wet.*, 80, 128, 1977.
133. **Bissett, J.**, *Paecilomyces breviramosus* sp. nov., in *Fungi Canadenses*, No. 159, National Mycological Herbarium, Biosystematics Research Institute, Research Branch, Agriculture, Ottawa, Ontario, Canada, 1981, 177.
134. **Liang, Z.**, Two new species of *Paecilomyces* from insects, *Acta Microbiol. Sinica*, 21, 31, 1981.
135. **Zimmerman, G.**, *Paecilomyces tenuipes* (Peck) Samson, ein seltener insehten pathogenes Pilz an Noctuiden, *Anz. Schaedlingsk., Pflanz. Umweltschutz.*, 53, 69, 1980.
136. **Carilli, A. and Pacioni, G.**, Occurrence of an entomogenous fungus on Odonata, *J. Invertebr. Pathol.*, 26, 259, 1975.
137. **Aoki, J.**, Mixed infection of the gypsy moth, *Lymantria dispar japonica* Mothschulsky (Lepidoptera: Lymantriidae), in a larch forest by *Entomophthora aulicae* (Reich.), Sorok. and *Paecilomyces canadensis* (Vuill.), Brown et Smith, *Appl. Entomol. Zool.*, 9, 185, 1974.
138. **Samson, R. A. and Evans, H. C.**, Notes on entomogenous fungi from Ghana. I. The genera *Gibellula* and *Pseudogibellula, Acta Bot. Neerl.*, 22, 522, 1973.
139. **Samson, R. A. and Evans, H. C.**, New species of *Gibellula* on spiders (Araneida) from South America, 1986.
140. **de Hoog, G. S.**, Notes on some fungicolous hyphomycetes and their relatives, *Persoonia*, 10, 33, 1978.
141. **Zaim, M., Baker, K. K., and Newson, H. D.**, Natural infection of the tree hole breeding mosquito, *Aedes triseriatus* (Diptera: Culicidae) with the fungus *Funicularius triseriatus, J. Invertebr. Pathol.*, 34, 199, 1979.
142. **Mains, E. B.**, Entomogenous species of *Hirsutella, Tilachlidium* and *Synnematium, Mycologia*, 43, 691, 1951.
143. **Evans, H. C. and Samson, R. A.**, *Cordyceps* species and their anamorphs pathogenic on ant (Formicidae) in tropical ecosystems. I. The Cephalotus (Myrmicinae) complex, *Trans. Br. Mycol. Soc.*, 79, 431, 1982.
144. **Pacioni, G.**, Some entomogenous fungi originally referred to *Isaria, Trans. Br. Mycol. Soc.*, 74, 239, 1980.

145. **Minter, D. W. and Brady, B. L. K.**, Mononematous species of *Hirsutella*, *Trans. Br. Mycol. Soc.*, 74, 271, 1980.
146. **Sturhan, D. and Schneider, R.**, *Hirsutella heteroderae*, ein neuer nematoden-parasitarer Pilz., *Phytopathol. Z. Schrift.*, 99, 105, 1980.
147. **Jaffee, B. A. and Zehr, E. I.**, Parasitism of the nematode *Criconemella xenoplax* by the fungus *Hirsutella rhossiliensis*, *Phytopathology*, 72, 1378, 1982.
148. **McCoy, C. W. and Kanavel, R. F.**, Isolation of *Hirsutella thompsonii* from the citrus rust mite, *Phyllocoptruta oleivora*, and its cultivation on various synthetic media, *J. Invertebr. Pathol.*, 14, 386, 1969.
149. **Samson, R. A., McCoy, C. W., and O'Donnell, K. L.**, Taxonomy of the acarine parasite *Hirsutella thompsonii*, *Mycologia*, 72, 359, 1980.
150. **McCoy, C. W., Stamper, D. H., and Tuveson, R. W.**, Conidiogenous cell differences among mutant and wild-type pathotypes of *Hirsutella thompsonii* var. *thompsonii*, *J. Invertebr. Pathol.*, 43, 414, 1984.
151. **Boucias, D. G. and McCoy, C. W.**, Isozyme differentiation among 17 geographical isolates of *Hirsutella thompsonii*, *J. Invertebr. Pathol.*, 39, 329, 1982.
152. **Prior, C. and Perry, C. H.**, Infection of *Promecotheca papuana* with *Synnematium jonesii*, *J. Invertebr. Pathol.*, 35, 14, 1980.
153. **Fabres, G.**, Intervention de *Synnematium jonesii* (Fungi-Imperfecti) et a'*Aphytis cocherauii* (Hym. Aphelinidae) lors d'une pullulation de *Lepidosaphes beckii* (Hom. Diaspididae) dans les habitats ombrage de la Nouvelle-Caledonie, *Ann. Zool. Ecol. Anim.*, 9, 601, 1977.
154. **Samson, R. A. and Evans, H. C.**, Notes on entomogenous fungi from Ghana. III. The genus *Hymenostilbe*, *Proc. K. Ned. Akad. Wet.*, 78, 73, 1975.
155. **Gams, W.**, *Cephalosporium — artige Schimmelpilze (Hyphomycetes). -G.*, Gustav Fisher, Stuttgart, 31, 1971.
156. **Samson, R. A. and Brady, B. L.**, *Paraisaria*, a new genus for *Isaria dubia*, the anamorph of *Cordyceps gracilis*, *Trans. Br. Mycol. Soc.*, 81, 285, 1983.
157. **Speare, A. T.**, Further studies of *Sorosporella uvella*, a fungus parasite of noctuid larvae, *J. Agric. Res.*, 18, 399, 1920.
158. **MacLeod, D. M.**, Investigations on the genera *Beauveria* Vuill. and *Tritirachium* Limber., *Can. J. Bot.*, 32, 818, 1954.
159. **de Hoog, G. S.**, The genera *Beauveria*, *Isaria*, *Tritirachium* and *Acrodontium* gen. nov., *Stud. Mycol.*, p. 1, 1972.
160. **Fargues, J., Duriez, T., and Popeye, R.**, Analyse serologique de deuz chapignons entomopathogenes, *Beauveria bassiana* (Bals.) Vuill. et *Beauveria tenella* (Delacr.) Siem., *C.R. Acad. Sci.*, 278, 2245, 1974.
161. **Alyeshina, O. A., Boyarskij, B. G., and Il'icheva, S. N.**, Production of a specific serum of the entomopathogenic fungus *Beauveria bassiana*, *Dokl. TSKhA*, 209, 121, 1975.
162. **Evans, H. C. and Samson, R. A.**, Two new *Beauveria* spp. from South America, *J. Invertebr. Pathol.*, 39, 93, 1982.
163. **Kish, L. P., Samson, R. A., and Allen, G. E.**, The genus *Nomuraea* Maublanc, *J. Invertebr. Pathol.*, 24, 154, 1974.
164. **Ignoffo, C. M.**, The fungus *Nomuraea rileyi*, as a microbial insecticide, in *Microbial Control of Pests and Plant Diseases 1970—1980*, Burges, H. D., Ed., Academic Press, New York, 1981, 514.
165. **Joslyn, D. J. and Boucias, D. G.**, Isozyme differentiation among three pathotypes of the entomogenous fungus *Nomuraea rileyi*, *Can. J. Microbiol.*, 27, 364, 1981.
166. **Samson, R. A. and Evans, H. C.**, personal communication, 1985.
167. **Tulloch, M.**, The genus *Metarhizium*, *Trans. Br. Mycol. Soc.*, 66, 407, 1976.
168. **Fargues, J., Duriez, T., Andrieu, S., Popeye, R., and Robert, P. H.**, Etude immunologique comparee de soyches de *Metarhizium anisopliae* (Delacr.) Siem., champignon hyphomycete entomopathogene, *C. R. Acad. Sci.*, 281, 1781, 1975.
169. **Balazy, S.**, A review of entomopathogenic species of the genus *Cephalosporium* corda, *Bull. Soc. Amis. Sci. Poznan, Ser. D*, 14, 101, 1973.
170. **Hall, R. A.**, The fungus *Verticillium lecanii* as a microbial insecticide of aphids and scales, in *Microbial Control of Pests and Plant Diseases 1970—1980*, Burges, H. D., Ed., Academic Press, New York, 1981, 483.
171. **Lightner, D. V. and Fontaine, C. T.**, A mycosis of the American lobster, *Homarus americanus*, caused by *Fusarium* sp., *J. Invertebr. Pathol.*, 25, 239, 1975.
172. **Barson, G.**, *Fusarium solani*, a weak pathogen of the larval stages of the elm bark beetle *Scolytus scolytus*, *J. Invertebr. Pathol.*, 27, 307, 1976.
173. **Couch, J. N., Romney, S. V., and Rao, B.**, A new fungus, which attacks mosquitoes and related Diptera, *Mycologia*, 66, 374, 1974.
174. **Sweeney, A. W.**, The insect pathogenic fungus *Culicinomyces* in mosquitoes and other hosts, *Aust. J. Zool.*, 23, 59, 1975.

175. **Sweeney, A. W., Couch, J. N., and Panter, C.,** The identity of an Australian isolate of *Culicinomyces, Mycologia,* 74, 162, 1982.
176. **Sweeney, A. W.,** Infection of aseptically reared mosquito larvae with *Culicinomyces* sp., *J. Invertebr. Pathol.,* 30, 273, 1977.
177. **Gams, W.,** *Tolypocladium,* eive neve Hyphomycetengattung mit geschwollene Phialiden., *Persoonia,* 6, 185, 1971.
178. **Federici, B. A., Lasko, J. F., Soares, G., and Tsao, P. W.,** Fungi show promise in biological control, *Calif. Agric.,* 34, 25, 1980.
179. **Soares, G. G.,** Pathogenesis of infection by the hyphomycetous fungus, *Tolypocladium cylindrosporum* Gams in *Aedes sierrensis* and *Culex tarsalis* (Diptera: Culicidae), *Entomophaga,* 27, 283, 1982.
180. **Samson, R. A. and Soares, G. G.,** Entomopathogenic species of the hyphomycete genus *Tolypocladium, J. Invertebr. Pathol.,* 43, 133, 1984.
181. **Mains, E. B.,** Notes concerning entomogenous fungi, *Bull. Torrey Bot. Club,* 78, 122, 1951.
182. **Samson, R. A., Evans, H. C., and van de Klashorst, G.,** Notes on entomogenous fungi from Ghana. V. The genera *Stilbella* and *Polycephalomyces, Proc. K. Ned. Akad. Wet. Ser. C,* 84, 289, 1981.
183. **Thaxter, R.,** On certain new or peculiar North American Hyphomycetes. II. *Helicocephalum, Gonatorrhodiella, Desmidiospora* nov. gen. and *Everhartia lignatillis* n. sp., *Bot. Gaz.,* 16, 201, 1891.
184. **Evans, H. C. and Samson, R. A.,** *Cordyceps* species and their anamorphs pathogenic on ants (Formicidae) in tropical forest ecosystems. II. The *Camponotus* (Formicinae) complex, *Trans. Br. Mycol. Soc.,* 82(1), 127, 1984.
185. **de Hoog, G. S.,** The genera *Blastobotrys, Sporothrix, Calcarisporium* and *Calcarisporiella* gen. nov., *Stud. Mycol.,* 7, 34, 1974.
186. **Samson, R. A., Gams, W., and Evans, H. C.,** *Pleurodesmospora,* a new genus for the entomogenous hyphomycete *Gonatorrhodiella coccorum, Persoonia,* 11, 65, 1980.
187. **Samson, R. A. and Evans, H. C.,** *Clathroconium,* a new helicosporous hyphomycete from spiders, *Can. J. Bot.,* 60, 1577, 1982.
188. **Mains, E. B.,** Species of *Aschersonia* (Sphaeropsidales), *Lloydia,* 22, 215, 1959.
189. **Mains, E. B.,** North American species of *Aschersonia* parasitic on Aleyrodidae, *J. Insect Pathol.,* 1, 43, 1959.
190. **Protsenko, E. P.,** The importance of the fungus *Aschersonia* in nature and its practical use by man in the biological control of insects, *Sb. Karantinu Rast.,* 19, 147, 1967.
191. **Samson, R. A. and McCoy, C. W.,** *Aschersonia aleyrodis,* a fungal pathogen of whitefly. I. Scanning electron microscopy of the development on the citrus whitefly, *Z. Angew. Entomol.,* 96, 380, 1983.
192. **Fawcett, H. S.,** An important entomogenous fungus, *Mycologia,* 2, 164, 1910.
193. **Petch, T.,** Studies in entomogenous fungi. IX. *Aegerita, Trans. Br. Mycol. Soc.,* 11, 50, 1926.
194. **Madelin, M. F.,** Fungal parasites of invertebrates, in *The Fungi: An Advanced Treatise,* Vol. 3, Ainsworth, G. C. and Sussman, A. S., Eds., Academic Press, New York, 1968, 227.
195. **Khan, S. R. and Aldrich, H. C.,** The haustoria of *Termitaria snyderi, J. Invertebr. Pathol.,* 25, 247, 1975.
196. **Broome, J. R., Sikorowski, P. P., and Norment, B. R.,** A mechanism of pathogenicity of *Beauveria bassiana* on larvae of the imported fire ant, *Solenopsis richteri, J. Invertebr. Pathol.,* 28, 87, 1976.
197. **Wasti, S. S. and Hartmann, G. C.,** Experimental parasitization of larvae of the gypsy moth, *Porthetria dispar* (L.) with entomogenous fungus *Beauveria bassiana* (Bals.) Vuill., *Parasitology,* 70, 341, 1975.
198. **Schabel, H. G.,** Oral infection of *Hylobius pales* by *Metarhizium anisopliae, J. Invertebr. Pathol.,* 27, 377, 1976.
199. **Hedlund, R. C. and Pass, B. C.,** Infections of the alfalfa weevil, *Hypera postica,* by the fungus, *Beauveria bassiana, J. Invertebr. Pathol.,* 11, 25, 1968.
200. **Clark, T. B., Keller, W. R., Fukuda, W. R., and Lindgren, J. E.,** Field and laboratory studies of the pathogenicity of the fungus *Beauveria bassiana* to three genera of mosquitoes, *J. Invertebr. Pathol.,* 11, 1, 1968.
201. **Ferron, P.,** Etude en laboratoire des conditions ecologiques favorisant le development de la mycose a *Beauveria tenella* du ver Blanc, *Entomophaga,* 12, 257, 1967.
202. **Weiser, J. and Muma, M. H.,** *Entomophthora floridana* n. sp. a parasite of the Texas citrus mite, *Eutetranychus banksi, Fla. Entomol.,* 49, 155, 1966.
203. **Selhime, A. G. and Muma, M. H.,** Biology of *Entomophthora floridana* attacking *Eutetranychus banksi, Fla. Entomol.,* 49, 161, 1966.
204. **Carner, G. R. and Canerday, T. D.,** Field and laboratory investigations with *Entomophthora fresenii,* a pathogen of *Tetranychus* spp., *J. Econ. Entomol.,* 61, 956, 1968.
205. **Anderson, S. O.,** Biochemistry of insect cuticle, *Annu. Rev. Entomol.,* 24, 29, 1979.
206. **Fargues, J.,** Adhesion of fungal spore to insect cuticle in relation to pathogenicity, in *Infection Processes of Fungi,* Aist, J. and Roberts, D. W., Eds., Rockefeller Foundation Study Center, Bellagio, Italy, 1984, 90.

207. **Grula, E. A., Woods, S. P., and Russell, H.,** Studies utilizing *Beauveria bassiana* as an entomopathogen, in *Infection Processes of Fungi,* Aist, J. and Roberts, D. W., Eds., Rockefeller Foundation Study Center, Bellagio, Italy, 1984, 147.
208. **Sweeney, A. W., Wright, R. G., and van der Lubbe, L.,** Ultrastructural observations on the invasion of mosquito larvae by the fungus *Culicinomyces, Micron,* 11, 487, 1980.
209. **Wallengren, H. and Johansson, R.,** On the infection of *Pyrasta nubilalis* by *Metarhizium anisopliae* (Metsch)., *Sor. Sci. Rep. Int. Corn Borer Invest.,* 2, 131, 1979.
210. **Michel, B.,** Recherches Experimentales sur la Penetration des Champignons Pathogens chez les Insectes, these 3 eme cycle, University Montpellier, 1981, 170.
211. **Al-Aidroos, K. and Seifert, A. M.,** Polysaccharide and protein degradation, germination, and virulence against mosquitoes in the entomopathogenic fungus *Metarhizium anisopliae, J. Invertebr. Pathol.,* 36, 29, 1980.
212. **Zacharuk, R. Y.,** Fine structure of the fungus *Metarhizium anisopliae* infecting three species of larval *Elateridae* (Coleoptera). I. Dormant and germinating conidia, *J. Invertebr. Pathol.,* 15, 63, 1970.
213. **Veen, K. H.,** Recherches sur la maladie due a *Metarhizium anisopliae* chez le cricket perlerin, *Meded. Landbouwhogesch. Wageningen,* 77, 43, 1968.
214. **Fargues, J.,** Specificite des Hyphomycetes Entomopathogenes et Resistance Interspecifique des Larvae d'Insectes, these Doct. es Sciences Naturelles, University of Pierre et Marie Curie, Paris, 1981, 252.
215. **Smith, R. J. and Grula, E. A.,** Nutritional requirements for conidial germination and hyphal growth of *Beauveria bassiana, J. Invertebr. Pathol.,* 37, 222, 1981.
216. **Hunt, D. W. A., Borden, J. H., Rahe, J. E., and Whitney, H. S.,** Nutrient-mediated germination of *Beauveria bassiana* conidia on the integument of the bark beetle *Dendroctonus ponderosae, J. Invertebr. Pathol.,* 44, 304, 1984.
217. **Woods, S. P. and Grula, E. A.,** Utilizable surface nutrients on *Heliothis zea* available for growth of *Beauveria bassiana, J. Invertebr. Pathol.,* 43, 259, 1984.
218. **Boucias, D. G. and Pendland, J. C.,** Nutritional requirements for conidial germination of several host range pathotypes of the entomopathogenic fungus, *Nomuraea rileyi, J. Invertebr. Pathol.,* 43, 288, 1984.
219. **Mohamed, A. K. A., Sikorowski, P. P., and Bell, J. V.,** Histopathology of *Nomuraea rileyi* in the larvae of *Heliothis zea* and *in vitro* enzymatic activity, *J. Invertebr. Pathol.,* 31, 345, 1978.
220. **Pekrul, S. and Grula, E. A.,** Mode of infection of the corn earworm, *Heliothis zea* by *Beauveria bassiana* as revealed by scanning electron microscopy, *J. Invertebr. Pathol.,* 34, 238, 1979.
221. **Boucias, D. G. and Pendland, J. C.,** Ultrastructural studies on the fungus, *Nomuraea rileyi,* infecting the velvetbean caterpillar, *Anticarsia gemmatalis, J. Invertebr. Pathol.,* 39, 338, 1982.
222. **Schabel, H. G.,** Percutaneous infection of *Hylobius pales* by *Metarhizium anisopliae, J. Invertebr. Pathol.,* 31, 180, 1978.
223. **Vey, A. and Fargues, J.,** Histological and ultrastructural studies of *Beauveria bassiana* infection in *Leptinotarsa decemlineata* during ecdysis, *J. Invertebr. Pathol.,* 30, 207, 1977.
224. **Pendland, J. C. and Boucias, D. G.,** Ultrastructural aspects of germination in the entomogenous hyphomycete, *Nomuraea rileyi, J. Invertebr. Pathol.,* 43, 432, 1984.
225. **Pendland, J. C. and Boucias, D. G.,** Use of labeled lectins to investigate cell wall surfaces of the entomogenous hyphomycete *Nomuraea rileyi, Mycopathologia,* 87, 141, 1984.
226. **David, W. A. L.,** The physiology of the insect integument in relation to the invasion of pathogens, in *Insects and Physiology,* Beament, J. W. L. and Treherne, J. E., Eds., Oliver and Boyd, Edinburg, 1967, 17.
227. **St. Leger, R. J., Cooper, R. M., and Charnley, A. K.,** Cuticle degrading enzymes of entomopathogenic fungi: regulation of production of chitinolytic enzymes, *J. Gen. Microbiol.,* 132, 1509, 1986.
228. **McCauley, V. J. E., Zacharuk, R. Y., and Tinline, R. D.,** Histopathology of green muscardine in larvae of four species of Elateridae (Coleoptera), *J. Invertebr. Pathol.,* 12, 444, 1968.
229. **Grula, E. A., Burton, R. L., Smith, R., Mapes, T. L., Cheung, P. Y. K., Pekrul, S., Champlin, F. R., Grula, M., and Abegaz, B.,** Biochemical basis for the entomopathogenicity of *Beauveria bassiana,* in Proc. First Joint US/USSR Conf. on Production, Selection, and Standardization of Entomopathogenic Fungi of the US/USSR Joint Working Group on the Production of Substances by Microbiological Means, Ignoffo, C. M., Ed., 1978, 192.
230. **Hanel, H.,** The life cycle of the insect pathogenic fungus *Metarhizium anisopliae* in the termite *Nasutitermes exitiosus, Mycopathologia,* 80, 137, 1982.
231. **Samsinakova, A., Misikova, S., and Leopold, J.,** Action of enzymatic systems of *Beauveria bassiana* on the cuticle of the greater wax moth larvae *(Galleria mellonella), J. Invertebr. Pathol.,* 18, 322, 1971.
232. **Smith, R. J. S., Pekrul, S., and Grula, E. A.,** Requirement for sequential enzymatic activities for penetration of the integument of the corn earworm *(Heliothis zea), J. Invertebr. Pathol.,* 38, 335, 1981.
233. **Leopold, J. and Samsinakova, A.,** Quantitative estimation of chitinase and several other enzymes in the fungus *Beauveria bassiana, J. Invertebr. Pathol.,* 15, 34, 1970.

234. **Rosata, Y. P., Messias, C. L., and Azevedo, J. L.,** Production of extracellular enzymes by isolates of *Metarhizium anisopliae, J. Invertebr. Pathol.,* 38, 1, 1981.
235. **Mohamed, A. K. A. and Turner, A. G.,** Proteolytic activity of *Nomuraea rileyi* on casein and host cuticle, *Mycopathologia,* 82, 13, 1983.
236. **Samsinakova, A., Bajan, C., Kalalova, S., Knitowa, K., and Wajciechowska, M.,** The effect of some entomophagous fungi on the Colorado potato beetle and their enzyme activity, *Bull. L'Acad. Pol. Sci. Ser. Sci. Biol.,* 25, 8, 1977.
237. **Smith, R. J. and Grula, E. A.,** Chitinase is an inducible enzyme in *Beauveria bassiana, J. Invertebr. Pathol.,* 42, 319, 1983.
238. **St. Leger, R. J., Charnley, A. K., and Cooper, R. M.,** Cuticle degrading enzymes of entomopathogenic fungi: synthesis in culture on cuticle, *J. Invertebr. Pathol.,* 48, 85, 1986.
239. **St. Leger, R. J., Cooper, R. M., and Charnley, A. K.,** Cuticle degrading enzymes of entomopathogenic fungi: cuticle degradation in vitro by enzymes from entomopathogens, *J. Invertebr. Pathol.,* 47, 167, 1986.
240. **Champlin, F. R., Cheung, P. Y. K., Pekrul, S., Smith, R. J., Burton, R. L., and Grula, E. A.,** Virulence of *Beauveria bassiana* mutants for the pecan weevil, *J. Econ. Entomol.,* 74, 617, 1981.
241. **Zacharuk, R. Y.,** Fine structure of the fungus, *Metarhizium anisopliae,* infecting three species of Elateridae (Coleoptera). III. Penetration of host integument, *J. Invertebr. Pathol.,* 15, 372, 1970.
242. **Kawakami, K. and Mikuni, T.,** Time required for the penetration of some muscardine fungi into the silkworm larvae, *Bombyx mori* L., *Sericult. Exp. Stn. Sansi-Kenkyu (Acta Sericolog.),* 56, 35, 1965.
243. **Roberts, D. W.,** Toxins of entomopathogenic fungi, in *Microbial Control of Pests and Plant Diseases 1970—1980,* Burges, H. D., Ed., Academic Press, New York, 1981, 41.
244. **Suzuki, A., Taguchi, H., and Tamura, S.,** Isolation and structure elucidation of three new insecticidal cyclodepsipeptides, destruxins C and D and dimethyldestruxin, produced by *Metarhizium anisopliae, Agric. Biol. Chem.,* 34, 813, 1970.
245. **Suzuki, A. and Tamura, S.,** Isolation and structure of prodestruxin from *Metarhizium anisopliae, Agric. Biol. Chem.,* 36, 896, 1972.
246. **Kucera, M.,** Proteases from the fungus *Metarhizium anisopliae* toxic for *Galleria mellonella, J. Invertebr. Pathol.,* 35, 304, 1980.
247. **Roberts, D. W.,** Toxins from the entomogenous fungi, *Metarhizium anisopliae:* isolation of destruxins from submerged cultures, *J. Invertebr. Pathol.,* 14, 82, 1969.
248. **Vey, A. and Quiot, J. M.,** Effect *in vitro* de substances toxiques produites par le champignon *Metarhizium anisopliae* (Metsch.) Sorok. sur la reaction hemocytaire du coleoptere *Oryctes rhinoceros, L., C. R. Acad. Sci. Paris,* 280, 931, 1975.
249. **Kaijiang, L. and Roberts, D. W.,** The production of destruxins by the entomogenous fungus *Metarhizium anisoplaie* var. *major, J. Invertebr. Pathol.,* 47, 120, 1986.
250. **Hanel, H.,** A bioassay for measuring the virulence of the insect pathogenic fungus, *Metarhizium anisopliae* (Metsch.) Sorok. (Fungi Imperfecti) against the termite *Nasutitermes exitiosus* (Hill) (Isoptera, Termitidae), *Z. Angew. Entomol.,* 92, 9, 1981.
251. **Zacharuk, R. Y.,** Fine structure of the fungus *Metarhizium anisopliae* infecting three species of larval elateridae (Coleoptera). IV. Development within the host, *Can. J. Microbiol.,* 17, 525, 1971.
252. **Vey, A., Quiot, S., Vago, C., and Fargues, J.,** Effet immuno depresseur de toxines fongiques: inhibition de al reaction d'ecapsulation multicellulaire par les destruxines, *C. R. Acad. Sci. Paris,* 300, 647, 1985.
253. **Kucera, M. and Samsinakova, A.,** Toxins of the entomophagous fungus *Beauveria bassiana, J. Invertebr. Pathol.,* 12, 316, 1968.
254. **Grove, J. F. and Pople, M.,** The insecticidal activity of beauvericin and the enniatin complex, *Mycopathologia,* 70, 103, 1980.
255. **West, E. J. and Briggs, J. D.,** *In vitro* toxin production by the fungus *Beauveria bassiana* and bioassay in greater wax moth, *J. Econ. Entomol.,* 61, 684, 1968.
256. **Champlin, F. R. and Grula, E. A.,** Noninvolvement of beauvericin in the entomopathogenicity of *Beauveria bassiana, Appl. Environ. Microbiol.,* 37, 1122, 1979.
257. **Kanaoka, M. A., Isogai, A., Murakoshi, S., Ichineo, M., Suzuki, A., and Tamura, S.,** Bassianolide, a new insecticidal cyclodepsipeptide from *Beauveria bassiana* and *Verticillium lecanii, Agric. Biol. Chem.,* 42, 629, 1978.
258. **Kodaria, Y.,** Biochemical studies on the muscardine fungi in the silkworms, *Bombyx mori, J. Fac. Text. Sci. Technol. Shinshu Univ. Ser. E:,* 5, 1, 1961.
259. **Pendland, J. C. and Boucias, D. G.,** Ultrastructural aspects of conidiogenesis in the entomogenous hyphomycete, *Nomuraea rileyi, Can. J. Bot.,* 60, 26, 1982.
260. **Pendland, J. C.,** Resistant structures in the entomogenous hyphomycete, *Nomuraea rileyi:* an ultrastructural study, *Can. J. Bot.,* 60, 1569, 1982.
261. **Wilding, N.,** Discharge of conidia of *Entomophthora thaxteriana* Petch. from the pea aphid *Acyrthosiphon pisum* Harris, *J. Gen. Microbiol.,* 69, 417, 1971.

262. **Yendol, W. G. and Paschke, J. D.**, Pathology of an *Entomophthora* infection in the eastern subterranean termite *Reticutitermis flavipes, J. Invertebr. Pathol.*, 7, 414, 1965.
263. **Gabriel, B. P.**, Histochemical study of the insect cuticle infected by the fungus *Entomophthora coronata, J. Invertebr. Pathol.*, 11, 82, 1968.
264. **Lambiase, J. T. and Yendol, W. G.**, The fine structure of *Entomophthora apiculata* and its penetration of *Trichoplusia ni, Can. J. Microbiol.*, 23, 452, 1977.
265. **Brobyn, P. J. and Wilding, N.**, Invasive and developmental processes of *Entomophthora* species infecting aphids, *Trans. Br. Mycol. Soc.*, 69, 346, 1977.
266. **Tomiyama, H. and Aoki, J.**, Infection of *Erynia blunckii* (Lak ex Zimm) Rem et Henn. (Entomophthorales: Entomophthoracae) in the diamond-black moth, *Plutella xylostella* L. (Lepidoptera: Yponomeutidae), *Appl. Entomol. Zool.*, 17, 375, 1982.
267. **Brobyn, P. J. and Wilding, N.**, Invasive and developmental processes of *Entomophthora muscae* infecting houseflies *(Musca domestica), Trans. Br. Mycol. Soc.*, 80, 1, 1983.
268. **Brey, P. T., Latge, J. P., and Prevost, M. C.**, Integumental penetration of the pea aphid, *Acyrthosiphon pisum*, by *Conidiobolus obscurus* (Entomophthoraceae), *J. Invertebr. Pathol.*, 48, 34, 1986.
269. **Bitton, S., Kenneth, R. G., and Ben-Ze'ev, I.**, Zygospore overwintering and sporulative germination in *Triplosporium fresenii* (Entomophthoraceae) attacking *Aphis spiraecola* on citrus in Israel, *J. Invertebr. Pathol.*, 34, 295, 1979.
270. **Nemato, H., Kobayashi, M., and Takizawa, Y.**, Scanning electron microscopy of *Entomophora* (Triplosporium) *floridana* (Zygomycetes: Entomophthorales) attacking the Sugi spider mite, *Oligonychus hondoensis* (Acarina: Tetranychidae), *Appl. Entomol. Zool.*, 14, 376, 1979.
271. **Carner, G. R.**, A description of the life cycle of *Entomophthora* sp. in the two spotted spider mite, *J. Invertebr. Pathol.*, 28, 245, 1976.
272. **Brey, P. T.**, Contribution a l'etude de le pathogenicite de *Conidiobolus obscurus* vis-a-vis du puceron *Acyrthosiphon pisum*, These 3 eme Cycle, Paris, 1982, 64.
273. **Latge, J. P., Monsigny, M., Prevost, M. C., Roche, A. C., Keida, C., and Fournet, B.**, Carbohydrate binding proteins in the entomogenous fungus *Conidiobolus obscurus, Biol. Cell*, 51, 52a, 1984b.
274. **Latge, J. P., Sampedro, L., Brey, P. T., and Oiaquin, M.**, Aggressiveness of *Conidiobolus obscurus* against the pea aphid. V. Influence of cuticular extracts on the germination behavior of aggressive and non-aggressive strains, *J. Gen. Microbiol.*, in press.
275. **Kerwin, J. L.**, Chemical control of the germination of asexual spores of *Entomophthora culicis*, a fungus parasitic on diperans, *J. Gen. Microbiol.*, 28, 2179, 1982.
276. **Jonsson, A. G.**, Protease production by species of *Entomophthora, Appl. Microbiol.*, 16, 450, 1967.
277. **Harion, N., Fromentin, H., and Keil, B.**, Proteolytic enzymes of *Entomophthora coronata*, characterization of a collagenase, *Comp. Biochem. Physiol.*, 56, 259, 1977.
278. **Tyrrell, D.**, Occurrence of protoplasts in the natural cycle of *Entomophthora egressa, Exp. Mycol.*, 1, 259, 1977.
279. **Butt, T. M., Beckett, A., and Wilding, N.**, Protoplasts in the *in vivo* life cycle of *Erynia neoaphidis, J. Gen. Microbiol.*, 127, 417, 1981.
280. **Dunphy, G. B. and Nolan, R. A.**, A study of the surface proteins of *Entomophthora egressa* protoplasts and of larval spruce budworm hemocytes, *J. Invertebr. Pathol.*, 38, 352, 1981.
281. **Dunphy, G. B. and Nolan, R. A.**, Cellular immune responses of spruce budworm larvae to *Entomophthora egressa* protoplasts and other test particles, *J. Invertebr. Pathol.*, 39, 81, 1982.
282. **Dunphy, G. B. and Nolan, R. A.**, Morphogenesis of protoplasts of *Entomophthora egressa* in simplified culture media, *Can. J. Bot.*, 55, 3046, 1977.
283. **Prasertphon, S.**, Mycotoxin production by species of *Entomophthora, J. Invertebr. Pathol.*, 9, 281, 1976.
284. **Yendol, W. G., Miller, E. M., and Behnke, C. H.**, Toxic substances from Entomophthoraceous fungi, *J. Invertebr. Pathol.*, 10, 313, 1968.
285. **Claydon, N.**, Insecticidal secondary metabolites from entomogenous fungi: *Entomophthora virulenta, J. Invertebr. Pathol.*, 32, 319, 1978.
286. **Dunphy, G. B. and Nolan, R. A.**, Mycotoxin production by the protoplast stage of *Entomophthora egressa, J. Invertebr. Pathol.*, 39, 261, 1982.
287. **Latge, J. P., Prevost, M. C., Perry, D. F., and Reisinger, O.**, Etude en microscope electronique de *Conidiobolus obscurus*. I. Formation et germination des azygospores, *Can. J. Bot.*, 60, 413, 1982.
288. **Matanomi, B. A. and Libby, J. L.**, The life cycle of *Entomophthora virulenta* (Entomophthorales: Entomophthoraceae), *Mycopathologia*, 56, 125, 1975.
289. **Perry, D. F. and Latge, J. P.**, Dormancy and germination of *Conidiobolus obscurus* azygospores, *Trans. Br. Mycol. Soc.*, 78, 221, 1982.
290. **Latge, J. P.**, Etude morphologique et cytologique de la formation de "spores durables" chez une espece d'Entomophthorale, *C. R. Sci. Acad. Ser. D:*, 282, 605, 1976.
291. **Krejzova, R.**, Germination process in resting spores of some *Entomophthora* species and pathogenicity of spore material for lepidopterous larvae, *Z. Angew. Entomol.*, 85, 42, 1978.

292. **Newman, G. G. and Carner, G. R.,** Factors affecting the spore form of *Entomophthora gammae, J. Invertebr. Pathol.,* 26, 29, 1975.
293. **Wilding, N. and Lauckner, F. B.,** *Entomophthora* infecting wheat bulb fly at Rothamsted, Hertfordshire, 1967—1971, *Ann. Appl. Biol.,* 76, 161, 1974.
294. **Shimazu, M.,** Resting spore formation of *Entomophthora sphaerosperma* Fresmois *(Entomophthorales: Entomophthoraceae)* in brown plant hopper, *Nilaparvata lugens* (Stol) (Hemiptera: Delphacidae), *Appl. Entomol. Zool.,* 14(4), 383, 1978.
295. **Whistler, H. C., Zebold, S. L., and Shemanchuk, J. A.,** Life history of *Coelomomyces psorophorae, Proc. Natl. Acad. Sci. U.S.A.,* 782, 693, 1975.
296. **Travland, L. B.,** Initiation of infection of mosquito larvae *(Culiseta inornata)* by *Coelomomyces psorophorae, J. Invertebr. Pathol.,* 33, 124, 1979.
297. **Travland, L. B.,** Structures of the motile cells of *Coelomomyces psorophorae* and function of the zygote in encystment on a host, *Can. J. Bot.,* 57, 1021, 1979.
298. **Zebold, S. L., Whistler, H. C., Shemanchuk, J. A., and Travland, L. B.,** Host specificity and penetration in the mosquito pathogen *Coelomomyces psorophorae, Can. J. Bot.,* 57, 2766, 1979.
299. **Kerwin, J. L.,** Biological aspects of the interaction between *Coelomomyces psorophorae* zygotes and the larvae of *Culiseta inornata:* host-mediated factors, *J. Invertebr. Pathol.,* 41, 224, 1983.
300. **Wong, T. L. and Pellai, J. S.,** *Coelomomyces opifexa* Pellai and Smith (Coelomomycetaceae: Blastocladiales). VI. Observations on the mode of entry into *Aedes australis* larvae, *N. Z. J. Zool.,* 7, 135, 1980.
301. **Martin, W. W.,** A morphological and cytological study of development in *Coelomomyces punctatus* parasitic in *Anopheles quadrimaculatus, J. Elisha Mitchell Sci. Soc.,* 85, 59, 1969.
302. **Powell, M. J.,** Ultrastructural changes in the cell wall surface of *Coelomomyces punctatus* infecting mosquito larvae, *Can. J. Bot.,* 54, 1419, 1976.
303. **Umphlett, C. J.,** Development of the resting sporangia of two species of *Coelomomyces, Mycologia,* 56, 488, 1964.
304. **Federici, B. A.,** Inviability of interspecific hybrids in the *Coelomomyces dodgei* complex, *Mycologia,* 74, 555, 1982.
305. **Umphlett, C. J.,** A note to identify a certain isolate of *Lagenidium* which kills mosquito larvae, *Mycologia,* 65, 970, 1973.
306. **Couch, J. N. and Romney, S. V.,** Sexual reproduction in *Lagenidium giganteum, Mycologia,* 65, 250, 1973.
307. **Kerwin, J. L. and Washino, R. K.,** Ground and aerial application of the sexual and asexual stages of *Lagenidium giganteum* (Oomycetes: Lagenidiales) for mosquito control, *J. Am. Mosq. Control Assoc.,* 2, 182, 1986.
308. **Domnas, A. J., Giebel, P. E., and McInnis, T. M.,** Biochemistry of mosquito infection: preliminary change in *Culex pipiens quinquefasciatus* following infection with *Lagendium giganteum, J. Invertebr. Pathol.,* 24, 293, 1974.
309. **Giebel, P. E. and Domnas, A. J.,** Soluble trehalases from larvae of the mosquito, *Culex pipiens* and the fungal parasite *Lagenidium giganteum, Insect Biochem.,* 6, 303, 1976.
310. **Dean, D. D. and Domnas, A. J.,** The extracellular proteolytic enzymes of the mosquito-parasitizing fungus *Lagenidium giganteum, Exp. Mycol.,* 7, 31, 1983.
311. **Roberts, D. W.,** Toxins of entomopathogenic fungi, in *Microbial Control of Pests and Plant Diseases 1970—1980,* Burges, H. D., Ed., Academic Press, New York, 1981, 442.
312. **Hamill, R. L., Higgens, C. E., Boaz, H. E., and Gorman, M.,** The structure of beauvericin, a new depsipeptide antibiotic toxic to *Artemia saliwa, Tetrahedron Lett.,* 49, 4255, 1969.
313. **Roberts, D. W.,** Toxins from the entomogenous fungus *Metarhizium anisopliae, J. Invertebr. Pathol.,* 8, 212, 1966.
314. **Roberts, D. W.,** Some effects of *Metarhizium anisopliae* and its toxins on mosquito larvae, in *Insect Pathology and Microbial Control,* van der Laan, P. A., Ed., North-Holland, Amsterdam, 1967, 243.
315. **Pais, M., Das, B. C., and Ferron, P.,** Depsipeptides from *Metarhizium anisopliae, Phytochemistry,* 20, 715, 1981.
316. **Kaijiang, L. and Roberts, D. W.,** The production of destruxins by the entomogenous fungus, *Metarhizium anisopliae* var. *major, J. Invertebr. Pathol.,* 47, 120, 1986.
317. **Burnside, C. E.,** Fungus diseases of the honeybee, *U.S. Dept. Agric. Tech. Bull.,* 149, 1930, 279.
318. **Myokei, R., Sakurai, A., Chang, C. F., Kodaira, Y., Takahashi, N., and Tamura, S.,** Aspochracin, a new insecticidal metabolite of *Aspergillus ochraceus, Agric. Biol. Chem.,* 33, 1491, 1969.
319. **Tascono, N. and Reeves, E. L.,** Effect of *Aspergillus flavus* mycotoxin of *Culex* mosquito larvae, *J. Invertebr. Pathol.,* 22, 55, 1973.
320. **Huber, J.,** Untersuchungen zur physiologie insektentotender, *Pilze. Arch. Mikrobiol.,* 29, 257, 1958.
321. **Fargues, J. and Remaudiere, G.,** Considerations on the specificity of entomopathogenic fungi, *Mycopathologia,* 62, 31, 1977.

322. **Fargues, J.,** Specificite des champignois pathogenes imparfaits (Hyphomycete) pour les larves des Coleopteres (Scarabaeidae et Chrysomelidae), *Entomophaga,* 21, 313, 1976.
323. **Ferron, P., Hurpin, B., and Robert, P. H.,** Sur las specificity de *Metarhizium anisopliae* (Metsch.) Sorokin, *Entomophaga,* 17, 165, 1972.
324. **Ignoffo, C. M., Puttler, B., Hostetter, D. L., and Dickerson, W. A.,** Susceptibility of the cabbage looper, *Trichoplusia ni* and the velvetbean caterpillar, *Anticarsia gemmatalis,* to several isolates of the entomopathogenic fungus, *Nomuraea rileyi, J. Invertebr. Pathol.,* 28, 259, 1976.
325. **Boucias, D. G., Schoborg, E. A., and Allen, G. E.,** The relative susceptibility of six noctuid species to infection by *Nomuraea rileyi* isolated from *Anticarsia gemmatalis, J. Invertebr. Pathol.,* 39, 238, 1982.
326. **Boucias, D. G., Bradford, D. L., and Barfield, C. S.,** Susceptibility of the velvetbean caterpillar and soybean looper (Lepidoptera: Nocluidae) to *Nomuraea rileyi:* effects of pathotype, temperature and host age, *J. Econ. Entomol.,* 77, 247, 1984.
327. **Feng, Z., Carruthers, R. I., Roberts, D. W., and Robson, D. S.,** Age-specific dose-mortality effects of *Beauveria bassiana* on the European corn borer, *Ostrinia nubilalis, J. Invertebr. Pathol.,* 46, 259, 1985.
328. **Papierok, B. and Wilding, N.,** Mise en evidence d'une difference de sensibilite entre 2 clones du puceron du pois *Acyrthosiphon pisum* Harr. (Homopteres: Aphididae) exposes 2 souche du champignon phycomycete: *Entomopththora obscura* Hall and Dunn, *C. R. Acad. Sci. Paris,* 288, 93, 1979.
329. **Milner, R. J.,** On the occurrence of pea aphids, *Acyrthosiphon pisum,* resistant to isolates of the fungal pathogen *Erynia neoaphidis,* Entomol. Exp. Appl., 32, 23, 1982.
330. **Ramoska, W. A. and Todd, T.,** Variation in efficacy and viability of *Beauveria bassiana* in the chinch bug as a result of feeding activity on selected host plants, *Environ. Entomol.,* 14, 146, 1985.
331. **Kawakami, K.,** Susceptibility of varieties of the silkworm, *Bombyx mori* L. to *Aspergillus* disease and germination of fungus spores in the hemolymph, *J. Sericult. Sci. Jpn.,* 44, 39, 1975.
332. **Riba, G., Katagiri, K., and Kawakami, K.,** Preliminary studies on the susceptibility of the silkworm, *Bombyx mori* (Lepidoptera: Bombycidae) to some entomogenous fungi, *Appl. Entomol. Zool.,* 17, 238, 1982.
333. **Al-Aidroos, K. and Roberts, D. W.,** Mutants of *Metarhizium anisopliae* with increased virulence towards mosquito larvae, *Can. J. Genet. Cytol.,* 20, 211, 1978.
334. **Koidsuma, K.,** Antifungal action of cuticular lipids in insects, *J. Insect Physiol.,* 1, 40, 1957.
335. **Evalakova, A. A. and Chekhourina, T. A.,** L'activite de defense de la cuticle de la punaise des cerales *(Eurygaster integriceps),* contre les microorganisms vegtaux, *Coll. In. Pathol. Insectes Paris,* p. 137, 1962.
336. **Smith, R. J. and Grula, E. A.,** Toxic components on the larvae surface of the corn earworm *(Heliothis zea)* and their effects on germination and growth of *Beauveria bassiana, J. Invertebr. Pathol.,* 36, 15, 1982.
337. **Saito, T. and Aoki, J.,** Toxicity of free fatty acids on the larvae surfaces of two lepidopterous insects towards *Beauveria bassiana* (Bals.) Vuill. and *Paecilomyces fumoso-rosea* (Wize) Brown et Smith (Deuteromycetes: Moniliales), *Ann. Entomol. Zool.,* 18(2), 225, 1983.
338. **Kerwin, J. L.,** Fatty acid regulation of the germination of *Erynia variabilis* conidia on adults and puparia of the lesser housefly, *Fannia canicularis,* Can. J. Microbiol., 30, 158, 1983.
339. **Woods, S. P. and Grula, E. A.,** Utilizable surface nutrients on *Heliothis zea* available for growth of *Beauveria bassiana, J. Invertebr. Pathol.,* 43, 259, 1984.
340. **Fargues, J.,** Etude des conditous d'infection des larvae de doryphore, *Leptinotarsa decemlincata* Say, par *Beauveria bassiana* (Bals.) Vuill. (Fungi Imperfecti), *Entomophaga,* 17, 319, 1972.
341. **Aoki, J. and Yanase, K.,** Phenoloxidase activity in the integument of the silkworm, *Bombyx mori* infected with *Beauveria bassiana* and *Spicaria fumoso-rosea, J. Invertebr. Pathol.,* 16, 459, 1970.
342. **Vey, A.,** Etude des reactions cellularies anticryptogamiques chez *Galleria mellonella* L.: structure and ultrastructure des granulomes a *Aspergillus niger* v. Teigh, *Ann. Zool. Ecol. Anim.,* 3, 17, 1971.
343. **Schmit, A. R. and Ratcliff, N. A.,** The encapsulation of foreign implants in *Galleria mellonella* larvae, *J. Insect Physiol.,* 23, 175, 1977.
344. **Unestam, T. and Soderhall, K.,** Soluble fragments from fungal cell wall elicit defense reactions in crayfish, *Nature (London),* 267, 45, 1977.
345. **Soderhall, K. and Unestam, T.,** Activation of serum prophenoloxidase in arthropod immunity. The specificity of cell wall glucan activation by purified fungal glycoproteins of crayfish phenoloxidase, *Can. J. Microbiol.,* 25, 406, 1979.
346. **Soderhall, K.,** Prophenoloxidase activating system and melanization — a recognition mechanism of arthropods? A review, *Dev. Comp. Immunol.,* 6, 601, 1982.
347. **Soderhall, K.,** B-1,3 glucan enhancement of protease activity in crayfish lysate, *Comp. Biochem. Physiol.,* 74B, 221, 1983.
348. **Smith, V. J. and Soderhall, K.,** Induction of degranulation and lysis of haemacytes in the freshwater crayfish, *Astacus astacus* by components of the prophenoloxidase activating system *in vitro, Cell Tissue Res.,* 233, 295, 1983.

349. **Ashida, M., Ishiziku, Y., and Iwahana, H.,** Activation of prophenoloxidase by bacterial cell walls or B-1,3 glucans in plasma of the silkworm, *Bombyx mori, Biochem. Res. Commun.,* 113, 562, 1983.
350. **Soderhall, K. and Ajaxon, R.,** Effect of quinones and melanin on mycelial growth of *Aphanomyces* spp. and extracellular protease of *Aphanomyces astaci,* a parasite on crayfish, *J. Invertebr. Pathol.,* 39, 105, 1982.
351. **Kucera, M.,** Inhibition of toxic proteases of *Metarhizium anisopliae* by extracts of *Galleria mellonella* larvae, *J. Invertebr. Pathol.,* 40, 299, 1982.
352. **Kucera, M.,** Partial purification and properties of *Galleria mellonella* larvae proteolytic inhibitors acting on *Metarhizium anisopliae* toxic protease, *J. Invertebr. Pathol.,* 43, 190, 1984.
353. **Hall, L. and Soderhall, K.,** Isolation and properties of protease inhibitors in crayfish *(Astacus astacus)* cuticle, *Comp. Biochem., Physiol., 76B, 699, 1983.*
354. **Cantwell, G. E. and Lehnert, T.,** Lack of effect of certain microbial insecticides on the honeybee, *J. Invertebr. Pathol.,* 33, 381, 1979.
355. **Kamat, M. N., Bagal, S. R., Thobbi, V. V., Rae, V. G., and Phadke, C. H.,** Biological control of castor semi-looper through use of entomogenous fungus *Nomuraea rileyi, Ind. J. Bot.,* 1, 69, 1978.
356. **King, E. G. and Bell, J. V.,** Interactions between a braconid, *Microplitis croceipes,* and a fungus, *Nomuraea rileyi,* in laboratory-reared bollworm larvae, *J. Invertebr. Pathol.,* 31, 337, 1978.
357. **Egerton, J. R., Hartley, W. J., Mulley, R. C., and Sweeney, A. W.,** Susceptibility of laboratory and farm animals and two species of duck to the mosquito fungus *Culicinomyces* sp., *Mosq. News,* 38, 260, 1978.
358. **Werth, R. J., Sabwe, Mubangu, Gatti, F., and Bastin, J. P.,** Second case of rhino entomophthora mycosis due to *Entomophthora coronata* observed in the Republic of Zaire, *Ann. Soc. Belg. Med. Trop.,* 52, 343, 1973.
359. **Lowe, R. E. and Kennek, E. W.,** Pathogenicity of the fungus *Entomophthora coronata* in *Culex pipiens quinquiefasciatus* and *Aedes taeniorhynchus, Mosq. News,* 32, 614, 1972.
360. **Soper, R. S. and Bryan, T. A.,** Mammalian safety of the aphid-attacking fungus *Entomophthora* nr. *thaxteriana, Environ. Entomol.,* 3, 346, 1974.
361. **Ignoffo, C. M.,** Effects of entomopathogens on vertebrates, *Ann. N.Y. Acad. Sci.,* 217, 141, 1973.
362. **Fromtling, R. A., Kosanke, S. D., Jensen, J. M., and Bulmer, G. S.,** Fatal *Beauveria bassiana* infection in a captive American alligator, *J. Am. Vet. Med. Assoc.,* 175, 934, 1979.
363. **Popov, A. I., Ignat'ev, V. I., Karpov, E. G., Khramova, T. V., and Sitchikhina, S. V.,** The effect of *Beauveria bassiana* 92-2K on mice, *Mikrobiol. Zh. (Kiev),* 44, 69, 1982.
364. **Ignoffo, C. M., Garcia, C., Kapp, R. W., and Coate, W. B.,** An evaluation of the risks to mammals of the use of an entomopathogenic fungus, *Nomuraea rileyi,* as a microbial insecticide, *Environ. Entomol.,* 8, 354, 1979.
365. **McCoy, C. W. and Heimpel, A. M.,** Safety of the potential mycoacaricide, *Hirsutella thompsonii* to vertebrates, *Environ. Entomol.,* 9, 47, 1980.
366. **Ignoffo, C. M., Barker, W. M., and McCoy, C. W.,** Lack of *per os* toxicity or pathogenicity in rats fed the fungus *Hirsutella thompsonii, Entomophaga,* 18, 333, 1973.
367. **Ignoffo, C. M. and Garcia, C.,** *In vitro* inactivation of conidia of the entomopathogenic fungus *Nomuraea rileyi* by human gastric juice, *Environ. Entomol.,* 7, 217, 1978.
368. **Hartmann, G. C., Wasti, S. S., and Hendrickson, D. L.,** Murine safety of two species of entomogenous fungi *Cordyceps militaris* and *Paecilomyces fumoso-rosea, Appl. Entomol. Zool.,* 14, 217, 1979.
369. **Wasti, S. S., Hartmann, G. C., and Rousseau, A. J.,** Gypsy moth *Lymantria dispar* mycoses by 2 species of entomogenous fungi and an assessment of their avian toxicity, *Parasitology,* 80, 419, 1980.
370. **Shadduck, J. A., Roberts, D. W., and Lause, L.,** Mammalian safety tests of *Metarhizium anisopliae:* preliminary results, *Environ. Entomol.,* 11, 189, 1982.
371. **Kmitowa, K., Bajan, C., and Wojciechowska, M.,** Differences in the pathogenicity of entomopathogenic fungi from France and Poland, *Pol. Ecol. Stud.,* 3, 115, 1977.
372. **Usenko, L. I. and Kirsanova, R. V.,** Genetics and selection of the entomopathogenic fungus *Beauveria bassiana* (Bals.) Vuill. II. The virulence of auxotrophic mutants of *Beauveria bassiana* on *Drosphila melanogaster, Genetika,* 9, 90, 1973.
373. **Paris, S. and Ferron, P.,** Study of the virulence of some mutants of *Beauveria brongniartii* (= *Beauveria tenella), J. Invertebr. Pathol.,* 34, 71, 1979.
374. **Tinline, R. D.,** Heterokaryosis in the entomogenous fungus, *Metarhizium anisopliae, Mycologia,* 63, 701, 1971.
375. **Tinline, R. D.,** Nuclear distribution in *Metarhizium anisopliae, Mycologia,* 63, 713, 1971.
376. **Yurchenko, L., Zakharov, I. A., and Levitin, M. M.,** Genetique et selection di champignon entomopathogene *Beauveria bassiana* (Bals.) Vuill., Etude de l'heterocaryose, *Genetika,* 10, 95, 1974.
377. **Paris, S.,** Heterocaryose chez *Beauveria tenella, Mycopathologia,* 61, 67, 1977.
378. **Riba, G.,** Combonasion apres heterocaryose chez le champignon entomopathogene *Paecilomyces fumoso-rosea* (Deuteromycete), *Entomophaga,* 23, 417, 1978.

379. **Wilding, N.,** Determinations of the infectivity of *Entomophthora* spp., *Proc. Int. Colloq. Invertebr. Pathol.*, 269, 1976.
380. **Vandenberg, J. S. and Soper, R. S.,** A bioassay technique for *Entomophthora sphaerosperma* on the spruce budworm, *Choristoneura fumiferana*, *J. Invertebr. Pathol.*, 33, 148, 1979.
381. **Milner, R. J. and Soper, R. S.,** Bioassay of *Entomophthora* against the spotted alfalfa aphid *Therioaphis trifolii* f. *maculata*, *J. Invertebr. Pathol.*, 37, 168, 1981.
382. **Hall, R. A.,** A bioassay of the pathogenicity of *Verticillium lecanii* conidiospores on the aphid, *Macrosiphoniella sanborni*, *J. Invertebr. Pathol.*, 27, 41, 1976.
383. **Hall, R. A.,** Pathogenicity of *Verticillium lecanii* conidia and blastospores against the aphid, *Macrosiphoniella sanborni*, *Entomophaga*, 24, 191, 1979.
384. **Ferron, P. and Robert, P. H.,** Virulence of entomopathogenic fungi (Fungi Imperfecti) for the adults of *Acanthoscelides obtectus* (Coleoptera: Bruchidae), *J. Invertebr. Pathol.*, 25, 379, 1975.
385. **Barson, G.,** Laboratory evaluation of *Beauveria bassiana* as a pathogen of the larval stage of the large elm bark beetle, *Scolytus scolytus*, *J. Invertebr. Pathol.*, 29, 361, 1977.
386. **Sinitsyna, L. P.,** Development of a biological method of evaluating entomopathogenic fungal preparations, in Proc. First Joint US/USSR Conf. on Production, Selection, and Standardization of Entomopathogenic Fungi of the US/USSR Joint Working Group on the Production of Substances by Microbiological Means, Ignoffo, C. M., Ed., 1978, 283.
387. **Puttler, B., Ignoffo, C. M., and Hostetter, D. L.,** Relative susceptibility of nine caterpillar species to the fungus *Nomuraea rileyi*, *J. Invertebr. Pathol.*, 27, 269, 1976.
388. **Ignoffo, C., Garcia, R., Alyoshina, O. A., and Lappa, N. V.,** Laboratory and field studies with boverin: a mycoinsecticidal preparation of *Beauveria bassiana* produced in the Soviet Union, *J. Econ. Entomol.*, 72, 562, 1979.
389. **Ignoffo, C. M., Garcia, C., Kroha, M., and Couch, T. L.,** Use of larvae of *Trichoplusia ni* to bioassay conidia of *Beauveria bassiana*, *J. Econ. Entomol.*, 75, 275, 1982.
390. **Ignoffo, C. M., Garcia, C., Kroha, M., Samsinakova, A., and Kalalova, S.,** A leaf surface treatment bioassay for determining the activity of conidia of *Beauveria bassiana* against *Leptinotarsa decemlineata*, *J. Invertebr. Pathol.*, 41, 385, 1983.
391. **Aoki, J.,** Pattern of conidial discharge of an *Entomophthora* species ("grylli" type) from infected cadavers of *Mamestra brassicae* L., *Appl. Entomol. Zool.*, 16, 216, 1981.
392. **Yamamoto, M. and Aoki, J.,** Periodicity of conidial discharge of *Erynia radicans*, *Trans. Mycol. Soc. Jpn.*, 24, 487, 1983.
393. **Milner, R. J.,** Patterns of primary spore discharge of *Entomophthora* spp. from the blue green aphid *Acyrthosiphon kondoi*, *J. Invertebr. Pathol.*, 38, 419, 1981.
394. **Roberts, D. W. and Campbell, A. S.,** Stability of entomopathogenic fungi, *Misc. Publ. Entomol. Soc. Am.*, 10, 19, 1977.
395. **Wallace, D. R., MacLeod, D. M., and Lyzer, D. M.,** Endogenous light action in germination of *Entomophthora aphidis* resting spores *in vitro*, *Bi-monthly Res. Notes*, 34, 24, 1978.
396. **Basith, M. and Madelin, M. F.,** Studies on the production of perithecial stromata by *Cordyceps militaris* in artificial culture, *Can. J. Bot.*, 46, 473, 1968.
397. **Ignoffo, C. M., Hostetter, D. L., Sikorowski, P. P., Sutter, G., and Brooks, W. M.,** Inactivation of representative species of entomopathogenic viruses, a bacterium, fungus and protozoan by an ultraviolet light source, *Environ. Entomol.*, 6, 411, 1977.
398. **Kreig, A., Groner, A., Huber, J., and Zimmerman, G.,** Inaktivierung von vershiedenen Insektenpathogenen durch ultraviolette Strahlen, *Z. Pflanzenkr. Pflanzenschutz*, 88, 38, 1981.
399. **Tuveson, R. W. and McCoy, C. W.,** Far-ultraviolet sensitivity and photoreactivation of *Hirsutella thompsonii*, *Ann. Appl. Biol.*, 101, 13, 1982.
400. **Gardner, W. A., Sutton, R. M., and Noblet, R.,** Persistence of *Beauveria bassiana*, *Nomuraea rileyi* and *Nosema nectrix* on soybean foliage, *Environ. Entomol.*, 6, 616, 1977.
401. **Ignoffo, C. M. and Batzer, O. F.,** Microencapsulation and ultraviolet protectants to increase sunlight stability of an insect virus, *J. Econ. Entomol.*, 64, 850, 1971.
402. **Zimmerman, G.,** Effect of high temperatures and artificial sunlight on the viability of conidia of *Metarhizium anisopliae*, *J. Invertebr. Pathol.*, 40, 36, 1982.
403. **Carruthers, R. I., Feng, Z., Robson, D. S., and Roberts, D. W.,** In vivo *temperature-dependent development of Beauveria bassiana* mycosis of the European corn borer, *Ostrinia nubilalis*, *J. Invertebr. Pathol.*, 46, 305, 1985.
404. **Soares, G. G. and Pinnock, D. E.,** Effect of temperature on germination, growth, and infectivity of the mosquito pathogen, *Tolypocladium cylindrosporium*, *J. Invertebr. Pathol.*, 43, 242, 1984.
405. **Wilding, N.,** Pest control by Entomophthorales, in *Microbial Control of Pests and Plant Diseases 1970—1980*, Burges, H. D., Ed., Academic Press, New York, 1981, 539.
406. **Payandeh, B., MacLeod, D. M., and Wallace, D. R.,** Germination of *Entomophthora aphidis* resting spores under constant temperatures, *Can. J. Bot.*, 56, 2328, 1978.

407. **Latge, J. P., Perry, D., Papierok, B., Coremans-Pelseneer, J., Remaudiere, G., and Reisinger, O.,** Germination des azygospores d' *Entomophthora obscura* Hall & Dunn, role du sol., *C. R. Acad. Sci. Paris,* 287, 943, 1978.
408. **Kawakami, K.,** Deep freeze storage of some entomogenous fungus cultures, *Sanshi Kenkyu (Acta Sericol.),* 76, 58, 1970.
409. **McCoy, C. W.,** Pest control by the fungus *Hirsutella thompsonii,* in *Microbial Control of Pests and Plant Diseases 1970—1980,* Burges, H. D., Ed., Academic Press, New York, 1981, 499.
410. **Bell, J. V. and Hamalle, R. J.,** Viability and pathogenicity of entomogenous fungi after prolonged storage on silica gel at $-20°C$, *Can. J. Bot.,* 20, 639, 1974.
411. **Walstad, J. E., Anderson, R. F., and Stambaugh, W. J.,** Effects of environmental conditions on two species of muscardine fungi *(Beauveria bassiana* and *Metarhizium anisopliae), J. Invertebr. Pathol.,* 16, 221, 1970.
412. **Steinhaus, E. A.,** The duration of viability and infectivity of certain insect pathogens, *J. Insect Pathol.,* 2, 255, 1960.
413. **Wilding, N.,** The survival of *Entomophthora* spp. in mummified aphids at different temperatures and humidities, *J. Invertebr. Pathol.,* 21, 309, 1973.
414. **Wilding, N.,** Effect of humidity on the sporulation of *Entomophthora aphidis* and *Entomophthora thaxteriana, Trans. Br. Mycol. Soc.,* 53, 126, 1969.
415. **Wilding, N.,** *Entomophthora* species infecting pea aphids, *Trans. R. Entomol. Soc.,* 127, 171, 1975.
416. **Wilding, N.,** *Entomophthora* conidia in the air-spora, *J. Gen. Microbiol.,* 62, 149, 1970.
417. **Milner, R. J. and Bourne, J.,** Influence of temperature and duration of leaf wetness on infection of *Acyrthosiphon kondoi* with *Erynia neoaphidis, Ann. Appl. Biol.,* 102, 19, 1983.
418. **Kish, L. P. and Allen, G. E.,** The biology and ecology of *Nomuraea rileyi* and a program for predicting its incidence on *Anticarsia gemmatalis* in soybean, *Fla. Agric. Exp. Stn. Bull.,* 795, 1, 1978.
419. **Garcia, C. and Ignoffo, C. M.,** Dislodgement of conidia of *Nomuraea rileyi* from cadavers of cabbage looper, *Trichoplusia ni, J. Invertebr. Pathol.,* 30, 114, 1977.
420. **Remaudiere, G. and Michel, M. F.,** Premiere experimentation ecologique sur les *Entomophthorales* (Phycomycetes) parasites de pucerons en vergers de pechers, *Entomophaga,* 16, 75, 1971.
421. **Papierok, B.,** Obtention *in vivo* des azygospores d'*Entomophthora thaxteriana* Petch, champignon pathogene de pucerons (Homopteres, Aphididae), *C. R. Acad. Sci. Paris,* 286, 1503, 1978.
422. **Daoust, R. A. and Roberts, D. W.,** Studies on the prolonged storage of *Metarhizium anisopliae* conidia: effect of temperature and relative humidity on conidial viability and virulence against mosquitoes, *J. Invertebr. Pathol.,* 41, 143, 1983.
423. **Ignoffo, C. M., Garcia, C., Hostetter, D. L., and Pinnell, R. E.,** Stability of conidia of an entomopathogenic fungus, *Nomuraea rileyi,* in and on soil, *Environ. Entomol.,* 7, 724, 1978.
424. **Reisinger, O., Fargues, J., Robert, P., and Arnold, M. F.,** Effect de L'argile sur la conservation des microorganismes. I. Etude ultrastructurale de la biodegradation dans le sol de L'hyphomycete entomopathogene *Beauveria bassiana* (Bals.) Vuill., *Ann. Microbiol. (Inst. Pasteur),* 128B, 271, 1977.
425. **Lingg, A. J. and Donaldson, M. D.,** Biotic and abiotic factors affecting stability of *Beauveria bassiana* conidia in soil, *J. Invertebr. Pathol.,* 38, 191, 1981.
426. **Shields, M. S., Lingg, A. J., and Heimsch, R. C.,** Identification of a *Penicillium urticae* metabolite which inhibits *Beauveria bassiana, J. Invertebr. Pathol.,* 38, 374, 1981.
427. **Wartenberg, H. and Freund, K.,** Der Konservierungseffekt anti-biotischer mikroorganismen an Konidien von *Beauveria bassiana* (Bals.) Vuill., *Zentrabl. Bakteriol. Parasitenk.,* 114, 718, 1961.
428. **Wojciechowska, M., Kmitowa, K., Fedorko, A., and Bajan, C.,** Duration of activity of entomopathogenic microorganisms introduced into the soil, *Pol. Ecol. Stud.,* 3, 141, 1977.
429. **Bell, J. V. and Hamalle, R. J.,** Three fungi tested for control of the cowpea curculio, *Chalcodermus aeneus, J. Invertebr. Pathol.,* 15, 447, 1970.
430. **Latteur, G.,** Sur la possibilite d'infection directe d'aphides par *Entomophthora* a partir de sols herbergeant un inoculum natural, *C. R. Acad. Sci.,* 284, 2253, 1977.
431. **Latge, J. P., Perry, D., Reisinger, O., Papierok, B., and Remaudiere, G.,** Induction de la formation des spores de resistance d' *Entomophthora obscura* Hall & Dunn, *C. R. Acad. Sci. Paris,* 288, 599, 1979.
432. **Ignoffo, C. M., Garcia, G., Hostetter, D. L., and Pinnell, R. E.,** Vertical movement of conidia of *Nomuraea rileyi* through sand and loam soils, *J. Econ. Entomol.,* 70, 163, 1977.
433. **Milner, R. J. and Lutton, G. G.,** *Metarhizium anisopliae:* survival of conidia in the soil, *Proc. 1st Int. Colloq. Invertebr. Pathol.,* Queen's University Press, Kingston, Canada, 1976, 428.
434. **Lord, J. C. and Roberts, D. W.,** Effects of salinity, pH, organic solutes, anaerobic conditions, and the presence of other microbes on production and survival of *Lagenidium giganteum* zoospores, *J. Invertebr. Pathol.,* 45, 331, 1985.
435. **Jaronski, S. T. and Axtell, R. C.,** Effects of organic water pollution on the infectivity of the fungus *Lagenidium giganteum* for larvae of *Culex quinquefasciatus* field and laboratory evaluation, *J. Med. Entomol.,* 19, 255, 1982.

436. **Merriam, T. L. and Axtell, R. C.,** Salinity tolerance of two isolates of *Lagenidium giganteum*, a fungal pathogen of mosquito larvae, *J. Med. Entomol.*, 19, 388, 1982.
437. **Roberts, D. W.,** Isolation and development of fungus pathogens of vectors, in *Biological Regulation of Vectors*, Briggs, J. D., Ed., U.S. Department of Health, Education and Welfare, Publ. No. 77-1180, Washington, D.C., 1977, 85.
438. **Burleigh, J. G.,** Comparison of *Heliothis* spp. larval parasitism and *Spicaria* infection in closed and open canopy cotton varieties, *Environ. Entomol.*, 4, 574, 1975.
439. **McCoy, C. W.,** Migration and development of citrus rust mite on the spring flush of Valencia orange, *Proc. Fla. State Hortic. Soc.*, 92, 48, 1979.
440. **Samson, R. A. and McCoy, C. W.,** A new fungal pathogen of the scavenger mite, *Tydeus gloveri*, *J. Invertebr. Pathol.*, 40, 216, 1982.
441. **Kish, L. P. and Allen, G. E.,** Conidial production of *Nomuraea rileyi* on *Pseudoplusia includens*, *Mycologia*, 68, 436, 1976.
442. **Ignoffo, G. M., Garcia, C., Hostetter, D. L., and Pinnell, R. E.,** Laboratory studies on the entomopathogenic fungus *Nomuraea rileyi*: soil-borne contamination of soybean seedlings and dispersal of diseased larvae of *Trichoplusia ni*, *J. Invertebr. Pathol.*, 29, 147, 1977.
443. **Zimmerman, G.,** Uber die wirkung systemischer fungizide auf verschiedene insektenpathogene Fungi Imperfecti, *Nachrichtenbl. Dtsch. Pflanzenschutz.*, 27, 113, 1975.
444. **Zimmerman, G.,** Uber die wirkung systemischer fungizide auf aphiden befallende Entomophthoraceen (Zygomycetes) *in vitro*, *Z. Pflanzenkr. Pflanzenschutz*, 83, 261, 1976.
445. **Zimmerman, G.,** Laborversuche uber den einflusz systemischer fungizide auf Verpilzung und Konidienbikdung durch aphidenpathogene Entomophthoraceen (Zygomycetes) bei Getreideblattlausen, *Z. Pflanzenkr. Pflanzenschutz*, 85, 513, 1978.
446. **Zimmerman, G. and Basedow, T.,** Freilanduntersuchungen zum Einflusz von Fungiziden auf die durch Entomophthoraceen (Zygomycetes) verursachte Mortalitat bei Getreideblattlausen, *Z. Pflanzenkr. Pflanzenschutz*, 87, 65, 1980.
447. **Delorme, R. and Fritz, R.,** Action de fongicides sur le development d'une mycose a *Entomophthora aphidis*, *Entomophaga*, 23, 389, 1978.
448. **Fritz, R.,** Action de quelques fongicides sur des entomophthorales pathogenes de pucerons, *Phytiatr. Phytopharm.*, 26, 193, 1977.
449. **Wilding, N.,** The effect of fungicides on field populations of *Aphis fabae* and on the infection of the aphids by Entomophthoraceae, *Ann. Appl. Biol.*, 100, 221, 1982.
450. **Livingston, J. M., Yearian, W. C., Young, S. Y., and Stacey, A. L.,** Effect of benomyl on an *Entomophthora* epizootic in a *Pseudoplusia includens* population, *J. Ga. Entomol. Soc.*, 16, 511, 1981.
451. **Oncuer, C. and Latteur, G.,** Etude de l'influence des 10 fongicides sur le pouvoir infectant des conidies d'*Entomophthora obscura* Hall & Dunn presentes a la surface d'un sol non sterile, *Parasitica*, 35, 3, 1979.
452. **Fritz, R.,** Action de quelques fongicides sur la croissance mycelieene de trois especes d'entomophthorales, *Entomophaga*, 21, 239, 1976.
453. **Keller, S.,** Untersuchungen uber den Einflusz von Dimilin (Diflubenzuron) auf Wachstum und Konidienkeimung einiger insektenpathogener Pilze, *Anz. Schaedlingskd. Pflanz. Umweltschutz*, 51, 81, 1978.
454. **McCoy, C. W.,** unpublished data, 1985.
455. **Nolan, R. A. and Dunphy, G. B.,** Effects of hormones on *Entomophthora egressa* morphogenesis, *J. Invertebr. Pathol.*, 33, 242, 1979.
456. **Sutton, R. M., Gardner, W. A., Kraus, D. W., and Noblet, R.,** Effects of ecdysone, juvenile hormone, and structurally-related compounds on *Nomuraea rileyi*, *Beauveria bassiana*, and *Tritirachium dependens* in vitro, *J. Ga. Entomol. Soc.*, 14, 364, 1979.
457. **Bajan, C., Kmitowa, K., and Wojciechowska, M.,** The effect of enolofos 50 and its active substance — chlorfenvinphos on growth and pathogenicity of entomopathogenic fungi, *Pol. Ecol. Stud.*, 3, 21, 1977.
458. **Wojciechowska, M., Kmitowa, K., and Bajan, C.,** The effects of carbamide herbicides, linuron and monolinuron, on three species of entomopathogenic fungi, *Pol. Ecol. Stud.*, 3, 43, 1977.
459. **Bajan, C., Kmitowa, K., and Wojciechowska, M.,** The effect of treatments used to control the Colorado beetle on the composition of the species of the fungi isolated from dead beetles, *Pol. Ecol. Stud.*, 3, 59, 1977.
460. **Clark, R. A., Casagrande, R. A., and Wallace, D. B.,** Influence of pesticides on *Beauveria bassiana*, a pathogen of the Colorado potato beetle, *Environ. Entomol.*, 11, 67, 1982.
461. **Ignoffo, C. M., Hostetter, D. L., Garcia, C., and Pinnell, R. E.,** Sensitivity to the entomopathogenic fungi *Nomuraea rileyi* to chemical pesticides used on soybeans, *Environ. Entomol.*, 4, 765, 1975.
462. **Johnson, D. W., Kish, L. P., and Allen, G. E.,** Field evaluation of selected pesticides on the natural development of the entomopathogen, *Nomuraea rileyi*, on the velvetbean caterpillar in soybean, *Environ. Entomol.*, 5, 964, 1976.
463. **Horton, D. L., Carner, G. R., and Turnipseed, S. G.,** Pesticide inhibition of the entomogenous fungus *Nomuraea rileyi* in soybeans, *Environ. Entomol.*, 9, 304, 1980.

464. **Garcia, C. and Ignoffo, C. M.,** Sensitivity of *Nomuraea rileyi* to antibiotics, sulfonamides, and fungicidal substances, *J. Invertebr. Pathol.,* 33, 124, 1979.
465. **Oho, N. and Satoh, Y.,** Studies on a fungus, *Aschersonia* sp. parasitic on whitefly, *Dialeurodes citri* Ashmead. I. Fundamental studies on cultural characters and some properties against pesticides, *Bull. Hortic. Res. Stn. Jpn.* A5, 179, 1966.
466. **Hall, R. A.,** Laboratory studies on the effects of fungicides, acaricides and insecticides on the entomopathogenic fungus, *Verticillium lecanii, Entomol. Exp. Appl.,* 29, 39, 1981.
467. **McCoy, C. W., Brooks, R. F., Allen, J. C., and Selhime, A. G.,** Management of arthropod pests and plant diseases in citrus agroecosystem, *Proc. Tall Timbers Conf. Ecol. Anim. Control Habitat Managem.,* 6, 10, 1976.
468. **McCoy, C. W., Brooks, R. F., Allen, J. C., Selhime, A. G., and Wardowski, W. F.,** Effect of reduced pest control programs on yield and quality of 'Valencia' orange, *Proc. Fla. State Hortic. Soc.,* 89, 74, 1976.
469. **Riehl, L. A., Brooks, R. F., McCoy, C. W., Fisher, T. W., and Dean, H. A.,** Accomplishments toward improving integrated pest management for citrus, in *New Technology of Pest Control,* Huffaker, Carl B., Ed., John Wiley & Sons, New York, 1980, 319.
470. **McCoy, C. W.,** Resurgence of citrus rust mite populations following applications of methidathion, *J. Econ. Entomol.,* 70, 748, 1977.
471. **Easwaramoorthy, S. and Jayaraj, S.,** Effect of certain insecticides and fungicides on the growth of the coffee green bug fungus, *Cephalosporium lecanii* Zimm., *Madras Agric. J.,* 64, 243, 1977.
472. **Shimazu, M.,** Factors affecting conidial germination of *Entomophthora delphacis* Hori (Entomophthorales: Entomophthora), *Appl. Entomol. Zool.,* 12, 260, 1977.
473. **Yendol, W. G.,** Factors affecting germination of *Entomophthora* conidia, *J. Invertebr. Pathol.,* 10, 116, 1968.
474. **Schneider, R.,** Untersuchungen uber Feuchtigkeitsanspruche parasitischer, *Pilze. Phytopathol. Z.,* 21, 63, 1953.
475. **Doberski, J. W.,** Comparative laboratory studies on three fungal pathogens of the elm bark beetle, *Scolytus scolytus:* effect of temperature and humidity on infection by *Beauveria bassiana, Metarhizium anisopliae* and *Paecilomyces farinosus, J. Invertebr. Pathol.,* 37, 195, 1981.
476. **Ferron, P.,** Influence of relative humidity on the development of fungal infection caused by *Beauveria bassiana* (Fungi Imperfecti, Moniliales) in images of *Acanthoscelides obtectus* (Coleoptera: Bruchidae), *Entomophaga,* 22, 393, 1977.
477. **Moore, G.,** Pathogenicity of three entomogenous fungi to the southern pine beetle at various temperatures and humidities, *Environ. Entomol.,* 2, 54, 1973.
478. **Ramoska, W. A.,** The influence of relative humidity on *Beauveria bassiana* infectivity and replication in the chinch bug, *Blissus leucopterous, J. Invertebr. Pathol.,* 43, 389, 1984.
479. **Gerson, U., Kenneth, R., and Muttath, T. I.,** *Hirsutella thompsonii,* a fungal pathogen of mites. II. Host-pathogen interactions, *Ann. Appl. Biol.,* 91, 29, 1979.
480. **Kenneth, R., Muttath, T. I., and Gerson, U.,** *Hirsutella thompsonii,* a fungal pathogen of mites. I. Biology of the fungus *in vitro, Ann. Appl. Biol.,* 91, 21, 1979.
481. **Getzin, L. W.,** *Spicaria rileyi* (Farlow) Charles, an entomogenous fungus of *Trichoplusia ni* (Hukner), *J. Insect Pathol.,* 3, 2, 1961.
482. **Sweeney, A. W.,** The effects of salinity on the mosquito pathogenic fungus *Culicinomyces, Aust. J. Zool.,* 26, 55, 1978.
483. **Wilding, N.,** The effect of temperature on the infectivity and incubation periods of the fungi *Entomophthora aphidis* and *Entomophthora thaxteriana* for the pea aphid *Acyrthosiphon pisum, Proc. 4th Int. Colloq. Insect Pathol.,* College Park, Md., 1970, 84.
484. **Diomande, T.,** Contribution a l'etude du developpement de la muscardine verte a *Metarhizium anisopliae* (Metsch) Sorokin des larves de *Oryctes monoceros* OL., *Bull. Inst. Fondam. Afr. Noire Ser. A,* 21, 1381, 1969.
485. **Voronina, E. G.,** Ecological features of strains of the fungus *Entomophthora thaxteriana* Petch. isolated from *Acyrthosiphon pisum* Harris and *Myzodes persicae* Sulz., *Tr. Vseross. Nauchno-Issled. Inst. Zashch. Rast.,* 31, 394, 1968.
486. **Callaghan, A. A.,** Effect of nutrient level, pH, and light on conidial germination in entomophthoraceous fungi, *Trans. Br. Mycol. Soc.,* 70, 271, 1978.
487. **Gabriel, B. P.,** Fungus infection of insects via the alimentary tract, *J. Insect Pathol.,* 1, 319, 1959.
488. **Samson, R. A.,** Laboratory culture and maintenance of entomopathogenic fungi, *Proc. 3rd Intern. Colloq. Invertebr. Pathol.,* Brighton, England, 1982, 409.
489. **Dulmage, H. T. and Rhodes, R. A.,** Production of pathogens in artificial media, in *Microbial Control of Insects and Mites,* Burges, H. D. and Hussey, N. W., Eds., Academic Press, New York, 1971, chap. 24.

490. **Kerwin, J. L. and Washino, R. K.**, Sterol induction of sexual reproduction in *Lagenidium giganteum*, *Exp. Mycol.*, 7, 109, 1983.
491. **Latge, J. P. and Saglier, J. J.**, Optimisation de la croissance et de al sporulation de *Conidiobolus obscurus* en milieu defini, *Can. J. Bot.*, 63, 68, 1985.
492. **Latge, J. P.**, Sporulation de *Entomophthora obscura* Hall & Dunn en culture liquide, *Can. J. Microbiol.*, 26, 1038, 1980.
493. **Kononova, E. V.**, Selection of commercial strains of the fungus, *Beauveria bassiana*, in Proc. First Joint US/USSR Conf. on Production, Selection, and Standardization of Entomopathogenic Fungi of the US/USSR Joint Working Group on the Production of Substances by Microbiological Means, Ignoffo, C. M., Ed., 1978, 173.
494. **Ignoffo, C. M. and Hink, W. F.**, Propagation of arthropod pathogens in living systems, in *Microbial Control of Insects and Mites*, Burges, H. D. and Hussey, N. W., Eds., Academic Press, New York, 1971, chap. 25.
495. **Ignoffo, C. M.**, Entomopathogens as insecticides, *Environ. Lett.*, 8, 23, 1975.
496. **Hall, I. M. and Dunn, P. H.**, Artificial dissemination of entomophthorous fungi pathogenic to the spotted alfalfa aphid, *J. Econ. Entomol.*, 51, 341, 1958.
497. **Laird, Marshall,** Microbiology and mosquito control, *Mosq. News*, 20, 127, 1960.
498. **Shapiro, M. and Roberts, D. W.**, Growth of *Coelomomyces psorophorae* mycelium *in vitro*, *J. Invertebr. Pathol.*, 27, 399, 1976.
499. **Nolan, R. A.**, Effects of plant hormones on germination of *Coelomomyces psorophorae* resistant sporangia, *J. Invertebr. Pathol.*, 21, 26, 1973.
500. **Castillo, J. M. and Roberts, D. W.**, *In vitro* studies of *Coelomomyces punctatus* from *Anopheles quadrimaculatus* and *Cyclops vernalis*, *J. Invertebr. Pathol.*, 35, 144, 1980.
501. **Federici, B. A. and Chapman, H. C.**, *Coelomomyces dodgei*: establishment of an *in vivo* laboratory culture, *J. Invertebr. Pathol.*, 30, 288, 1977.
502. **Federici, B. A.**, Production of the mosquito-parasitic fungus *Coelomomyces dodgei* through synchronized infection and growth of the intermediate copepod host, *Cyclops vernalis*, *Entomophaga*, 25, 209, 1980.
503. **Sweeney, A. W.**, The effects of temperature on the mosquito pathogenic fungus *Culicinomyces*, *Aust. J. Zool.*, 26, 47, 1978.
504. **Jaronski, S., Axtell, R. C., Fagan, S. M., and Domnas, A. J.**, *In vitro* production of zoospores by the mosquito pathogen *Lagenidium giganteum* on solid media, *J. Invertebr. Pathol.*, 41, 305, 1983.
505. **Domnas, A. J., Fagan, S. M., and Jaronski, S.**, Factors influencing zoospore production in liquid cultures of *Lagenidium giganteum*, *Mycologia*, 74, 820, 1982.
506. **Kerwin, J. L. and Washino, R. K.**, Cyclic nucleotide regulation of oosporogenesis by *Lagenidium giganteum* and related fungi, *Exp. Mycol.*, 8, 215, 1984.
507. **Kerwin, J. L. and Washino, R. K.**, Oosporogenesis by *Lagenidium giganteum*: induction and maturation are regulated by calcium and calmodulin, *Can. J. Microbiol.*, 32, 663, 1986.
508. **Latge, J. P.**, Production of entomophthorales, *Proc. 3rd Int. Colloq. Invertebr. Pathol.*, Brighton, England, 1982, 164.
509. **Sawyer, W. H.**, Observations on some entomogenous members of the *Entomophthoraceae* in artificial culture, *Am. J. Bot.*, 16, 87, 1928.
510. **Dunphy, G. B., Nolan, R. A., and MacLeod, D. M.**, Comparative growth and development of protoplast isolates of *Entomophthora egressa*, *J. Invertebr. Pathol.*, 31, 267, 1978.
511. **MacLeod, D. M., Tyrrell, D., and Welton, M. A.**, Isolation and growth of the grasshopper pathogen *Entomophthora grylli*, *J. Invertebr. Pathol.*, 36, 85, 1980.
512. **Latge, J. P. and de Bievre, C.**, Influence des lipides et acides gras du jaune d'oeuf sur la croissance et la sporulatoin des Entomophthorales, *Ann. Microbiol.*, 127, 261, 1976.
513. **Gustafsson, M.**, On species of the genus *Entomophthora* Fres. in Sweden. II. Cultivation and physiology, *Lantbrukshoegsk. Ann.*, 31, 405, 1965.
514. **Gustafsson, M.**, On species of the genus *Entomophthora* Fres. in Sweden. III. Possibility of usage in biological control, *Lantbrukshoegsk. Ann.*, 35, 235, 1969.
515. **Latge, J. P.**, Croissance et sporulation de 6 especes d'Entomophthorales. I. Influence de la nutrition carbonee, *Entomophaga*, 20, 201, 1975.
516. **Latge, J. P.**, Croissance et sporulation de 6 especes d'Entomophthorales. II. Influence de diverses sources d'azote, *Mycopathologia*, 57, 53, 1975.
517. **Latge, J. P. and Remaudiere, G.**, Croissance et sporulation de six especes d'Entomophthorales. III. Influence des concentrations de carbonee et d'azote et du rapport C/N, *Rev. Mycol.*, 34, 239, 1975.
518. **Voronina, E. G., Gindina, G. M., and Mitsyganov, V. A.**, *Entomophthora thaxteriana* spore formation depending on the sources of nitrogen nutrition, *Mikol. Fitopatol.*, 15, 92, 1981.
519. **Muller-Kogler, E.**, On mass cultivation, determination of effectiveness and standardization of insect pathogenic fungi, in *Insect Pathology and Microbial Control, Proc. Int. Colloq. Insect Pathol. and Microbiol. Control*, North-Holland, Amsterdam, 1967, 330.

520. **Kmitowa, K.,** Effect of culture media on entomogenous fungi, *Pol. Ecol. Stud.,* 4, 1, 1978.
521. **Goral, V. M. and Lappa, N. V.,** Deep and surface culture of the green muscardine fungus, *Zashch. Rast.,* 1, 19, 1973.
522. **Bajan, C., Kmitowa, K., and Wojciechowska, M.,** A simple method of obtaining the infectious material of entomogenous fungi, *Bull. Acad. Pol. Sci.,* 13, 45, 1975.
523. **Fargues, J., Robert, P. H., and Reisinger, O.,** Formulation des productions de masse de l'hyphomycete entomopathogene *Beauveria* en vue des applications phytosanitaries, *Ann. Zool. Ecol. Anim.,* 11, 247, 1979.
524. **Bertatlief, Z.,** Studies on the entomopathogenic fungus *Beauveria bassiana* (Bals.) Vuill. and its action on the Colorado potato beetle *(Leptinotarsa decemlineata* Say) and the sugar-beet weevil *(Bothynoderes punctiventris* Germ.), *An. Inst. Cercet. Prot. Plant. Bucarest,* 15, 233, 1979.
525. **de Aquino, M., Cavalcanti, V. A., Sena, R. C., and Queiroz, G. F.,** Nova tecnologia de multiplicacao do fungo *Metarhizium anisopliae, Bol. Tec. CODE-CAP, Recife,* 4, 1, 1975.
526. **Aquino, de M., Vital, A. E., Cavalcanti, V. A., and Nascimento, M. G.,** Cultura de *Metarhizium anisopliae* (Metsch.) Sorokin em sacos de polipropileno, *Bol. Tec. Com. Exec. Defesa Fitossanitaria da Lavoura Canavieira de per Nambuco,* 5, 11, 1977.
527. **Bell, J. V.,** Production and pathogenicity of the fungus *Spicaria rileyi* from solid and liquid media, *J. Invertebr. Pathol.,* 26, 129, 1975.
528. **Phadke, C. H. and Rao, V. G.,** Mass production of spores by *Nomuraea rileyi* for biological control of insect pest, *Biovigyanam* 3, 125, 1977.
529. **Galani, G.,** Studies on the variation of pathogenicity of *Verticillium lecanii* (Zimm.) Viegae to larvae of *Trialeurodes vaporariorum* Westw., *An. Inst. Cercet. Prot. Plant. Bucarest,* 15, 244, 1979.
530. **Ponomarenk, N. G., Prilepskaya, H. A., Murvanidze, M. Y., and Stolyarova, L. A.,** *Aschersonia* against whiteflies, *Zashch. Rast.,* 6, 44, 1975.
531. **Iren, Z. and Soran, H.,** The study of the mass production of the fungus *Aschersonia aleyrodis,* a pathogenic agent for *Dialeurodes citri, Univ. Ankara Yearb. Fac. Agric.,* 25, 281, 1975.
532. **Vezina, C., Singh, K., and Seghal, S. N.,** Sporulation of filamentous fungi in submerged culture, *Mycologia,* 57, 722, 1965.
533. **Goral, V. M.,** Morphological characteristics of development of the entomopathogenic fungus *Beauveria bassiana* (Bals.) Vuill. in deep cultures, *Mikol. Fitopathol.,* 9, 989, 1975.
534. **Van Winkelhoff, A. J. and McCoy, C. W.,** Conidiation of *Hirsutella thompsonii* var. *synnematosa* in submerged culture, *J. Invertebr. Pathol.,* 43, 59, 1984.
535. **Adamek, L.,** Submersed cultivation of the fungus *Metarhizium anisopliae* (Metsch.), *Folia Microbiol.,* 10, 255, 1965.
536. **Samsinakova, A.,** Growth and sporulation in submerged culture of the fungus *Beauveria bassiana* in various media, *J. Invertebr. Pathol.,* 8, 395, 1966.
537. **Catroux, G., Calvez, J., Ferron, P., and Blachere, H.,** Mise au point d'uno preparation entompathogene a base de blastospores de *Beauveria tenella* (Delacr.) Siemaszko pour la lutte microbiologique contre le ver blanc *(Melolontha melolontha* L.), *Ann. Zool. Ecol. Anim.,* 2, 281, 1970.
538. **Samsinakova, A., Kalalova, S., Vlcek, V., and Kybal, J.,** Mass production of *Beauveria bassiana* for regulation of *Leptinotarsa decemlineata* populations, *J. Invertebr. Pathol.,* 38, 169, 1981.
539. **Barnes, G. L., Boethel, D. J., Eikenbary, R. D., Criswell, J. T., and Gentry, C. R.,** Growth and sporulation of *Metarhizium* and *Beauveria bassiana* on media containing various peptone sources, *J. Invertebr. Pathol.,* 25, 301, 1975.
540. **Campbell, R. K., Perring, T. M., Barnes, G. L., Eikenbary, R. D., and Gentry, C. R.,** Growth and sporulation of *Beauveria bassiana* and *Metarhizium anisopliae* on media containing various amino acids, *J. Invertebr. Pathol.,* 31, 289, 1978.
541. **Campbell, R. K., Barnes, G. L., Cartwright, B. O., and Eikenbary, R. D.,** Growth and sporulation of *Beauveria bassiana* and *Metarhizium anisopliae* in a basal medium containing various carbohydrate sources, *J. Invertebr. Pathol.,* 41, 117, 1983.
542. **Belova, R. N.,** Development of the technology of boverin production by the submersion method, in Proc. First Joint US/USSR Conf. on Production, Selection, and Standardization of Entomopathogenic Fungi of the US/USSR Joint Working Group on the Production of Substances by Microbiological Means, Ignoffo, C. M., Ed., 1978, 102.
543. **McCoy, C. W., Hill, A. J., and Kanavel, R. F.,** Large-scale production of the fungal pathogen *Hirsutella thompsonii* in submerged culture and its formulation for application in the field, *Entomophaga,* 20, 229, 1975.
544. **McCoy, C. W., Hill, A. J., and Kanavel, R. F.,** A liquid medium for the large-scale production of *Hirsutella thompsonii* in submerged culture, *J. Invertebr. Pathol.,* 19, 370, 1972.
545. **McCoy, C. W., Couch, T. L., and Weatherwax, R.,** A simplified medium for the production of *Hirsutella thompsonii, J. Invertebr. Pathol.,* 31, 137, 1978.

546. **Latge, J. P., Soper, R. S., and Madore, C. D.,** Media suitable for industrial production of *Entomophthora virulenta* zygospores, *Biotechnol. Bioeng.,* 19, 1269, 1977.
547. **Latge, J. P., Remaudiere, G., and Diaguin, M.,** Un nouveau milieu pour la croissance et la sporulation d'Entomophthorales pathogenes d'aphides, *Ann. Microbiol.,* 129, 463, 1978.
548. **Matanomi, B. A. and Libby, J. L.,** The production and germination of resting spores of *Entomophthora virulenta, J. Invertebr. Pathol.,* 27, 279, 1976.
549. **Yegina, K. Y.,** Technology of the growing of *Entomophthora,* in Proc. First Joint US/USSR Conf. on Production, Selection, and Standardization of Entomopathogenic Fungi of the US/USSR Joint Working Group on the Production of Substances by Microbiological Means, Ignoffo, C. M., Ed., 1978, 243.
550. **Latge, J. P., Remaudiere, G., Soper, R. S., Madore, C. D., and Diaquin, M.,** Growth and sporulation of *Entomophthora virulenta* on semi-defined media in liquid culture, *J. Invertebr. Pathol.,* 31, 225, 1978.
551. **Perry, D. F. and Latge, J. P.,** Chemically defined media for growth and sporulation of *Entomophthora virulenta, J. Invertebr. Pathol.,* 35, 43, 1980.
552. **Soper, R. S., Holbrook, F. R., Majchrowicz, I., and Gordon, C. C.,** Production of *Entomophthora* resting spores for biological control of aphids, *Maine Agric. Exp. Stn. Bull.,* 76, 1, 1975.
553. **Latge, J. P.,** Sporulation de *Entomophthora obscura* en culture liquide, *J. Can. Microbiol.,* 26, 1038, 1980.
554. **Perry, D. F. and Latge, J. P.,** Dormancy and germination of *Conidiobulus obscurus* azygospores, *Trans. Br. Mycol. Soc.,* 78, 221, 1982.
555. **Kerwin, J. L., Simmons, C. A., and Washino, R. K.,** Oosporogenesis by *Lagenidium giganteum* in liquid culture, *J. Invertebr. Pathol.,* 47, 258, 1986.
556. **Couch, T. L.,** Standardization of entomogenous fungi, in Proc. First Joint US/USSR Conf. on Production, Selection and Standardization of Entomopathogenic Fungi of the US/USSR Joint Working Group on the Production of Substances by Microbiological Means, Ignoffo, C. M., Ed., 1978, 138.
557. **McCabe, D. and Soper, R. S.,** U.S. Patent No. 4,530,834, 1985.
558. **Blachere, H., Calvez, J., Ferron, P., Corrieu, G., and Peringer, P.,** Etude de la formulation et de la conservation d'une preparation entomopathogene a base de blastospores de *Beauveria tenella* (Delace. Siemaszko), *Ann. Zool. Ecol. Anim.,* 5, 67, 1973.
559. **Fargues, J., Reisinger, O., Robert, P. H., and Aubart, C.,** Biodegradation of entomopathogenic hyphomycetes: influence of clay coating on *Beauveria bassiana* blastospore survival in soil, *J. Invertebr. Pathol.,* 41, 131, 1983.
560. **Daoust, R. A., Ward, M. G., and Roberts, D. W.,** Effect of formulation on the viability of *Metarhizium anisopliae* conidia, *J. Invertebr. Pathol.,* 41, 151, 1983.
561. **Daoust, R. A., Ward, M. G., and Roberts, D. W.,** Effect of formulation on the virulence of *Metarhizium anisopliae* on mosquito larvae, *J. Invertebr. Pathol.,* 40, 228, 1982.
562. **Gruner, L. and Abud-Antun, A.,** Etude des conditions de sporulation et de conservation d une souche de *Metarhizium anisopliae* Sorokin isolee de *Phyllophaga pleei* en Guadeloupe (Coleoptera: Scarabaeidae), *Turrialba,* 26, 241, 1976.
563. **Hall, R. A.,** Effect of repeated subculturing on agar and passage through an insect host on pathogenicity, morphology, and growth rate of *Verticillium lecanii, J. Invertebr. Pathol.,* 36, 216, 1980.
564. **Samsinakova, A. and Kalalova, S.,** The influence of a single-spore isolate and repeated subculturing on the pathogenicity of conidia of the entomophagous fungus *Beauveria bassiana, J. Invertebr. Pathol.,* 42, 156, 1983.
565. **Hall, I. M., Dulmage, H. T., and Arakawa, K. Y.,** Laboratory tests with entomogenous bacteria and the fungus *Beauveria bassiana* against the little house fly species *Fannia canicularis* and *F. femoralis, Environ. Entomol.,* 1, 105, 1972.
566. **Hartman, G. C. and Wasti, S. S.,** Infection of the gypsy moth, *Porthetria dispar* with the entomogenous fungus, *Conidiobolus coronatus, Entomophaga,* 19, 353, 1974.
567. **Krejzova, R.,** Experimental infections of several species of aphids by specimens of the genus *Entomophthora exitialis, Vestn. Cesk. Spol. Zool.,* 37, 21, 1973.
568. **Fargues, J. and Robert, P.,** Pathologie des invertebres — adaptabilite de deux pathotypes de *Metarhizium anisopliae* (Metsch.) Sor. (Fungi Imperfecti: Hyphomycetes) par culture sur milieu artificiel et par passage sur insecte-hoted origine, *C. R. Acad. Sci. Paris,* 287, 165, 1978.
569. **Goral, V. M.,** Effect of cultivation conditions on the entomopathogenic properties of muscardine fungi, in Proc. First Joint US/USSR Conf. on Production, Selection, and Standardization of Entomopathogenic Fungi of the US/USSR Joint Working Group on the Production of Substances by Microbiological Means, Ignoffo, C. M., Ed., 1978, 217.
570. **Voronina, E. G.,** Epizootics of Entomophthorosis of *Acyrthosiphon pisum, Entomol. Obozr.,* 50, 780, 1971.
571. **Ignoffo, C. M., Marston, N. L., Hostetter, D. L., Puttler, B., and Bell, J. V.,** Natural and induced epizootics of *Nomuraea rileyi* in soybean caterpillars, *J. Invertebr. Pathol.,* 27, 191, 1976.

572. **Ignoffo, C. M.,** Possibilities of mass-producing insect pathogens in *Insect Pathology and Microbial Control, Proc. Int. Colloq. Insect Pathol. and Microbiol Control,* North-Holland, Amsterdam, 1967, 91.
573. **Miller, L. K., Lingg, A. J., and Bulla, L. A., Jr.,** Bacterial, viral and fungal insecticides, *Science,* 219, 715, 1983.
574. **Roberts, D. W.,** Means for insect regulation: fungi, *Ann. N.Y. Acad. Sci.,* p. 76, 1973.
575. **Couch, T. L.,** Potential for commercial production of entomogenous fungi, in *Proc. 1st Int. Colloq. on Invertebr. Pathol.,* Queen's University Press, Kingston, Canada, 1976, 305.
576. **Sweeney, A. W.,** The time-mortality response of mosquito larvae infected with the fungus *Culicinomyces, J. Invertebr. Pathol.,* 42, 162, 1983.
577. **Sweeney, A. W. and Panter, C.,** The pathogenicity of the fungus *Culicinomyces* to mosquito larvae in natural field habitats, *J. Med. Entomol.,* 14, 495, 1977.
578. **Sweeney, A. W.,** Preliminary field tests of the fungus *Culicinomyces* against mosquito larvae in Australia, *Mosq. News,* 41, 470, 1981.
579. **Sweeney, A. W., Cooper, R., Medcraft, B. E., Russell, R. C., O'Donnell, M., and Panter, C.,** Field tests of the mosquito fungus *Culicinomyces clavisporus* against the Australian encephalitis vector *Culex annulirostris, Mosq. News,* 43, 290, 1983.
580. **Merriam, T. L. and Axtell, R. C.,** Evaluation of the entomogenous fungi *Culicinomyces clavosporus* and *Lagenidium giganteum* for control of the salt marsh mosquito *Aedes taeniorhynchus, Mosq. News,* 42, 594, 1982.
581. **McCray, E. M., Womeldrof, D. J., Husbands, R. C., and Eliason, D. A.,** Laboratory observations and field tests with *Lagenidium* against California mosquitoes, *Proc. Pap. Annu. Conf. Calif. Mosq. Control Assoc.,* 41, 123, 1973.
582. **Jaronski, S. and Axtell, R. C.,** Persistence of the mosquito fungal pathogen *Lagenidium giganteum* after introduction into natural habitats, *Mosq. News,* 43, 332, 1983.
583. **Kerwin, J. L. and Washino, R. K.,** Ground and aerial application of the sexual and asexual stages of *Lagenidium giganteum* for mosquito control, *J. Am. Mosq. Control Assoc.,* 2, 182, 1986.
584. **Pickford, R. and Riegert, P. W.,** The fungus disease caused by *Entomophthora grylli* Fres., and its effects on grasshopper populations in Saskatchewan in 1963, *Can. Entomol.,* 96, 1158, 1964.
585. **Hamm, J. J.,** Epizootics of *Entomophthora aulicae* in lepidopterous pests of sorghum, *J. Invertebr. Pathol.,* 36, 60, 1980.
586. **Latge, J. P., Remaudiere, G., and Papierok, B.,** Un exemple de recherche en lutte biologique: les champignons *Entomophthora* pathogenes de pucerous, *Bull. Soc. Pathol. Exot.,* 71, 196, 1978.
587. **Milner, R. J., Soper, R. S., and Lutton, G. G.,** Field release of an Israeli strain of the fungus *Zoophthora radicans* for biological control of *Therioaphis trifolii* f. *maculata, J. Aust. Entomol. Soc.,* 21, 113, 1982.
588. **Dedryver, C. A.,** Initiating an epizootic in a glasshouse with *Entomophthora fresenii* on *Aphis fabae* through inoculum introduction and control of the relative humidity, *Entomophaga,* 24, 443, 1979.
589. **Wilding, N.,** The effect of introducing aphid-pathogenic Entomophthoraceae into field populations of *Aphis fabae, Ann. Appl. Biol.,* 99, 11, 1981.
590. **Chudare, Z. P.,** Application of Entomophthora, in Proc. First Joint US/USSR Conf. on Production, Selection, and Standardization of Entomopathogenic Fungi of the US/USSR Joint Working Group on the Production of Substances by Microbiological Means, Ignoffo, C. M., Ed., 1978, 254.
591. **Latge, J. P. and Perry, D. F.,** The utilization of an *Entomophthora obscura* resting spore preparation in biological control experiments against cereal aphids, *IOBC Bull.,* 111, 19, 1980.
592. **Soper, R. S.,** Development of *Entomophthora* species as possible microbial insecticides, in Proc. First Joint US/USSR Conf. on Production, Selection, and Standardization of Entomopathogenic Fungi of the US/USSR Joint Working Group on the Production of Substances by Microbiological Means, Ignoffo, C. M., Ed., 1978, 270.
593. **Ferron, P.,** Pest control by the fungi *Beauveria* and *Metarhizium,* in *Microbial Control of Pests and Plant Diseases 1970—1980,* Burges, H. D., Ed., Academic Press, New York, 1981, 465.
594. **Pospelov, V. P.,** Microbiological method of controlling agricultural pests, *Proc. Leningrad Acad. Agric. Sci.;* abstracted in *Rev. Appl. Entomol. Ser. A,* 29, 52, 1944.
595. **Telenga, N. A., Sikura, A. I., Smetnik, A. I.,** The use of the biological material boverin in conjunction with insecticides to control *Leptinotarsa decemlineata* Say, *Zashch. Rast. (Kiev),* 4, 3, 1964.
596. **Pristavko, V. P.,** Processes pathologiques consecutifs a l'action de *Beauveria bassiana* (Bals.) Vuill. associe a de fibles doses de DDT chez *Leptinotarsa decemlineata* Say, *Entomophaga,* 11, 311, 1966.
597. **Fedorko, A., Bajan, C., Kmitowa, K., and Wojciechowska, M.,** Effect of a joint introduction into the soil of several entomopathogenic microorganisms on the level of reduction of the Colorado beetle, *Pol. Ecol. Stud.,* 3, 135, 1977.
598. **Lakhidov, A. N.,** Biopreparations for the control of the Colorado beetle, *Zashch. Rast. (Kiev),* 11, 44, 1979.
599. **Drapatyi, N. A. and Beilakh, G.,** A mixture of beauverin and despirol tested against the Colorado beetle, *Kartofel Ovoshchi,* 5, 37, 1979.

600. **Fargues, J., Cugier, J. P., and van de Weghe, P.,** Field experiments *Beauveria bassiana* hyphomycete against *Leptinotarsa decemlineata* Coleoptera a Chrysomelidae, *Acta Oecol. Appl.*, 1, 49, 1980.
601. **Fargues, J., Cugier, J. P., and van de Weghe, P.,** Experimentation en parcelles du champignon *Beauveria bassiana* contre *Leptinotarsa decemlineata*, *Acta Ecol.*, 1, 49, 1980.
602. **Beratlief, Z.,** Investigations on the entomopathogenic fungus *Beauveria bassiana* and its action on the Colorado beetle and the beet weevil, *An. Inst. Cercet. Prot. Plant. Inst. Ont. Cercet. Agric. Bucharest,* 15, 233, 1979.
603. **Watt, B. A. and Lebrun, R. A.,** Soil effects of *Beauveria bassiana* on pupal populations of the Colorado potato beetle, *Environ. Entomol.,* 13, 15, 1984.
604. **Sikura, A. I. and Smietnik, A. I.,** The effectiveness of beauverin in conjunction with some insecticides to control *Carpocapsa pomonella, Zashch. Rast. (Kiev),* 5, 20, 1967.
605. **Lappa, N. V., Goral, V. M., and Drozda, V. F.,** Effectiveness of boverin and pecilomin in the control of the codling moth, *Zasch. Rast. (Kiev),* 6, 24, 1977.
606. **Ferron, P. and Vincent, J. J.,** Preliminary experiments on the use of *Beauveria bassiana* against *Carpocapsa pomonella, Mitt. Biol. Bundesanst. Land Forstwirtsch. Berlin-Dahlem,* 180, 120, 1978.
607. **Chiang, H. C. and Huffaker, C. B.,** Insect pathology and microbial control of insects in the People's Republic of China, in *Proc. 1st Int. Colloq. on Invertebr. Pathol.,* Queen's University Press, Kingston, Canada, 1976, 42.
608. **Riba, G.,** Application en essais parcellaires de plein champ d'un mutant artificiel du champignan entomopathogene *Beauveria bassiana* contre la pyrale du mais, *Ostrinia nubilalis, Entomophaga,* 29, 41, 1984.
609. **Dunn, P. H. and Mechalas, B. J.,** The potential of *Beauveria bassiana* (Balsamo) Vuillemin as a microbial insecticide, *J. Insect Pathol.,* 5, 451, 1963.
610. **Muller-Kogler, E. and Stein, W.,** Gewachshausversuche mit *Beauveria bassiana* (Bals.) Vuill. zur infection von *Sitona lineatus* (L.) (Coleopt., Curcul.) im Boden, *Z. Angew. Entomol.,* 65, 59, 1970.
611. **Ayala, J. L. and Monzon, S.,** Test on different doses of *Beauveria bassiana* for the control of the banana weevil *(Cosmopolites sordidus), Cent. Agric.,* 4, 19, 1979.
612. **Gottwald, T. R. and Tedders, W. L.,** Colonization, transmission and longevity of *Beauveria bassiana* and *Metarhizium anisopliae* on pecan weevil larvae, *Curculio caryae* in the soil, *Environ. Entomol.,* 13, 557, 1984.
613. **McCoy, C. W., Beavers, G. M., and Tarrant, C. A.,** Susceptibility of *Artipus floridanus* to different isolates of *Beauveria bassiana, Fla. Entomol.,* 68, 402, 1985.
614. **Ferron, P.,** Modification of the development of *Beauveria tenella* mycosis in *Melolontha melolontha* larvae by means of reduced doses of organophosphorus insecticides, *Entomol. Exp. Appl.,* 14, 457, 1971.
615. **Keller, S.,** The importance of fungal diseases in the regulation of cockchafer populations, *Mitt. Schweiz. Entomol. Ges.,* 55, 392, 1982.
616. **Ferron, P.,** Lutte microbiologique contre le hanneton commun *Melolontha melolontha, Meded. Fac. Landouwwet. Rijksuniv. Gent.,* 42, 1323, 1977.
617. **Keller, S., Keller, E., and Ramser, E.,** Results of a field trial for the microbial control of the cockchafer *Melolontha melolontha* with the fungus *Beauveria tenella, Mitt. Schweiz. Entomol. Ges.,* 52, 35, 1979.
618. **Latch, G. C. M.,** *Metarhizium anisopliae* strains in New Zealand and their possible use for controlling pasture-inhabiting insects, *J. Agric. Res.,* 8, 384, 1965.
619. **Latch, G. C. M. and Falloon, R. E.,** Studies on the use of *Metarhizium anisopliae* to control *Oryctes rhinoceros, Entomophaga,* 21, 39, 1976.
620. **Peterson, G. D.,** Research on the control of the coconut palm rhinoceros beetle in Fiji, Tonga, Western Samoa, review of project results, *U. N. Dev. Programme FAO,* 21, 109, 1977.
621. **Guagliumi, P., Marques, E. J., and Vilas Boas, A. M.,** Contribucaco oa estudo da cultura e applicacao de *Metarhizium anisopliae* no controle da "cigarrinha da folha," *Mahanarva posticata* (Stal.) no Nordeste do Brasil, *Bol. Tec. CODECAP Recife,* 3, 45, 1974.
622. **Ramiro, Z. A.,** Field experiments with different levels of the fungus *Metarhizium anisopliae* (Metsch.) Sorokin in the control of the spittle bugs *Deois flavopicta* and *Zulia entreriana* in grasslands, *Biologico,* 45, 199, 1979.
623. **Coles, R. B. and Pinnock, D. E.,** Control of the pasture cockchafer with the fungal pathogen *Metarhizium anisopliae,* Proc. 3rd Aust. Conf. Grasslands, 1981, 191.
624. **Xiong, D. Z. and Wu, G.,** The application of *Metarhizium anisopliae* to control *Colasposoma metallicum, Zhiwu Baohu,* 3, 36, 1981.
625. **Gottwald, T. R. and Tedders, W. L.,** Suppression of pecan weevil populations with entomopathogenic fungi, *Environ. Entomol.,* 12, 471, 1983.
626. **Risco, B.,** Development of a multiplication program for the fungus *Metarhizium anisopliae* for biological control of *Mahanarva posticata* in the states northeast of Brazil, *Int. Soc. Sugarcane Technol. Entomol. Newsl.,* 5, 17, 1978.
627. **Allen, G. E., Greene, G. L., and Whitcomb, W. H.,** An epizootic of *Spicaria rileyi* on the velvet bean caterpillar, *Anticarsia gemmatalis* in Florida, *Fla. Entomol.,* 54, 189, 1971.

628. **Sprenkel, R. K., Brooks, W. M., van Duyn, J. W., and Deitz, L. L.,** The effect of three cultural variables on the incidence of *Nomuraea rileyi*, phytophagous lepidoptera and their predators on soybeans, *Environ. Entomol.*, 8, 334, 1979.
629. **Ignoffo, C. M., Marston, N. L., Puttler, B., Hostetter, D. L., Thomas, G. D., Biever, K. D., and Dickerson, W. A.,** Natural biotic agents controlling insect pests of Missouri soybeans, *World Soybean Res. Rep.*, 561, 1976.
630. **Sprenkel, R. K. and Brooks, W. M.,** Artificial dissemination and epizootic initiation of *Nomuraea rileyi* an entomogenous fungus of lepidopterous pests of soybeans, *J. Econ. Entomol.*, 68, 847, 1975.
631. **Samsinakova, A. and Kalalova, S.,** Artificial infection of scale insects with entomophagous fungi, *Verticillium lecanii* and *Aspergillus candidus*, *Entomophaga*, 20, 361, 1875.
632. **Hall, R. A.,** Aphid control by a fungus, *Verticillium lecanii*, within an integrated programme for chrysanthemum pests and diseases, *Proc. 8th Br. Insect Fungi. Conf.*, 1975, 93.
633. **Hall, R. A. and Burges, H. D.,** Control of aphids in glasshouses with the fungus *Verticillium lecanii*, *Ann. Appl. Biol.*, 93, 235, 1979.
634. **Hall, R. A.,** *Verticillium lecanii* on the aphid, *Macrosiphoniella sanborni*, *J. Invertebr. Pathol.*, 28, 389, 1976.
635. **Hall, R. A.,** Control of aphids by the fungus, *Verticillium lecanii*. I. Effect of spore concentration, *Entomol. Exp. Appl.* 27, 1, 1980.
636. **Kanagaratnan, P., Hall, R. A., and Burges, H. D.,** Control of glasshouse whitefly, *Trialeurodes vaporariorum*, by an "aphid" strain of the fungus *Verticillium lecanii*, *Ann. Appl. Biol.*, 100, 213, 1982.
637. **Hall, R. A.,** Control of whitefly, *Trialeurodes vaporariorum* and cotton aphid, *Aphis gossypii* in glasshouses by two isolates of the fungus *Verticillium lecanii*, *Ann. Appl. Biol.*, 101, 1, 1982.
638. **Ekbom, B. S.,** Investigations on the potential of a parasitic fungus for biological control of the greenhouse whitefly, *Swed. J. Agric. Res.*, 9, 129, 1979.
639. **Solovei, E. F. and Sogoian, L. N.,** Fungus against the glasshouse whitefly, biological control of *Trialeurodes vaporariorum*, *Zashch. Rast. (Kiev)*, 5, 28, 1982.
640. **Solovey, Y. F. and Koltsov, P. D.,** Effect of the entomopathogenic fungus *Aschersonia* on the orange whitefly, *Mikol. Fitopatol.*, 10, 425, 1976.
641. **Primak, T. A. and Chizhik, R. I.,** The basis for possible use of *Aschersonia aleyrodis* in the control of glasshouse whitefly, *Zakhyst Rosl. Resp. Mizhvid. Temat. Nauk. Zb.*, 22, 53, 1975.
642. **Spasova, P.,** Use of the entomopathogenic fungus *Aschersonia* against *Trialeurodes vaporariorum*, *Gradinarstvo*, 16, 13, 1974.
643. **Uchida, M.,** Studies on the use of the parasitic fungus *Aschersonia* sp. for controlling citrus whitefly, *Dialeurodes citri*, *Bull. Kanagawa Hortic. Exp. Stn.*, 18, 66, 1970.
644. **Ramaker, P. M. C.,** *Aschersonia aleyrodis*, a selective biological insecticide, *IOBC/WPRS Bull.*, VI, 167, 1983.
645. **Ramaker, P. M. J. and Samson, R. A.,** *Aschersonia aleyrodis*, a fungal pathogen of whitefly. II. Application as a biological insecticide in glasshouses, *Z. Angew. Entomol.*, 97, 1, 1984.
646. **McCoy, C. W., Selhime, A. G., Kanavel, R. F., and Hill, A. J.,** Suppression of citrus rust mite populations with application of fragmented mycelia of *Hirsutella thompsonii*, *J. Invertebr. Pathol.*, 17, 270, 1971.
647. **McCoy, C. W. and Selhime, A. G.,** The fungus pathogen, *Hirsutella thompsonii* and its potential use for control of the citrus rust mite in Florida, *Proc. Int. Citrus Congr.*, 2, 521, 1977.
648. **Yen, H.,** Isolation of the filamentous fungus, *Hirsutella thompsonii* from *Phyllocoptruta oleivora*, *Acta Entomol. Sinica*, 17, 225, 1974.
649. **Chen, D. M., Yan, S. X., Li, Y. O., and Chen, J. S.,** Liquid culture of *Hirsutella thompsonii* mycelium and its effects on citrus rust mite control in the field, *Weishengwuxue Tongbao*, 5, 204, 1981.
650. **van Brussel, E. W.,** Interrelations between citrus rust mite, *Hirsutella thompsonii* and greasy spot on citrus in Surinam, *Landbouwproefstn. Suriname Bull.*, 98, 43, 1975.
651. **Gardner, W. A., Oetting, R. D., and Storey, G. K.,** Susceptibility of the two-spotted spider mite, *Tetranychus urticae* to the fungal pathogen *Hirsutella thompsonii*, *Fla. Entomol.*, 65, 458, 1982.
652. **Litvinova, M. N., Aleshina, O. A., and Deysadze, T. A.,** Report on testing of an entomopathogenic preparation based on *Hirsutella thompsonii* obtained from the U.S.A., *Sci. Res. Rep. All Union Sci. Res. Inst. Bacterial Prep.*, 6, 1979.
653. **Almeida, S. L., Corte, C. R., Morais, A. A., Galhardo, L. C. S., Fekete, T. J., and Mariconi, F. A. M.,** Defensivos quimicos e o fungo *Hirsutella thompsonii* pulverizados contra *Phyllocoptruta oleivora*, *Solo*, 73, 11, 1981.
654. **Urueta, E. S.,** Control del acaro *Retracrus elaeis* mediante el hongo *Hirsutella thompsonii* e inhibicion de este por dos fungicidas, *Rev. Augura*, 6, 25, 1980.
655. **McCoy, C. W.,** personal observations.
656. **Humber, R. A.,** personal communication.

657. **Boucias, D. G. and Pendland, S. L.,** unpublished observations.
658. **Latge, J. P.,** personal communication.

INDEX

A

ABG-6178, 206
Acaracides, 196
Acremonium, 172
Acrodontium, 171
Aegerita, 176
Aegerita webberi, 166
Akanthomyces, characteristics of, 167, 168
Amblyospora, transovarian transmission of, 54
Ameson pulvis, early studies of, 2
Amoebae
   biocontrol with, 100—101
   characteristics of, 4
   taxonomy of, 5
   in vivo specificity of, 75
Amoebida, 4, 49, 94—95
Anisogamous gametes, 9
Anopheline mosquitoes, *N. algerae* as pathogen in, 31, also Mosquitoes
Anthonomus grandis, early studies of, 2
Antibiotics, for control of amoebic disease in grasshoppers, 66
Ants, 174
Aphids
   biocontrol of, 208, 210
   entomegenous fungi reported from, 166, 171, 174
   killed by entomogenous fungi, 186
   microbial control agents of, 172
Arthropod pests, biocontrol of
   with entomogenous fungi, 205—207
      Aschersonia, 211
      Beauveria, 209
      Culicinomyces, 207
      Entomophthoraceae, 208
      Hirsutella, 211—212
      Lagenidium, 208
      Metarhizium, 209
      Nomuraea, 210
      Verticillium, 210
   with entomogenous protozoa
      Amoebae, 100
      Coccidia, 104
      Eugregarines, 101—102
      flagellates, 100
      with microsporidia, 104—119
      N. algerae, 80, 81, 104
      Neogregarines, 102—104
Apanosporoblastina, classifications of, 10
Apicomplexa, 4, 8
Application techniques, 121, see also Field efficacy
Aquatic flea, Neogregarine specificity for, 78
Aquatic habitats, persistence of spores in, 68
Army work, *V. necatrix* in, 44
Aschersonia, 163, 174—176, 211
Ascogregarina, 8
Ascogregarina culicis

classification of, 19
description of, 20
general characteristics of, 15
hosts susceptible to, 77
life cycle of, 20—21
as potential microbial agent, 101—102
prevalence of infection in mosquitoes of, 21
Aseronija, 206
Ascomycotina, taxonomy of, 162—165
Azygospores, in entomogenous fungal infection, 187

B

Bacteria, as microbial control agents, 151
Barkbeetle, specificity of *Amobae* for, 75, see also Beetle
Basidiomycotina, taxonomy of, 165—166
Beauveria, 171, 172, 209
Bees, attacked by entomogenous fungi, 174, see also Honeybee; Leaf-cutting bees; Pollinator species
Beetle, controlled with *M. trogodermae*, 103—104, see also Barkbeetle; Colorado potato beetle; Dermestid beetle; Dytiscid beetle; Flour beetle; Red flour beetle; Rhinoceros beetle; Scarabeid beetle
Benomyl, as antimicrosporidian agent, 65—66
Biocontrol
   with entomogenous fungi, 205—207
      Aschersonia, 211
      Beauveria, 209
      Culicinomyces, 207
      Entomophthoraceae, 208
      Hirsutella, 211—212
      Lagenidium, 208
      Metarhizium, 209—210
      Nomuraea, 210
      Verticillium, 210—211
   with entomogenous protozoa
      Amoebae, 100
      Coccidia, 104
      Eugregarines, 101—102
      flagellates, 100
      microsporidia, 104—119
      N. algerae, 80, 104
      Neogregarines, 102—104
Biotechnology, 120
Biotrol FBB®, 153, 206
Biotrol FMA, 206
Black cutwork, Neogregarine specificity for, 78
Black fly, susceptibility to ciliates of, 92
Boll weevil, see also Cotton boll weevil; Weevils
   biocontrol of
      with *M. grandis*, 102—103
      with *N. gasti*, 108—109
   mass production of pathogens of, 97
Bombyx mori, early studies of, 2

Boverin, 206
Budworm, infected by *N. fumiferanae,* 33—34, see also Spruce budworm
Butterflies, see also Lepidoptera
 attacked by entomogenous fungi, 174
 entomogenous fungi found in, 171
 infected with *P. schubergi,* 115

C

Capillispores, of entomogenous fungi, 185
Carmine spider mite, controlled by *Hirsutella,* 212
Cellular immunity, to entomogenous protozoa, 48—50
Chalkbrood disease, 162—163
Chemicals, effect on entomogenous fungi of, 196—198
Chlamydospores, in entomogenous fungal infections, 187
Chrodata, susceptibility to *N. algerae* of, 81
Choristonedia fumiferane, infected by *N. fumiferanae,* 34
Ciliates
 biocontrol with, 119—120
 classification of, 12
 host range for, 92—93
 taxonomy of, 8
Citrus rust mite, controlled with *Hirsutella,* 211—212
Citrus whitefly, 153, 211, see also Whitefly
Clathroconium, 174
Clavicipitales, 163
Coccidae, entomogenous fungi specific to, 163
Coccidia
 biocontrol with, 104
 sporozoa of, 9
 taxonomy of, 6—8
 in vivo specificity with, 78—79
Codling moth, controlled with *Beauveria,* 209
Coelomomyces, 156, 157, 187—188
Coelomycetes, 166, 174—176
Coelomycidium, 156
Coleoptera, *N. gasti* specificity for, 83
Colonization, after penetration of germ tube, 182
Colorado potato beetle, controlled with *Beauveria,* 209, see also Beetle
Conidia, death point of, 194
Conidial adhesion process, 177—178
Conidiobolus, 161
Conidiogenesis, of *N. rileyi,* 178, 179
Conidiophores, 187
Conidiophores, of hyphomycetes, 167, 168
Cordycepioideus, 163
Cordyceps, 163—165
Corn borer, see also European corn borer
 susceptibility to *N. pyrausta* of, 84, 113—114
 susceptibility to *V. necatrix* of, 118
Corn meal, *V. necatrix* used to control pests of, 117—118
Cotton boll weevil, 23, 78, see also Boll weevil; Weevils

Crane fly, *Coccidia* specificity for, 79
Crustacea, susceptibility to *N. algerae* of, 81
Culicids, susceptibility to *V. culcis* of, 91—92
Culicinomyces, 174, 175, 177, 207
Cytopathology, induced by entomogenous protozoa, 54

D

Dermestid beetle, 78, 103—104
Desmidiospora, 174
Deuteromycetes, 166, 177—185, see also specific topics
Deuteromycotina, 166—176, see also specific topics
Diet, of hosts in whole organism technology, 95, 97
Diphasic fermentation, of entomogenous fungi, 203
Diptera, susceptibility to *V. culicis* of, 92
Dytiscid beetle, *Coccidia* specificity for, 79

E

Ecdysone, 197
Ectoparasites, 177
Emerald shiner, Neogregarine specificity for, 78
Endamoebidae, 4
Endoparasitic infections, 177
Engyodontium, 169
Entomophthora, 158—160
Entomophthoraceae, 160, 208
Entomophthorales, 185—187
Environmental Protection Agency, U.S., 13
Epizootics, fungal, 151, 213
Eriophid mite, controlled by *Hirsutella,* 212
Eucoccidiida, inflammatory response of insects to, 49
Eugregarines
 A. culicis, 19—21
 biocontrol with, 101—102
 characteristics of, 8
 inflammatory response of insects to, 49
 taxonomy of, 5
 in vivo production of, 94—95
 in vivo specificity of, 75
European corn borer, controlled with *Beauveria,* 209

F

Fermentation culture, of entomogenous fungi, 202—203
Fermentation technology, entomogenous protozoa in, 99—100
Field efficacy
 aerial applications, 112
 against *A. simplex,* 113
 of Bt and insecticide combinations, 114
 determination of, 112—113
 of spore bait and insecticide combinations, 112
 trials in Africa, 112—113
 with *V. necatrix,* 119
Field persistence, of entomogenous protozoa, 86
Filobasidiella, 166

Flagellates, in vivo specificity of, 74—75
Flour beetle, biocontrol of, 104
Flour moth, Neogregarine specificity for, 78
Foliage, effect on entomogenous fungi of, 196
Forest pests, *P. schubergi* used against, 115, 116
Fumagillin, effectiveness of, 56
Fumidil B., 64, 65
Fungal inhibitor, plant-produced, 190
Fungi, 151, 155
Fungi, entomogenous, 151
   biocontrol with
      Aschersonia, 211
      Hirsutella, 211—212
      Nomuraea, 210
      Verticillium, 210—211
   effect on stability and persistence
      of chemicals, 196—198
      of foliage, 196
      of host, 196
      of light, 194
      of moisture, 195
      of soil, 195—196
      of substrates, 195
      of temperature, 194—195
      of water, 196
   historical background of, 152—154
   infection cycle for
      under adverse environmental conditions, 185
      Coelomomyces, 187—188
      colonization in, 182—183
      Deuteromycetes, 177—185
      effect of physical factors on, 198—199
      Entomophthorales, 185
      and insect behavior, 184
      Lagenidium giganteum, 188
      successful penetration in, 180—182
   major taxa of, 155
   mass production of, 200, 201
      changes in virulence during, 204—205
      fermentation culture, 202—203
      formulation and stability of preparations for, 204
      in living systems, 200
      semisolid culture, 203
      submerged culture, 202—203
      surface culture, 200—202
   as microbial control agents, 212
   mode of parasitism for, 177
   pathogenicity of, 193
   as potential microbial insecticides, 206
   research in, 213
   specificity of, 189—192
   storage information on, 201
   taxonomy of
      Ascomycotina, 162—165
      Basidiomycotina, 165—166
      Deuteromycotina, 166—176
      Mastigiomycotina, 154, 156, 158
      Zygomycotina, 157—161
   toxic action of, 188—189
   virulence of, 192—193
Fungicides, organic, 196
Fungi Imperfecti, classification of, 166—176
Fungistatic agents, 195
Fusarium, 174

G

Gametes, of *M. grandis*, 25
Genetic engineering, 120
Germ tube, 180, 181, 190
Gibellula, 167
Grain pests, Neogregarine specificity for, 77
Grasshoppers
   attacked by entomogenous fungi, 174
   infected with *M. locustae*, 100—101
   in mass production of entomogenous protozoa, 95
   mass production of pathogens of, 97
   prevalence of infection with *N. locustae*, 37
   susceptibility to *N. locustae* of, 84, 109—112
   in vivo specificity of *Amoebae* for, 75
Greenhouse whitefly, controlled by *Aschersonia*, 211, see also Whitefly
Green muscardine disease, 173, 209
Growth regulators, insect, 197

H

Haplosporidia, classification of, 12
Heliscosporidia, classification of, 31
Heliothis virescens, V. necatrix used to control, 117
Heliothis zea, V. necatrix used to control, 117
Herbicides, effect on entomogenous fungi of, 197
Hirsutella
   biocontrol with, 211
   characteristics of, 167—170
   infection cycle for, 177
Honey bee
   amoebic disease in, 4
   chalkbrood disease in, 162—163
   Neogregarine specificity for, 78
Host
   effect on entomogenous fungi of, 196
   environmental factors and, 48
   methods of transmission for entomogenous protozoa and, 48, 50—51
   susceptibility of
      cellular immunity, 48—50
      to entomogenous protozoa, 46
      humoral immunity, 47—48
Humidity, and infective cycle of entomogenous fungi, 198
Humoral immunity, to entomogenous protozoa, 47—48
Hymenoptera, 84, 86, 90, see also Bees
Hymenostilbe, 169
Hymenostomatida, 12
Hyphomycetes, 166
Hypocreales, 163

I

Immunity, 47—50

Immunosuppressants, in infection cycle of entomogenous fungi, 182
Infective processes, of entomogenous protozoa
   characteristics of, 51—54
   signs and symptoms of, 1, 55—56
Insecticides
   entomogenous fungi as potential, 206
   effect on entomogenous fungi, 196—197
   fungal microbial, 207
   N. pyrausta used in combination with, 114
   sensitivity of protozoa to, 61, 64
Insect larvae, conidial germination on, 179
Insects
   fungi attacking, 152
   natural immunity of, 47
   susceptibility to *N. algerae* of, 80, 81
Instar nymphs, susceptibility to *N. locustae* of, 109
Integument, host, penetration of, 51
Invertebrates, entomogenous fungi specificity for, 189—192
Isaria, 171
Isogamous gametes, 8

L

Lagenidium, 156—158, 208
Lagenidium giganteum, infection cycle for, 188
Lambornella, 12, 13
Leaf-cutting bees, 162—163
Leafhoppers, entomogenous fungi on, 174
Lepidoptera, see also Butterfly; Moth
   N. fumiferanae specificity for, 83
   N. gasti specificity for, 83
   N. pyrausta specificity for, 90
   P. schubergi specificity for, 84, 86, 90
   susceptibility to *V. culicis* of, 92
   susceptibility to *V. necatrix* of, 86, 91, 116
Light, effect on entomogenous fungi of, 194
Loculoascomycetes, 163—164
Lygus bugs, controlled with *Beauveria,* 209
Lyophilization, of protozoan spores, 58, 60

M

Mammalia, susceptibility to *N. algerae* of, 81
Malameba locustae, 13
   cysts of, 14, 18
   early studies of, 2
   general characteristics of, 15
   host specificity of, 75
   hosts susceptible to, 76
   life cycle of, 18—19
   as microbial control agent, 100—101
   natural occurrence of, 19
   trophozoites of, 14, 18
Mastigiomycotina, taxonomy of, 154, 156, 158
Mattesia grandis, 21
   biocontrol with, 102—103
   description of, 22
   early studies of, 2
   general characteristics of, 15
   host range of, 78
   life cycle of, 22—25
   mass production of, 93
   morphometrics of, 23
   natural occurrence of, 62
Mattesia trogodermae
   biocontrol with, 103—104
   description of, 26—28
   general characteristics of, 15
   host range of, 78
   life cycle of, 26—28
   natural occurrence of, 28
Meronts, of *N. algerae,* 29
Metaquino, 206
Metarhizium, 171—173, 209
Metarhizium anisopliae, 153
Methoprene, 67
Methyl-p-hydroxybenzoate, 67
Microascus, 163
Microbial inhibitors, 67
Microspora, 9—11
Microsporea, classification of, 10
Microsporida
   classification of, 10
   dried spores of, 57
   inflammatory response of insects to, 49—50
   survival at subfreezing temperatures, 59
   in vivo production of, 94—95
Microsporidia
   biocontrol with
      N. algerae, 104—106
      N. fumiferanae, 106—107
      N. gasti, 108—109
      N. locustae, 109—113
      N. pyrausta, 113—115
      P. schubergi, 115, 116
      V. culicis, 119
      V. necatrix, 115—119
   host range for, 83—84
   N. algerae, 28—32
   N. fumiferanae, 31
   N. gasti, 34—35
   N. locustae, 35—37
   N. necatrix, 41—44
   N. pyrausta, 37—39
   P. schubergi schubergi, 39—40
   spores of, 52
   taxonomy of, 6—8
   V. culicis, 44—46
   virulence of, 55
   in vitro specificity of, 69—74
   in vivo specificity of
      N. algerae, 79—83
      N. locustae, 84
      N. pyrausta, 84
      P. schubergi, 84
      V. culicis, 91—92
      V. necatrix, 86, 91
Microsporidiology, biotechnological advancement in, 102
Mite, 174, 211—212, see also Carmine spider mite;

Citrus rust mite; Eriophid mite
Moisture, effect on entomogenous fungi of, 195
Monoclonal antibodies, production of, 120
Mormon cricket, susceptibility to *N. locustae* of, 113
Mosquito breeding, controlled with *Lagenidium*, 208
Mosquito larvae, entomogenous fungus of, 175
Mosquitoes, see also Tree hole mosquito
   biocontrol of, 105, 119
   susceptibility of
      to ciliates, 92, 119
      to Eugregarines, 75—77
Moth, see Flour moth; Noctuid moth; Wax moth
Muscardine disease, 152
   green, 173, 209
   white, 171, 209
Mycar, 153, 206
Mycoses, 151
Mycotal, 153, 206
Mycelia Sterilia, 166, 176
Myiophagus, 156

N

Nectria, 163, 165
Neogregarines
   biocontrol with
      M. grandis, 102—103
      M. trogodermae, 103—104
   characteristics of, 9
   classifications of, 9
   dried spores of, 57
   inflammatory response of insects to, 94
   M. grandis, 21—26
   M. trogoderma, 26
   survival at subfreezing temperatures, 59
   taxonomy of, 5—6
   in vivo production of, 95
   in vivo specificity of, 77—78
Neozygites, 160
Nephridiophaga blattellae, 12
Noctuid caterpillar, controlled with *Nomuraea*, 210
Noctuid moth, killed by entomogenous fungi, 171
Nosema, 12
Nosema algerae
   classification of, 28—29
   biocontrol with, 104—106
   description of, 29—32
   general characteristics of, 51
   infective process of, 53
   life cycle of, 29—32
   natural occurrence of, 31
   per os susceptibility of hosts to, 80
   virulence of, 55
   in vitro cultivation of, 70
   in vivo specificity of, 79—83
Nosema bombycis, early studies of, 2
Nosema fumiferanae
   biocontrol with, 106—107
   classification of, 31
   description of, 32—33
   general characteristics of, 16

host range of, 83
life cycle of, 32—33
natural occurrence of, 33
prevalence of infection by, 34
Nosema gasti
   biocontrol with, 108—109
   description of, 34—35
   early studies of, 2
   general characteristics of, 16
   host range of, 83—84
   life cycle of, 34—35
   mass production of, 93
   natural occurrence of, 35
Nosema locustae, 13
   biocontrol with, 109—113
   classification of, 35
   description of, 35—36
   general characteristics of, 16
   life cycle of, 35—36
   mass production of, 93, 96—97
   natural occurrence of, 36
   orthopteran hosts of, 84—86
   prevalence of infection of, 37, 110
   sensitivity to insecticides of, 64
Nosema pyrausta
   biocontrol with, 113—115
   classification of, 37
   description of, 37—38
   general characteristics of, 16
   host range of, 90
   life cycle of, 37—38
   natural occurrence of, 39

O

Ovipositors, contaminated, 51
Pathogenicity, 55—56, 193
Pest control, use of fungi in, 212, see also Arthropod
      pests
Pesticides, biorational, 121
Pest management program, with *N. locustae*, 112
pH balance, effect on entomogenous fungi of, 196
Plathelminthes, susceptibility to *N. algerae* of, 80
Plectomycetes, 162—163
Pleistophora, in vitro cultivation of, 72
Pleistophora schubergi
   biocontrol with, 115, 116
   classification of, 39—40
   description of, 40
   general characteristics of, 17
   host range of, 90
   life cycle of, 40
   natural occurrence of, 40—41
Pleurodesmospora, 174
Polar tubes, 39, 41
Polar tube activation, process of, 52
Pollinator species, chalkbrook disease in, 162—163
Pollution, effect on entomogenous fungi of, 196
Production
   of entomogenous fungi, 200, 201
      fermentation culture, 202—203

in living systems, 200
semisolid culture, 203
submerged culture, 202—203
surface culture, 200—202
entomogenous protozoa
fermentation technology, 99—100
tissue culture systems, 98—99
whole organism technology, 93—98
Protozoa, 48—55, 67
Protozoa, entomogenous, see also Spores
field persistence, 67—69
historical background for, 1—2
host susceptibility to, 46—48
as potential microbial insecticides, 13—17
production of
fermentation technology, 99—100
tissue culture systems, 98—99
in vivo, 93
taxonomy of
principle groups, 5—8
phyla, 3
revision of, 2—13
virulence of, 121
in vitro specificity of, 69—74
in vivo specificity of
ciliates, 92—93
Coccidia, 78—79
Eugregarines, 75, 77
flagellates, 74—75
microsporidia, 79—92
Neogregarines, 77—78
Pseudaletia unipuncta, *V. necatrix* in, 44
Purple scale, natural control of, 169, see also Scale insects
Pyroryclor, 197

R

Rain forests, entomogenous fungi from, 163, see also Forest pests
Radiation, sensitivity of microsporidian spores to, 62—63
Red flour beetle, Neogregarine specificity for, 77
Rhinosceros beetle, controlled with *Metarhizium*, 209
Rodentia, susceptibility to *N. algerae* of, 81
Rust fungi, 166

S

Salinity, effect on entomogenous fungi of, 196
Sarcomastigorphora, 3—4
Sawflies, *P. schubergi* specificity for, 84
Scale insects
attacked by engomogenous fungi, 169
controlled with *Verticilium*, 210
entomogenous fungi specific to, 163, 166, 174
microbial control agents of, 172
Scarabeid beetle, controlled with *Beauveria*, 209
Scarabeid grub, biocontrol of, 104
Semisolid culture, of entomogenous fungi, 203—204

Sensitivity
of entomogenous protozoa
to insecticides, 61, 64
to sunlight, 61
to therapeutic agents, 64—66
to UV radiation, 61—63
Septobasidium, 165
Silkworm, 152, 171
Silverfish, in vivo specificity of *Amoebae* for, 75
Soil, 68, 195
Sorghum, *V. necatrix* used to control pests of, 117—118
Sorbic acid, 67
Soybean, *V. necatrix* used to control pests of, 117—118
Specificity
of entomogenous fungi
for invertebrates, 189—192
for vertebrates, 192
of entomogenous protozoa
in vitro, 69—74
in vivo, 74—93
Spiders, 166, 167, 171
Spores, of entomogenous protozoa, 56
longevity in aqueous media of, 58
N. algerae, 30, 31, 106
N. fumiferanae, 33, 107
N. locustae, 36, 96—97
N. pyrausta, 38—39
of *P. schubergi*, 40—41
sensitivity of
to radiation, 61—63
to therapeutic agents, 64—66
to ultraviolet light, 61—63
stability of, 56
survival of
at above-freezing temperatures, 56—58
at subfreezing temperatures, 58, 59
held dry, 57
after lyophilization, 58, 60
toxic effect of chemicals on, 66—67
V. culicis, 45—46
of *V. necatrix*, 43
Sporonts, of *N. algerae*, 29
Sporoplasm, in infective process, 54
Sporothrix, 174
Spruce budworm, 83, 106—107
Stability
of entomogenous fungi
effect of chemicals on, 196—198
effect of foliage on, 196
host and, 196
light and, 194
moisture and, 195
in soil, 195—196
substrates and, 195
temperature and, 194—195
water and, 196
of entomogenous protozoa, 56—59
at above-freezing temperatures, 56—58
at subfreezing temperatures, 58—59

after lyophilization, 58, 60
Sterols, in production of entomogenous fungi, 200
Stibella, 174
Storage, of infective stages of protozoa, 1
Strongwellsea, 161
Subkingdom, protozoa as, 2
Submerged culture, of entomogenous fungi, 202—203
Substrates, effect on entomogenous fungi of, 195
Sunlight, 61, 94
Surface culture, of entomogenous fungi, 200—202
Synnematium, 169

T

Tabanomyces, 161
Temperature
  effect on entomogenous fungi of, 194—195
  spore survival of entomogenous fungi and, 199
  spore survival of entomogenous protozoa and, 56—59
Tetrahymena, 12
Tetrahymena pyriformis, 13
Tetranacrium, 175
Thipyrimeth, 66
Tilachlidiopsis, 174
Tilachlidium, characteristics of, 171
Tissue culture systems, for entomogenous protozoa, 98—99, see also Production
Tobacco, *V. necatrix* used to control pests of, 117—118
Tolypocladium, 174
Torrubiella, 163—165
Toxins, 186, 188—189
Transmission, methods of
  for entomogenous protozoa, 48, 50—51
  transovum vs. transovarian, 1, 49
Tree hole mosquito, entomogenous fungi in, 174
Trees, forest, susceptibility of pests of to *P. schubergi,* 115, 116
Trematoda, susceptibility to *N. algerae* of, 80
Trichomycetes, 161
Trogodermae, prevalence of infection of *M. trogodermae* in, 28
Trypanosomatids, 5—6, 75

U

Uredinella, 166

V

Vairimorpha necatrix
  biocontrol with, 115—119
  classification of, 41
  description of, 41, 45—46
  geimsa-stained stages of, 42
  general characteristics of, 17
  host range for, 86, 91
  Hyphantria cunea colonies and larvae infected with, 45
  in larvae of *Pseudaletia unipuncta,* 44
  life cycle of, 41, 45—46
  natural occurrence of, 44
  in vitro cultivation of, 73
Vairimorpha plodiae, 43
Vavraia culicis
  biocontrol with, 119
  classification of, 44—45
  general characteristics of, 17
  host range of, 92
  natural occurrence of, 46
  in vitro cultivation of, 73
Velvetbean caterpillar
  controlled with *Nomuraea,* 210
  entomogenous fungi of, 179
  susceptibility to *V. necatrix* of, 118
Vertalec, 153, 206
Vertebrates
  host specificity of entomogenous fungi for, 192
  possible development of *N. algerae* in, 81, 83
  possible hazards of *N. locustae* to, 87—89
Verticillium, 172, 173
Virulence
  of entomogenous fungi
    on artificial media, 204—205
    naturally occurring, 190—193
  of entomogenous protozoa, 55—56

W

Water, 196, 198
Water mold parasites, 156
Wax moth, Neogregarine specificity for, 78
Weevils, 174, 209
Whitefly, see also Greenhouse whitefly
  attacked by entomogenous fungi, 169
  controlled by *Aschersonia,* 211
  controlled with *Verticillium,* 210
  entomogenous fungi specific to, 163, 174, 176
White muscardine disease, 171, 209
Whole organism technology, 1, 93—98
Wood cockroach, *Coccidia* specificity for, 79

Z

Zygomycetes, characteristics of, 157—158